Studies in Organic Chemistry 5

COMPREHENSIVE CARBANION CHEMISTRY

PART B

Studies in Organic Chemistry

Volume 1 Complex Hydrides, by A. Hajós
Volume 2 Proteoglycans — Biological and Chemical Aspects in Human Life, by J.F. Kennedy
Volume 3 New Trends in Heterocyclic Chemistry, edited by R.B. Mitra, N.R. Ayyangar, V.N. Gogte, R.M. Acheson and N. Cromwell
Volume 4 Inositol Phosphates: Their Chemistry, Biochemistry and Physiology, by D.J. Cosgrove
Volume 5 Comprehensive Carbanion Chemistry. Part A. Structure and Reactivity, edited by E. Buncel and T. Durst
Comprehensive Carbanion Chemistry. Part B. Selectivity in Carbon-Carbon Bond Forming Reactions, edited by E. Buncel and T. Durst
Volume 6 New Synthetic Methodology and Biologically Active Substances, edited by Z. Yoshida
Volume 7 Quinonediazides, by V.V. Ershov, G.A. Nikiforov and C.R.H.I. de Jonge
Volume 8 Synthesis of Acetylenes, Allenes and Cumulenes: A Laboratory Manual, by L. Brandsma and H.D. Verkruijsse
Volume 9 Electrophilic Additions to Unsaturated Systems, by P.B.D. de la Mare and R. Bolton
Volume 10 Chemical Approaches to Understanding Enzyme Catalysis: Biomimetic Chemistry and Transition-State Analogs, edited by B.S. Green, Y. Ashani and D. Chipman
Volume 11 Flavonoids and Bioflavonoids 1981, edited by L. Farkas, M. Gábor, F. Kallay and H. Wagner
Volume 12 Crown Compounds: Their Characteristics and Applications by M. Hiraoka
Volume 13 Biomimetic Chemistry, edited by Z.-I. Yoshida and N. Ise
Volume 14 Electron Deficient Aromatic- and Heteroaromatic-Base Interactions. The Chemistry of Anionic Sigma Complexes, by E. Buncel, M.R. Crampton, M.J. Strauss and F. Terrier
Volume 15 Ozone and its Reactions with Organic Compounds, by S.D. Razumovskii and G.E. Zaikov

Studies in Organic Chemistry 5

COMPREHENSIVE CARBANION CHEMISTRY

Edited by

E. Buncel
Department of Chemistry, Queen's University, Kingston, Ontario, Canada K7L 3N6

T. Durst
Department of Chemistry, University of Ottawa, Ottawa, Ontario, Canada K1N 6N5

PART B
SELECTIVITY IN CARBON-CARBON BOND FORMING REACTIONS

ELSEVIER
Amsterdam — Oxford — New York — Tokyo 1984

7117 - 6408

ELSEVIER SCIENCE PUBLISHERS B.V.
Molenwerf 1
P.O. Box 211, 1000 AE Amsterdam, The Netherlands

Distributors for the United States and Canada:

ELSEVIER SCIENCE PUBLISHING COMPANY, INC.
52 Vanderbilt Avenue
New York, NY, 10017

ISBN 0-444-42267-6 (Vol. 5B)
ISBN 0-444-41737-0 (Series)

Printed in The Netherlands

FOREWORD

The impetus to the study of the structure, stability and reactivity of
carbanions was provided to a large extent by Professor Donald J. Cram, through
his own work, and in the publication in 1965 of the classic monograph "Funda-
mentals of Carbanion Chemistry". Since that time, the study of $_{13}$carbanions has
chemistry has intensified, in part as new techniques such as ^{13}C and hetero-
nuclear magnetic resonance spectroscopy have become available, and in part
as new theoretical concepts have come to light. A particularly interesting
development is that of ion cyclotron resonance and related techniques for
the gas phase investigation of carbanion reactivity. Such studies have
increased our awareness of medium effects and have caused a re-evaluation
of the theoretical basis of strucural and electronic effects.

Carbanion chemistry plays a central role in modern synthetic organic
chemistry. This relationship has developed at an extremely rapid rate since
the 1960's due to the commercial availability of a variety of alkyllithiums,
and the development of a number of potent non-nucleophilic bases such as
lithium diisopropylamide, lithium 2,2,6,6-tetramethylpiperidide and sodium
hexamethyldisilazide. The availability of these reagents has made accessible
many structural types of carbanions which are extremely valuable as synthetic
units since they generally undergo efficient carbon-carbon bond formation
with typical electrophiles.

In this series, which has been entitled "Comprehensive Carbanion Chemistry"
it is hoped to discuss in detail key aspects of carbanion chemistry, via
contributed chapters written by active investigators in the field. Several
volumes are currently planned, each devoted to related aspects of carbanion
chemistry. Thus, Part A, already published, is concerned with the physical
aspects of carbanions, namely the acidity of carban acids both in solution
and in the gas phase, the structure of mono- and dianions as determined by
proton and carbon nuclear magnetic resonance, ultraviolet, infrared and Raman
spectroscopy, and, the involvement of electron transfer processes in carbanion
processes in carbanion reactions. Part B, the current volume, emphasizes
regio- and stereoselective processes, for example in aldol condensations,
the use of enolate equivalents in synthesis, and some reactions of allenic-
alpha-acetylenic carbanions. Other topics discussed include the formation
of carbon-carbon bonds via π-complexes of transition metals, as well as
techniques used in carbanion chemistry. A future volume will be devoted to
carbanions stabilized by silicon, phosphorus, sulfur, etc., functionalities.

We would like to thank the authors of the indvidual chapters in the
present volume for their excellent contribution and the extra effort which
they undertook in preparing camera-ready copy.

June,1983

E.B
T.D.

CONTENTS

CHAPTER 1

FORMATION OF CARBON-CARBON BONDS VIA π-COMPLEXES OF TRANSITION METALS.

LOUIS S. HEGEDUS
Department of Chemistry, Colorado State University, Fort Collins, CO   80523

CONTENTS

I.   INTRODUCTION

Unsaturated organic compounds, such as simple olefins, dienes, and arenes, which are not substituted with electron withdrawing groups are normally unreactive toward carbanions, since both reactants represent electron rich systems. However, complexation of these unsaturated compounds to the appropriate transition metal can alter their reactivity significantly, and in many cases permit direct alkylation with carbanions. Unsaturated organic compounds coordinate (bond) to transition metals by overlap of their filled π-bonding molecular orbitals with vacant "dsp" hybrid orbitals on the metal. Concurrently, filled d orbitals on the metal have appropriate symmetry and energy to overlap with vacant π* antibonding orbitals on the unsaturated substrate. When these interactions between the metal and the organic ligand are substantial, the chemistry of the unsaturated organic ligand is perturbed. The transition metal can act as an "electron sink", both withdrawing electron density from the complexed organic ligand to facilitate nucleophilic attack upon it, and stabilizing the negatively charged species resulting from nucleophilic addition to the unsaturated organic ligand (Figure 1).

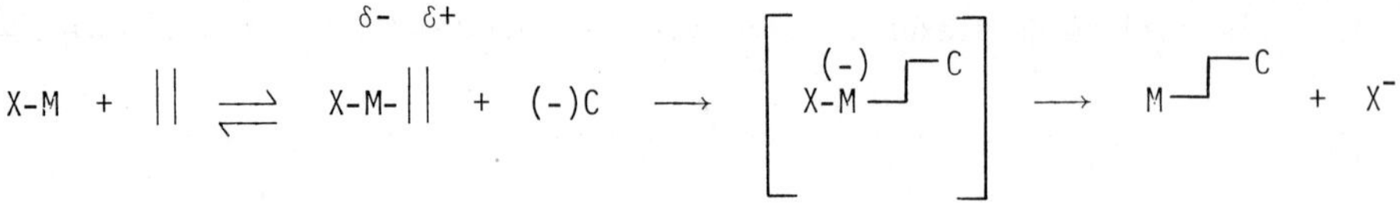

$$\delta- \quad \delta+$$

$$X\text{-}M + \; || \; \rightleftharpoons \; X\text{-}M\text{-}|| \; + \; (\text{-})C \; \longrightarrow \; \left[\begin{array}{c} (\text{-}) \quad \begin{array}{c} C \\ \end{array} \\ X\text{-}M \end{array} \right] \; \longrightarrow \; M \begin{array}{c} C \\ \end{array} \; + \; X^-$$

Figure 1

This is a sensitive reaction, even under the best of circumstances, in that reducti
of the metal or displacement of the complexed unsaturated ligand by the carbanion
are often major competing reactions. However, a great deal of progress has been
made in this area of research, and the direct alkylation of transition metal-
complexed unsaturated compounds is now a viable synthetic method. This review
will treat the current status of this subject by considering transition-metal
assisted alkylations of chelating olefins, simple olefins, diene and dienyl
systems, allylic complexes, and arenes via their reaction with carbanions.

II. DIRECT ALKYLATION OF OLEFINS WITH CARBANIONS

A. Chelating Olefins and Diolefins

Among the first successful alkylations of π-olefin complexes with carbanions
was the reaction of cyclooctadiene complexes of platinum (II) and palladium (II)
with stabilized carbanions such as diethyl malonate anion and acetylacetonate
anion (equation 1).[1-4] The reaction resulted in the formation of a carbon-carbon

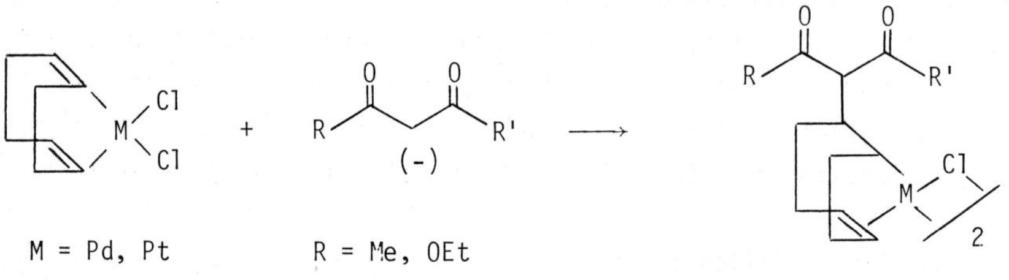

$$M = Pd, Pt \qquad R = Me, OEt \qquad R' = Me, OEt \tag{1}$$

bond between the carbanion and one terminus of the olefin, and a metal-carbon
bond at the other olefin terminus. The other double bond of the cyclooctadiene
resisted further alkylation, and the σ-alkylmetal complex was quite stable. A
careful study of the reaction of dichloro (endo-dicyclopentadiene)platinum (II)
and palladium (II) complexes with acetylacetone, ethyl acetoacetate, and diethyl
malonate showed that this alkylation indeed occurred in a trans exo fashion at
the 5,6-double bond without skeletal rearrangement of the dicyclopentadiene

group (equation 2). This reaction is quite sensitive to both the nature of the chelating diolefin and the nature of the carbanion. For example, the corresponding norbornadiene complexes of both palladium (II) and platinum (II) react with

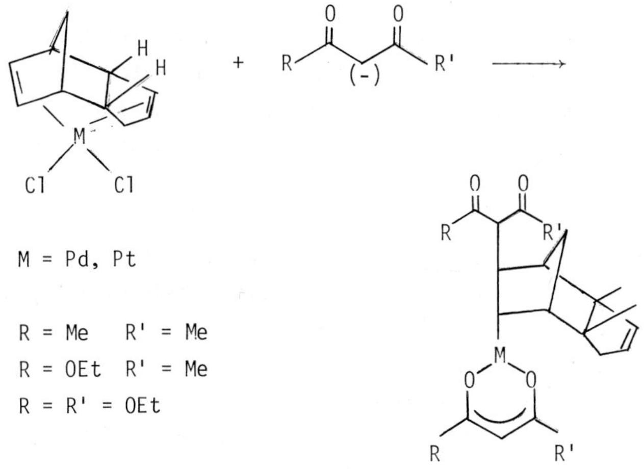

$$(2)$$

M = Pd, Pt

R = Me   R' = Me
R = OEt   R' = Me
R = R' = OEt

acetylacetone, ethyl acetoacetate, or diethyl malonate in the presence of sodium carbonate to produce only metallic palladium or platinum.[5] Apparently, the complexes of norbornadiene are easily reduced. In contrast, the cyclooctadiene complexes of palladium (II) and platinum (II) react with methyl- or aryl Grignard reagents to give relatively stable bis-alkylmetal complexes (equation 3).[6] With chelating diolefin complexes, displacement of the olefin by the carbanion (the major competing reaction with simple monoolefin complexes) has rarely been observed.

Although the alkylation of chelating diolefin complexes of palladium and platinum is a general process, very little use of the reaction has been made in organic synthesis. In contrast, the alkylation of palladium-coordinated allyl amines and allyl sulfides produces complexes which have been converted to useful organic products. N,N-Dimethylallylamine reacted with lithium chloropalladate and stabilized carbanions to produce stable palladiacycles in high yield (equation 4). Isopropyl allyl sulfide reacted in a similar fashion (equation 5).[7] In all cases, alkylation of the coordinated olefin occurred at the internal carbon of the olefin producing a five membered palladiacycle (the most stable ring size for palladiacycles). The regiospecificity of this reaction was clearly

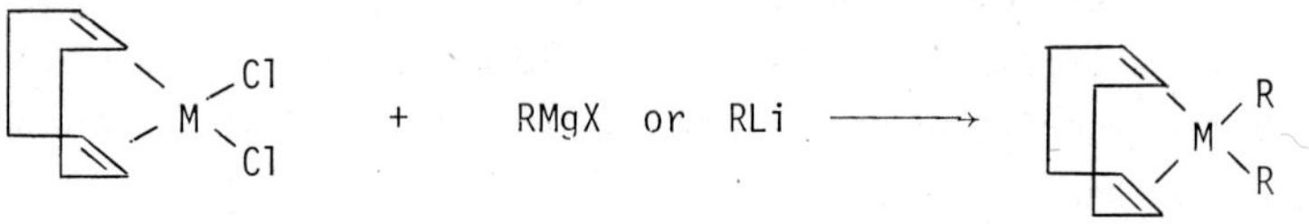

(3)

M = Pd   R = Me
M = Pt   R = Me, Et, Ph, O-Tolyl

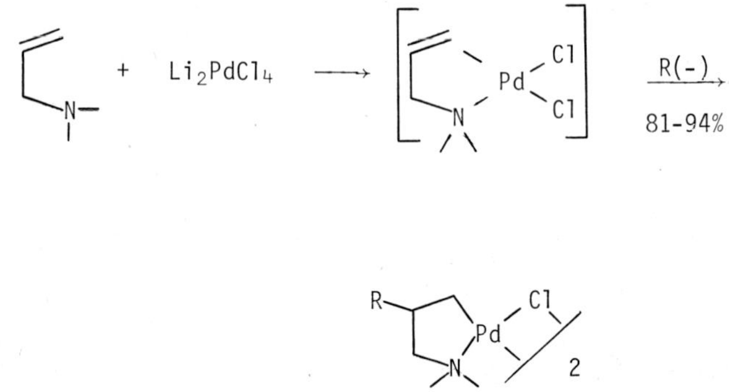

81-94%

(4)

R = CH(CO$_2$Et)$_2$, CH(COC$_6$H$_5$)$_2$, CHCO$_2$Et,
                                         COCH$_3$

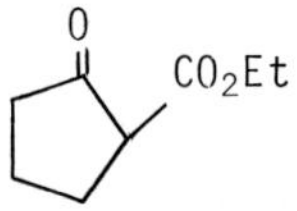

CH–COCH$_3$, C(CO$_2$Et)$_2$
COC$_6$H$_5$ C$_2$H$_5$

(5)

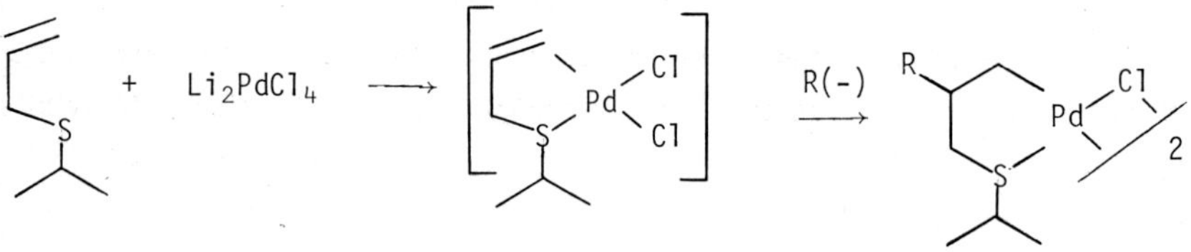

dictated by the formation of a stable (relative to the four membered ring) five
membered ring. This reaction was restricted to _stabilized_ carbanions. Simple
ketone enolates, or organocopper, or Grignard reagents produced intractable mixtures of organic products along with metallic palladium. The olefinic substrate
must be able to coordinate to palladium in a bidentate fashion, since allyl
alcohol, allyl phenyl ether, ethyl acrylate, and 1-octene (none of which coordinate
as strongly to palladium (II) as do nitrogen or sulfur-containing olefins) all
failed to undergo this olefin alkylation.

The real synthetic utility of this process lies in the further exceptional
chemistry of the product σ-alkylpalladium complexes. They undergo a facile reduction with sodium borohydride or hydrogen to produce compounds corresponding to a
formal addition of the stabilized carbanion across the olefin double bond
(equation 6).[7] The intermediate σ-alkyl palladium complexes need not be isolated
in this high yield preparation of γ-amino or γ-thio esters and ketones (equation 7).

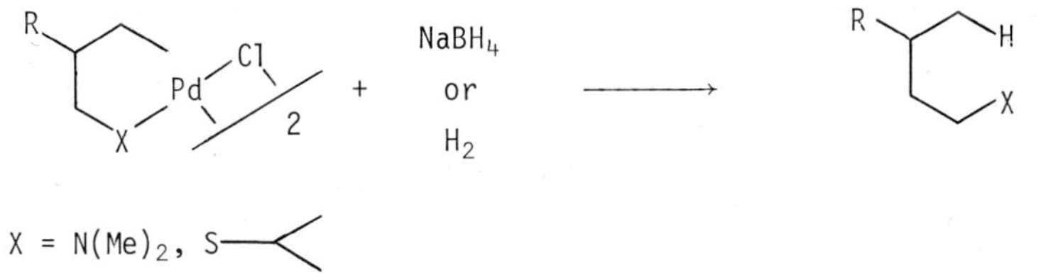

$$\text{(6)}$$

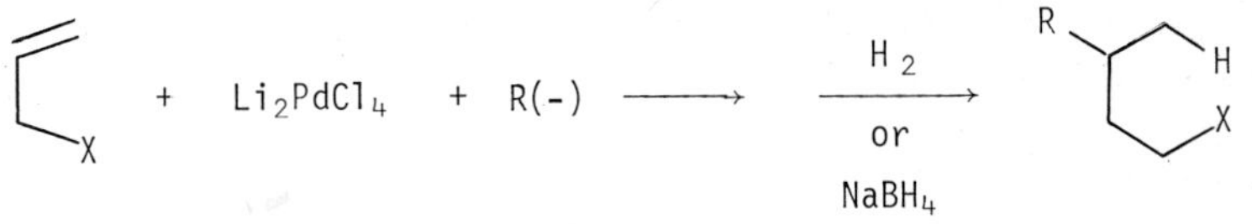

$$X = N(Me)_2, \quad S-\!\!<$$

$$\text{(7)}$$

Even more useful is the insertion of conjugated enones into the carbon-palladium
σ bond, a reaction characteritstic of stable σ-alkylpalladium (II) complexes
(equation 8).

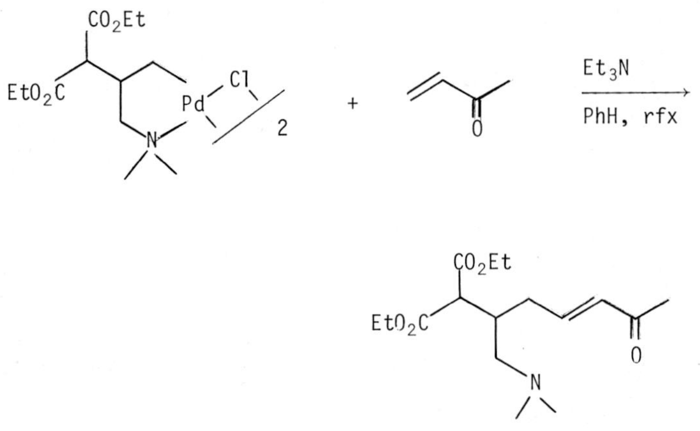

$$(8)$$

This chemistry has been used in the elegant synthesis of a key prostaglandin precursor described in equation 9.[8] The first step of this remarkable process was the alkylation of the cyclic allylamine in which palladium (II) not only activated the olefin for reaction with the carbanion, but also permitted the alkylation to proceed regiospecifically, to form the most stable five membered palladiacycle, and stereospecifically _trans_ to the palladium substituent. The olefin was regenerated by heating the complex to force a β-hydride elimination

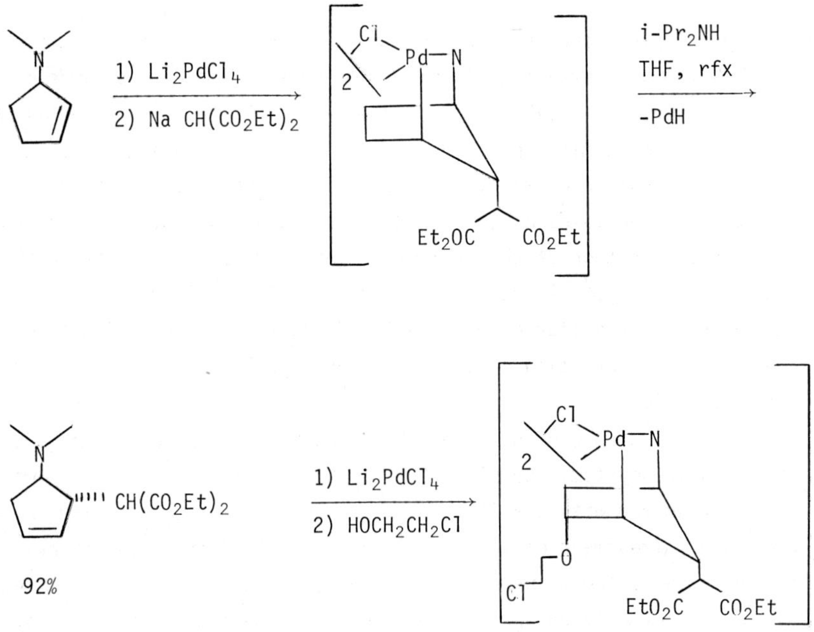

$$(9)$$

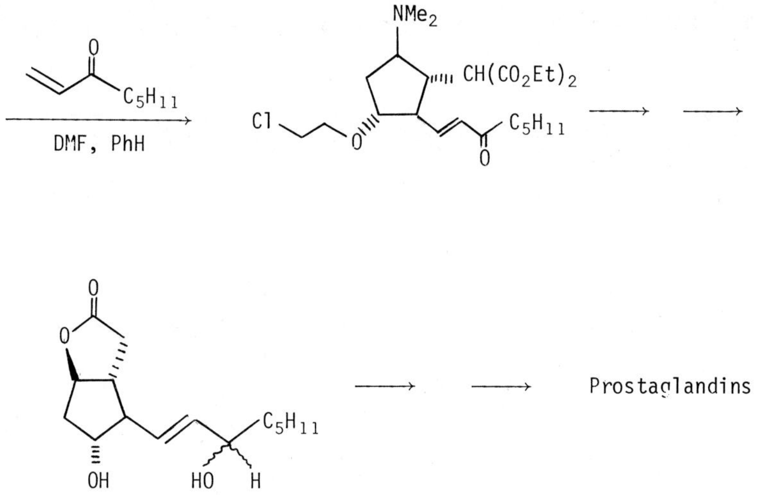

(9)

⟶   ⟶   Prostaglandins

reaction to ensue. The resulting homoallylic amine was alkoxypalladated (wherein
the 2-chloroethanol behaved as a nucleophile towards the complexed olefin), again
regio- and stereospecifically. The resulting σ-alkylpalladium complex inserted
oct-1-ene-3-one to give a tetrasubstituted cyclopentene which was converted to
"Corey's lactone". This compound has already been converted to a number of prostaglandins.

Homoallylic amines and sulfides underwent a similar "carbo-palladation" when
treated with the same stabilized carbanions and lithium chloropalladate.[9] In
direct contrast to the allylic systems, these homoallylic systems alkylated
exclusively at the terminal position of the olefin. The driving force, however,
was again the formation of the very stable five membered palladiacycle. These
σ-alkylpalladium complexes also reduced easily, again producing compounds resulting
from the formal addition of the carbanion precursor across the double bond
(equations 10 and 11).[9] Amines and sulfides having unsaturation further removed
from the nitrogen or sulfur atom did not undergo this "carbopalladation" reaction,
presumably because formation of a five membered palladium containing ring was
no longer possible.

A further chelating olefin system which has been "carbopalladated" is the
allylalkoxy amine system. Complexation of these compounds with $PdCl_2(MeCN)_2$
followed by reaction with stabilized carbanions again produced a five membered
palladiacycle. Treatment of these complexes with trimethylsilyl chloride
removed both the palladium and the ether function to produce the alkylated olefin.
The process is formally an S_N2' displacement of dimethylhydroxylamine by the

carbanion (equation 12).[10]

(10)

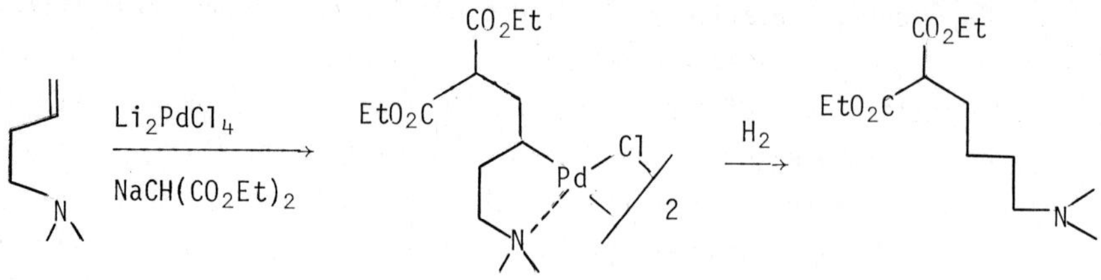

(11)

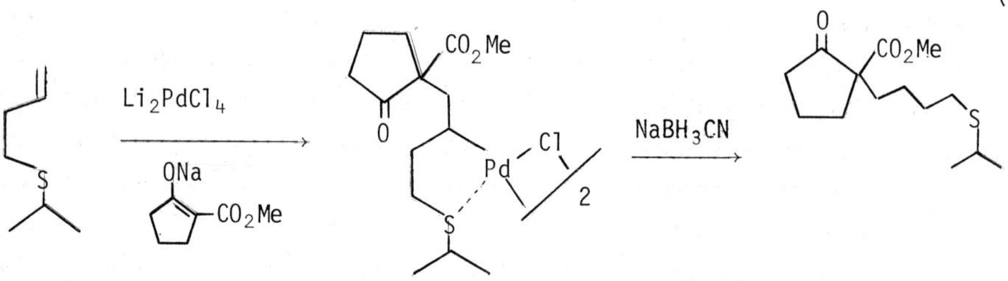

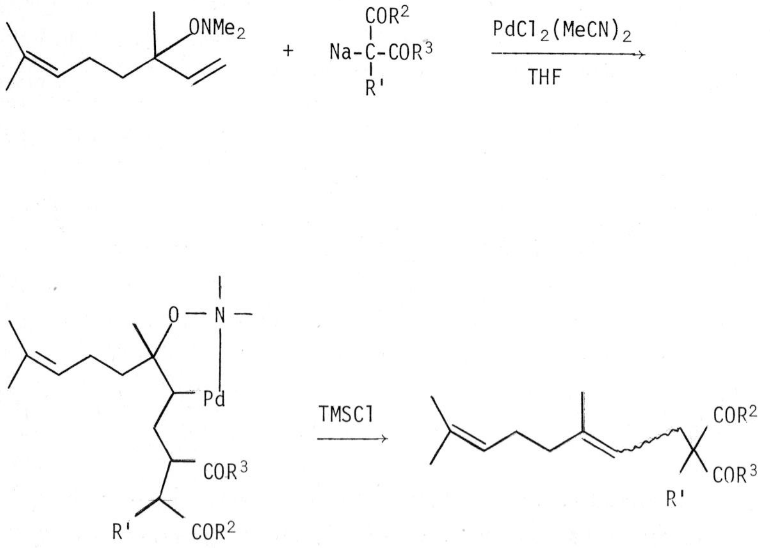

(12)

R^1 = H, Me, n-Bu
R^2 = MeO, Me, Et
R^3 = MeO, EtO

B. Simple Monoolefins

Although alkylation of chelating olefin complexes by carbanions is a fairly general process, relatively little use of this process has been made in organic synthesis. In contrast, the alkylation of simple olefins by carbanions would be a synthetically useful process, but, until recently, has not been successful. In this case, the major difficulty to overcome appeared to be reduction of the metal by the carbanion, or conversely, oxidative coupling of the carbanion by the metal. Methyllithium was found to react with styrene in the presence of palladium (II) acetylacetonate to give trans-β-methylstyrene in 90% yield.[11] In contrast, the use of palladium (II) acetate gave 75% yield, and palladium (II) chloride only 3% yield of β-methylstyrene. Addition of phosphines or phosphites as additional ligands, a ploy frequently successful in stabilizing delicate organopalladium intermediates, in this instance, completely retarded the alkylation. By the use of specifically deuterated cis- and trans-β-deuterio-styrene, the process was shown to proceed by a cis-addition, cis-elimination mechanism, in which the carbanion first attacked the metal, and the coordinated olefin then inserted into the methylpalladium σ-bond (equation 13). This reac-tion also displayed the regiospecificity (alkylation at the least hindered posi-tion) characteristic of insertion of a coordinated olefin into a metal-carbon

$$(13)$$

σ-bond. This is in marked contrast to the regioselectivity displayed by external nucleophilic attack of carbanions on complexed olefins (see below).

In contrast to the highly reactive methyllithium, the stabilized carbanion of diethyl malonate reacted with styrene in the presence of palladium (II) acetylacetonate giving alkylation product in very low yield (18%). In addition, the regiochemistry was opposite that observed with methyllithium, the product

being diethyl α-methylbenzylidenemalonate. In this case, the regiochemistry is
that characteristic of external nucleophilic attack on a metal-complexed olefin
(equation 14).[11]

$$(14)$$

A much more efficient and general alkylation of olefins by carbanions using
palladium (II) chloride-acetonitrile complexes to activate the olefin has recently
been developed.[12] The key to the success of this process derives from a related
study of the palladium assisted amination of olefins, wherein it was observed
that two equivalents of amine in excess of the one acting as the nucleophile
was required for reasonable yields. This suggested that either the step involving nucleophilic attack on the olefin required a palladium-olefin complex containing two amine ligands, or that two equivalents of amine were required to
produce a sufficient equilibrium concentration of the reactive species. Whatever
the case, reaction of palladium-olefin complexes with carbanions in the absence
of added triethylamine resulted in no alkylation but rather reduction of the
palladium complex to palladium (0). In contrast, in the presence of two equivalents of triethylamine, olefin-palladium (II) complexes reacted with stabilized
carbanions in good to excellent yields (equation 15).[12] Triethylamine was the
most efficient for promoting this reaction (yields of 90%).
    Trimethylamine (77%) and N,N-dimethylamine (65%) were somewhat less effective,
and pyridine (28%), diethylamine (5%) and strongly coordinating ligands such as
triphenylphosphine (0%) and tetramethylethylene-diamine (0%) were ineffective.
The reaction is quite sensitive to the nature of the olefin. Terminal monoolefins reacted in high (almost quantitative) yields, with alkylation at the
most substituted carbon of the olefin predominating. The electron rich olefin,
N-vinylacetamide also was alkylated in high yield, but the electron poor olefin

(15)

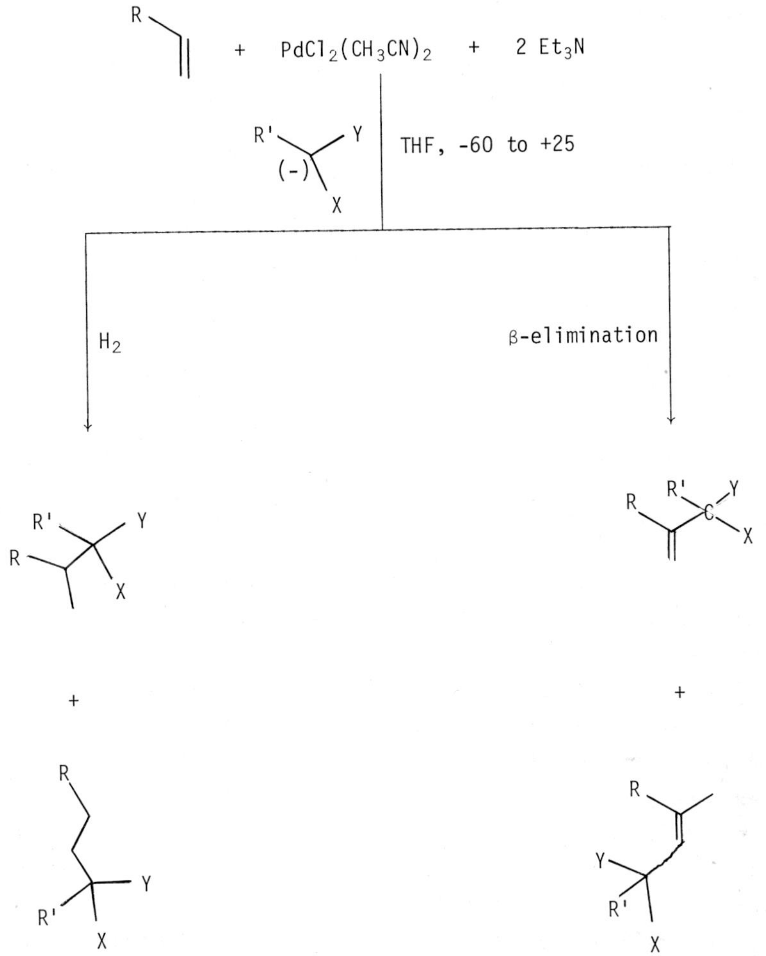

R = H, Me, Et, n-Bu, NHAc
R' = H, Me, n-hexyl

X = CO_2Et, CO_2Me, CO_2-t-Bu, COMe
Y = CO_2Et, CO_2Me, COMe, Ph

methyl acrylate was not alkylated under these conditions. Styrene reacted in only modest (20-50%) yield, undergoing alkylation exclusively at the benzylic carbon. Internal olefins such as cis and trans-2-butene were alkylated in only 30-40% yield: cyclohexene and isobutene were unreactive. An intramolecular alkylation (equation 16) also proceeded in modest yield. This reaction has been used to prepare α-amino acid derivatives using acetamidomalonate anion as the nucleophile (equation 17).[13]

$$\text{(olefin with CO}_2\text{Me)} \quad \xrightarrow[\text{THF}]{\text{PdCl}_2(\text{CH}_3\text{CN})_2,\ 2\ \text{Et}_3\text{N}} \quad \xrightarrow{\text{H}_2} \quad \text{(cyclopentane with CO}_2\text{Me, CO}_2\text{Me)} \qquad (16)$$

$$\text{olefin} \quad + \quad \text{PdCl}_2(\text{MeCN})_2 \quad + \quad \text{Et}_3\text{N} \quad + \quad (-)\underset{\text{AcHN}}{\text{C}}\overset{\text{CO}_2\text{Et}}{\underset{\text{CO}_2\text{Et}}{}} \quad \rightarrow \quad \left[\underset{\text{Pd}}{\overset{\text{AcNH}}{\underset{}{\text{C}}}}\overset{\text{CO}_2\text{Et}}{\underset{\text{CO}_2\text{Et}}{-\text{CO}_2\text{Et}}} \right]$$

$$\xrightarrow{\text{H}_2} \quad \underset{\text{NHAc}}{\text{C}}\overset{\text{CO}_2\text{Et}}{\underset{\text{CO}_2\text{Et}}{}} \qquad (17)$$

The above alkylation of simple olefins is restricted to stabilized carbanions, having pK_a's ranging from 10-17. With more basic carbanions, reduction of the olefin complexes to produce metallic palladium is always observed. However, addition of HMPA [$(\text{Me}_2\text{N})_3\text{PO}$, 10-20 equiv./equiv. of Pd (II)] permitted this olefin alkylation reaction to occur with considerably less stabilized carbanions. Thus, ketone and ester enolates, oxazoline anions (carboxylic acid anion equivalents), protected cyanohydrin anions (acyl anion equivalents) and even benzylmagnesium chloride alkylated olefins in fair to excellent yields. With nonstabilized carbanions, propene reacted almost exclusively at the 2-position, 1-hexene at the 1-position, and styrene at both positions, the product distribution depending on the specific carbanion.

These olefin-alkylation reactions proceed via σ-alkylpalladium (II) complexes which, in principle, permits further elaboration of the olefin by insertion reactions of small molecules into the palladium-carbon bond. In this fashion, a palladium-assisted carboacylation of olefins was possible (equation 18).[14] The first step of the reaction was the palladium assisted alkylation of olefins discussed above, and it had all of the scope and the limitations of those reactions. Although the intermediate σ-alkylpalladium (II) complexes lacked sufficient

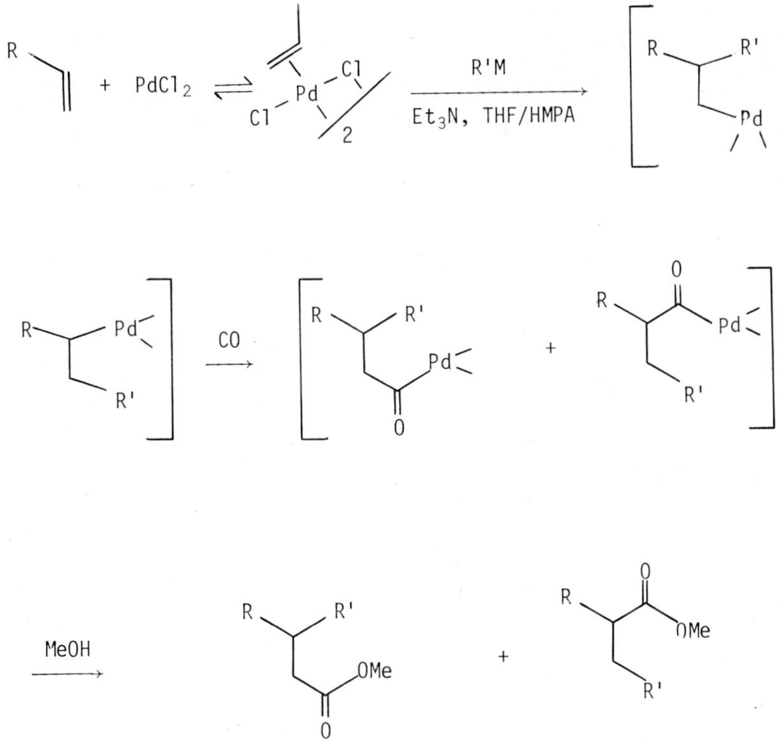

R'M

Et$_3$N, THF/HMPA

CO

(18)

MeOH

stability to permit their isolation, they did readily insert carbon monoxide, giving acylpalladium (II) complexes. These too were unstable and readily cleaved when treated with methanol to produce esters. The overall process corresponds to the alkylation of one carbon of an olefin and the carboxylation of the other. This carboacylation was quite efficient in cases in which the olefin underwent alkylation predominantly at the secondary carbon to produce a primary σ-alkyl-palladium complex. Whenever secondary σ-alkylpalladium complexes were the major products, extensive migration of palladium (by Pd-H elimination - readdition processes) occurred prior to carbon monoxide insertion, and mixtures of positional esters isomers were obtained (equation 19).

14

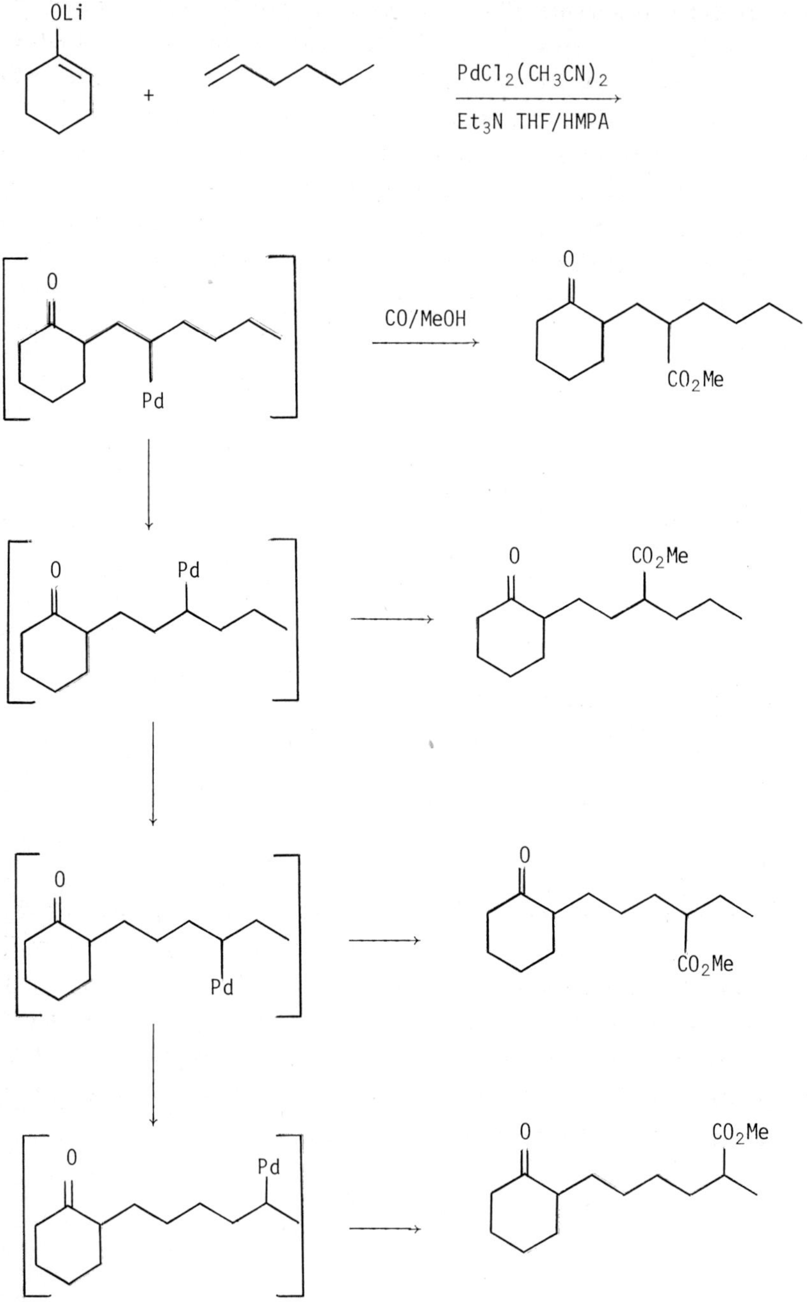

OLi

+

$PdCl_2(CH_3CN)_2$
Et$_3$N THF/HMPA

CO/MeOH

CO_2Me

CO_2Me

Pd

CO_2Me

Pd

CO_2Me

Pd

CO_2Me

O

O

O

O

O

O

O

Pd

An attractive feature of the palladium-assisted alkylation reactions discussed above is that the intermediate σ-alkylmetal complexes are unstable, and the organic product is easily freed from the metal once the reaction is complete. A number of iron complexes activate olefins toward alkylation reactions with carbanions. In these processes, the resulting σ-alkyliron complexes are rather stable, and the metal must be removed by an additional chemical step. However, these iron promoted alkylation reactions are still quite useful synthetically. Olefin complexes of iron tetracarbonyl reacted with stabilized carbanions to result in olefin alkylation after oxidative removal of the iron carbonyl residue from the product (equation 20).[15] Because simple olefin complexes of iron tetracarbonyl

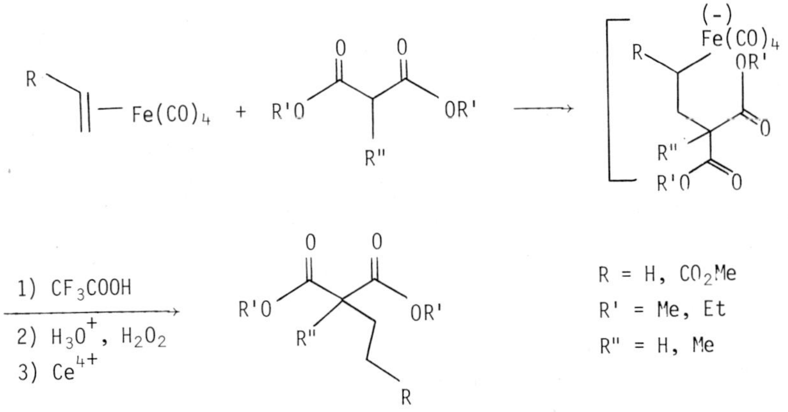

$$(20)$$

R = H, CO_2Me

R' = Me, Et

R" = H, Me

1) CF_3COOH

2) H_3O^+, H_2O_2

3) Ce^{4+}

are rather unstable, this process is of limited use for alkylation of simple monoolefins. However, complexes of α,β unsaturated carbonyl systems are considerably more stable, and hence more easily alkylated by this process. For example, use of α-halo-α,β- unsaturated esters in this system led to dialkylation of the olefinic substrate (equation 21).[16] Difunctionalization also resulted from treating the σ-alkyliron complex resulting from alkylation of complexed acrylates with organic halides (equation 22).[17]

(21)

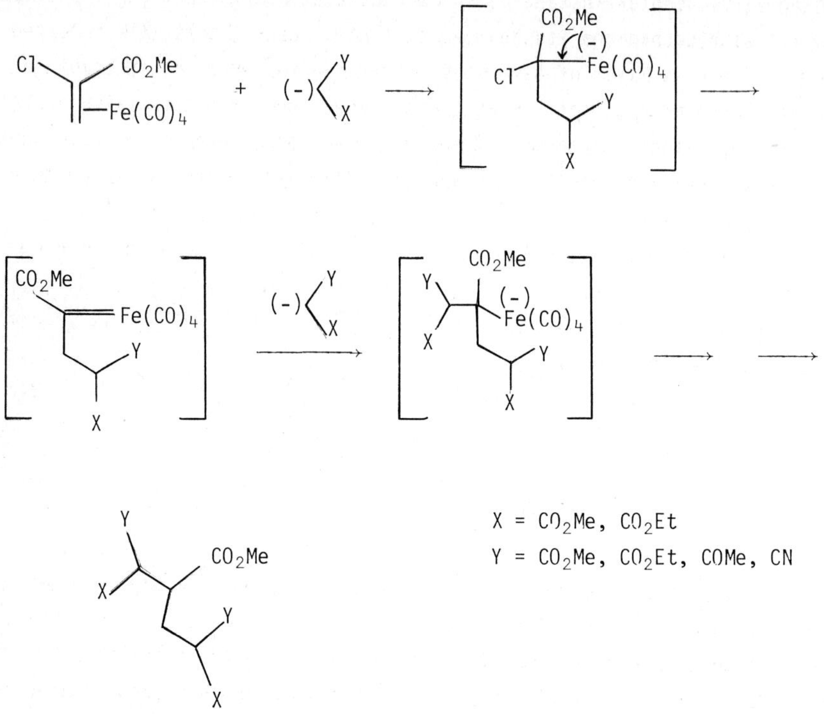

X = CO_2Me, CO_2Et
Y = CO_2Me, CO_2Et, $COMe$, CN

(22)

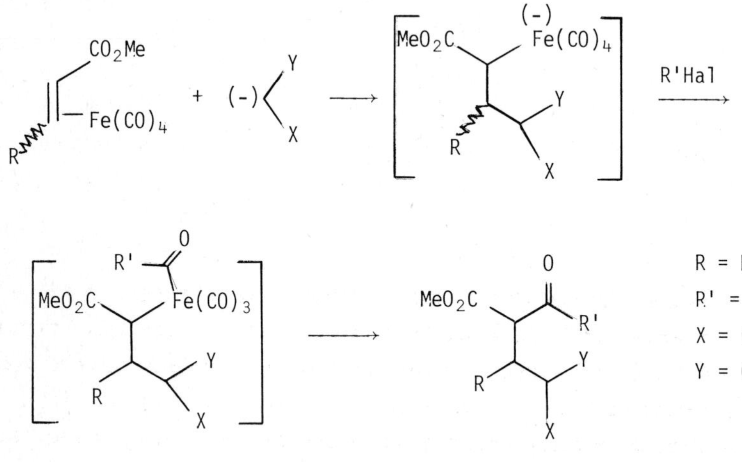

R = H, Me
R' = Me, Et, n-Pr
X = CO_2Et, CO_2Me
Y = CO_2Et, $COMe$, CN

40-60%

Another iron complex that has been extensively used to activate olefins toward alkylation with carbanions is the η^5-$C_5H_5Fe(CO)_2^+$ group, hereafter referred to as the F_p group. The F_p complex of methyl vinyl ketone reacted with the lithium enolate of cyclohexanone at $-78°$ to give an adduct, which upon treatment with basic alumina in refluxing methylene chloride gave the corresponding Robinson annellation product (equation 23).[18] This same methyl vinyl ketone - F_p^+ complex also

(23)

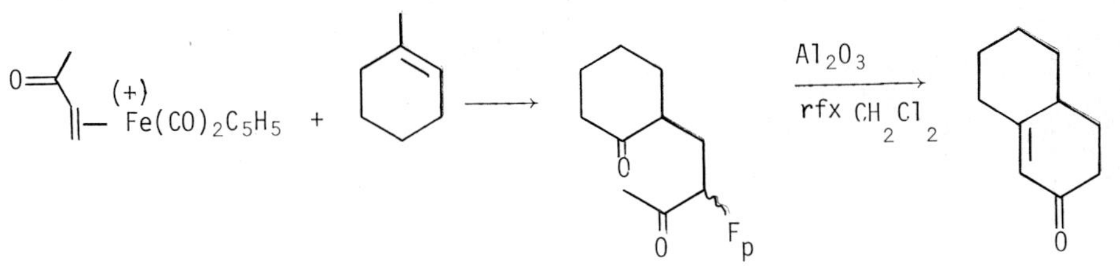

reacted rapidly at $0°$ with both cyclohexanone and cyclopentanone enamines to form the corresponding Michael adducts. In a similar fashion, silyl enol ethers of ketone enolates reacted regiospecifically under very mild conditions.

The alkylation of F_p^+ olefin complexes with carbanions has been studied in great detail.[19] The general process is presented in equation 24. The carbanions

(24)

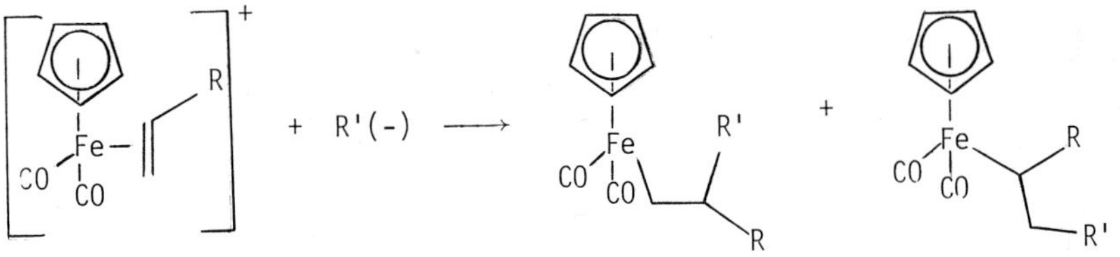

studied in this process were malonates, acetoacetates, cyanoacetates, nitromethane, and the enamines of isobutyraldehyde, cyclopentanone, and cyclohexanone. The olefins examined included ethene, propene, butadiene, cyclopentene, cyclohexene, and allene. The regiospecificity of the reaction depended on both the olefin and the anion. With propene, regioselectivity was relatively low, but styrene reacted primarily at the benzylic carbon. The butadiene complex underwent a 1,2-alkylation with malonate anion to produce a σ-alkyl iron complex which spontaneously inserted carbon monoxide (equation 25).

(25)

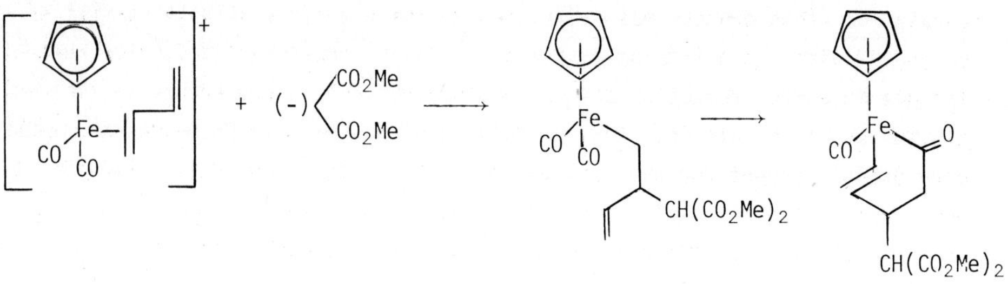

Allene complexes alkylated cleanly at the terminal carbon (equation 26).

(26)

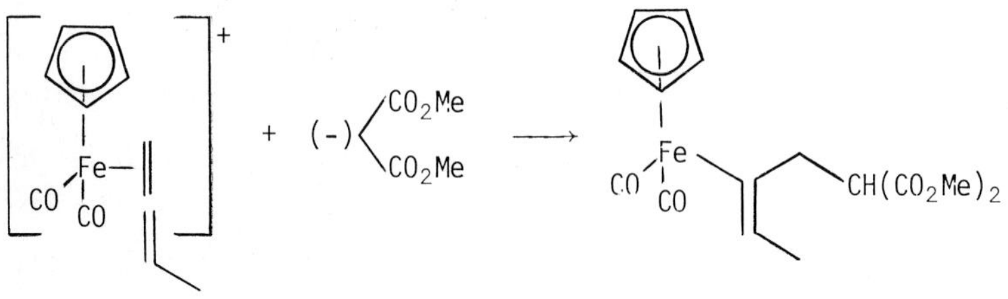

Oxidative removal of iron from the cyclohexanone adduct of styrene gave the corresponding methyl ester (<u>via</u> an oxidatively driven carbon monoxide insertion reaction) in high yield (equation 27).

(27)

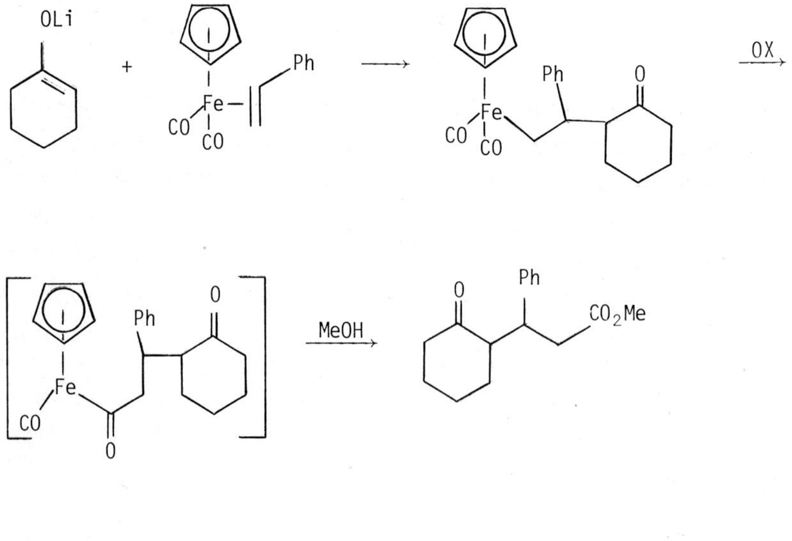

As is often the case in metal assisted alkylation reactions of olefin by carbanions, this process was restricted to the use of relatively stabilized carbanions. With organolithium reagents or most Grignard reagents, displacement of the olefin and/or reduction of the complex occurred in preference to the desired alkylation of the olefin. In contrast, lithium dimethylcuprate was a somewhat more useful reagent for the methylation of F_p-olefin$^+$ complexes. Finally, the phosphonium ylid of ethyl bromoacetate reacted with the F_p-ethene$^+$ complex in good yield. The resulting adduct reacted with base and benzaldehyde in a typical Wittig reaction (equation 28).

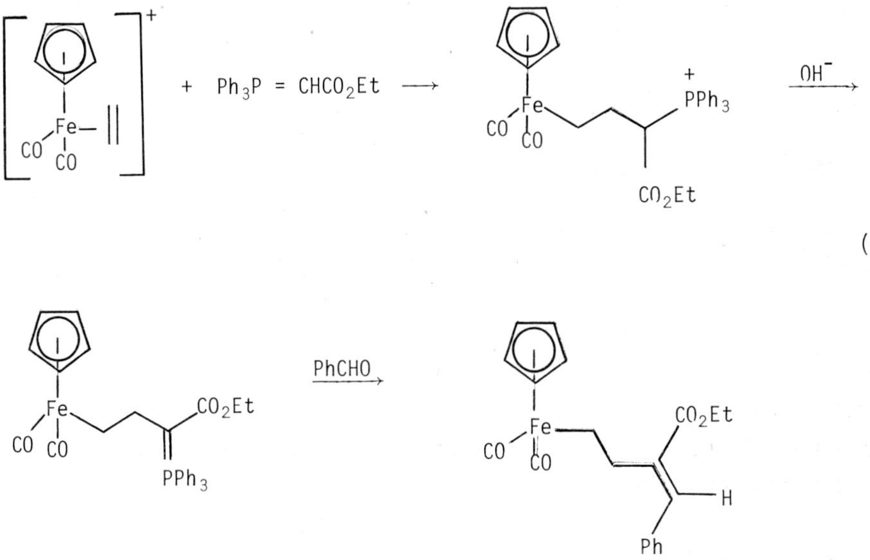

$$+ \quad Ph_3P = CHCO_2Et \longrightarrow$$

$$\xrightarrow{OH^-}$$

(28)

$$\xrightarrow{PhCHO}$$

The F_p^+ complexes of enol ethers also react with carbanions. This chemistry has been used to introduce vinyl groups into ketone enolates. The F_p^+-enol ether complexes may therefore be viewed as vinyl cation equivalents (equation 29).[20]

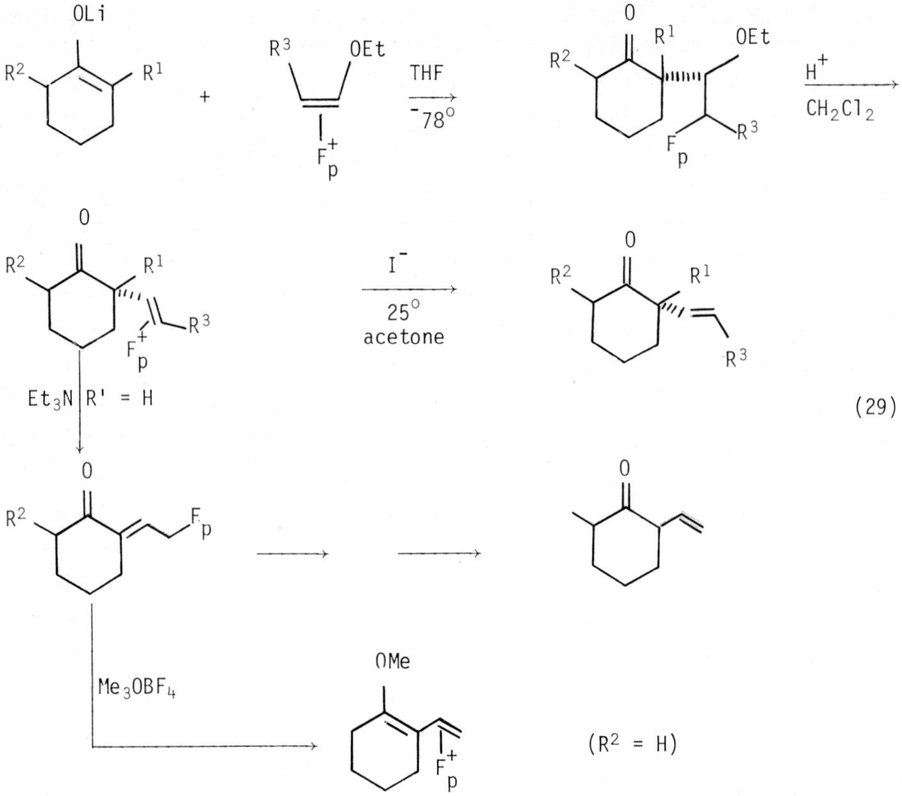

A cationic olefin complex of palladium (II) similar to the iron complexes discussed above has been shown to react with malonate anion in a clean _trans_ addition (equation 30).[21, 22] This is likely to be the stereochemical course of all the alkylation reactions discussed above.

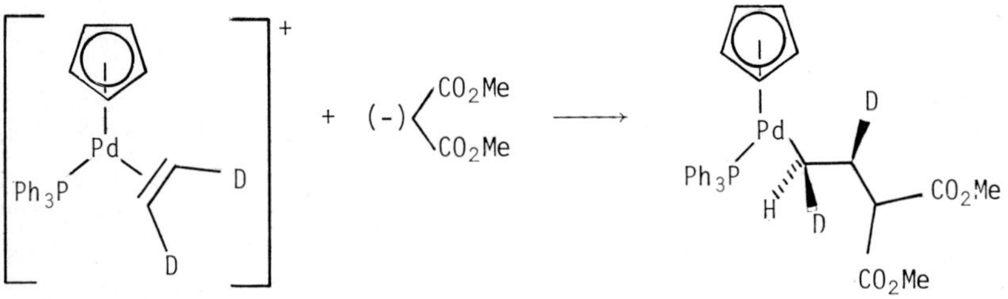

III. ALKYLATION OF METAL-STABILIZED CATIONS

A. Cationic Dienyl Complexes

Iron carbonyls react with conjugated dienes to produce remarkably stable dieneiron carbonyl complexes. When complexed to iron in this fashion, the diene becomes extremely inert chemically. For example the diene in η^4-butadieneiron tricarbonyl is inert to hydrogenation, and does not undergo Diels-Alder reactions with maleic anhydride. In fact, this diene complex is so stable that it can be made to undergo Friedel-Crafts acylation without suffering decomplexation from the iron. However, diene complexes of this type which have allylic hydrogens undergo a hydride abstraction process to produce cationic dienyl complexes which are quite reactive toward nucleophiles, including carbanions.[23]

Cyclohexadienyl complexes have been most extensively studied. They are usually prepared by the reaction of neutral cyclohexadiene complexes of iron with triphenylmethyl cation. Cyanide ion reacted with the methoxycyclohexadienyl complex of iron exclusively from the face opposite the iron to generate a stable cyclohexadiene complex. Treatment with palladium on carbon produced p-cyanoanisole (equation 31).[24] The 4-methyl-1-methoxycyclohexadienyl complex reacted with cyanide at the 4-position exclusively to produce the exo-cyanide adduct (equation 32).[25] This same process was used to introduce an angular nitrile group into a decalin ring system (equation 32a).

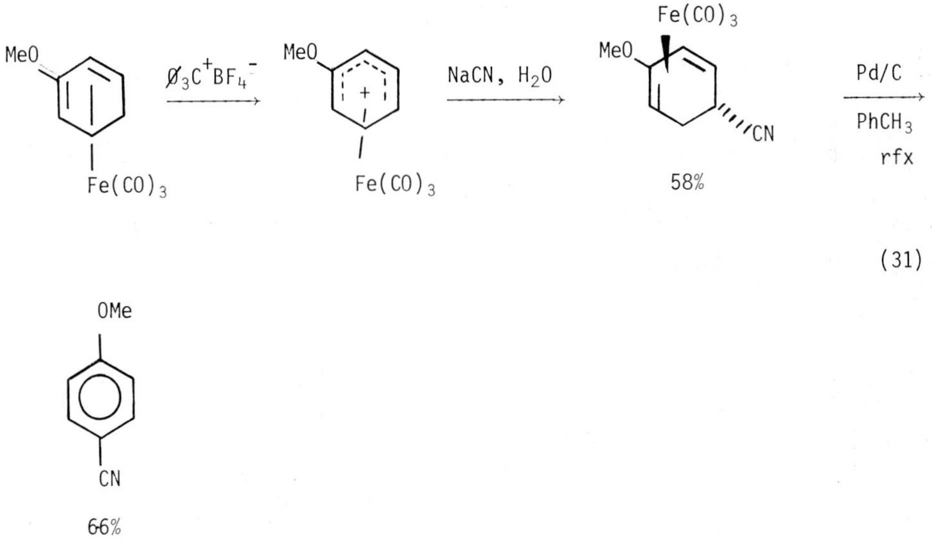

58%

$$\emptyset_3C^+BF_4^- \qquad NaCN, H_2O \qquad Pd/C \quad PhCH_3 \quad rfx$$

(31)

66%

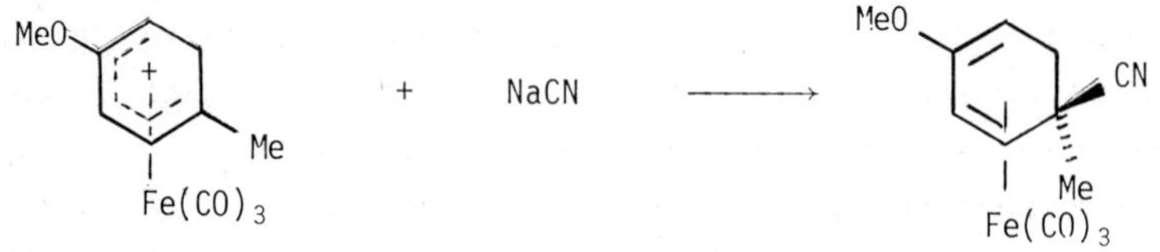

(32)

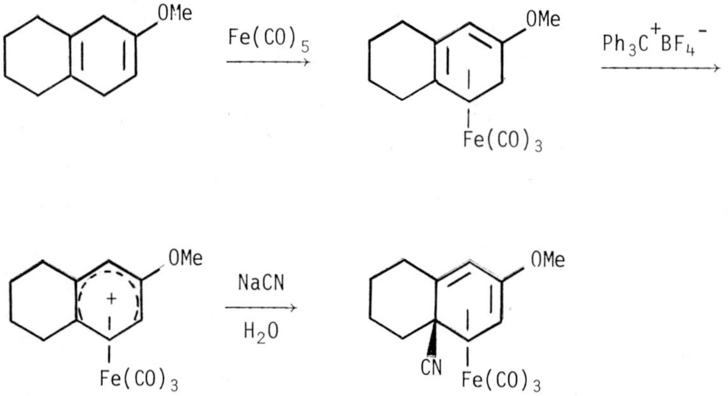

(32a)

Although Grignard reagents reductively dimerized these cationic dienyl complexes, less reducing organometallics such as dialkylcadmium and dialkylzinc
reagents alkylated them cleanly (equation 33).[26] Similarly, both dimethyl

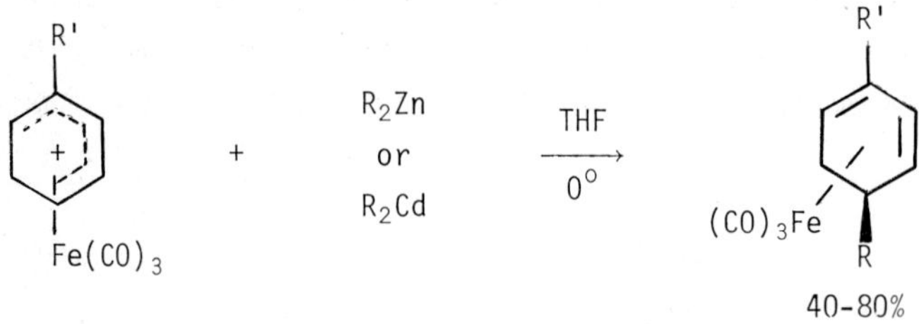

(33)

40-80%

cuprates[27] and mixed cuprates[28] alkylated cationic cyclohexadienyliron complexes.
In all cases, the alkylation occurred from the face opposite the metal to give
a stable diene complex. At the time when much of this work was carried out,

there was no general way to free the diene from the metal without causing at least some degradation. Fortunately, it has since been discovered that amine oxides are selective oxidants for the iron carbonyl system, and readily release coordinated organic ligands from the $Fe(CO)_3$ fragment without attacking and degrading the desired product. Hence, these alkylations of cationic dienyliron complexes should soon find increased use in organic synthesis.

Perhaps the most synthetically useful reactions of cationic dienyl complexes are those with stabilized carbanions. An early example was the synthesis of a dihydrobenzofuran derivative via the reaction of a cyclohexane-1,3-dione with a methoxy substituted dienyl complex (equation 34).[29] Again, in this case, the product was not freed from the metal, but with currently available procedures it surely could be.

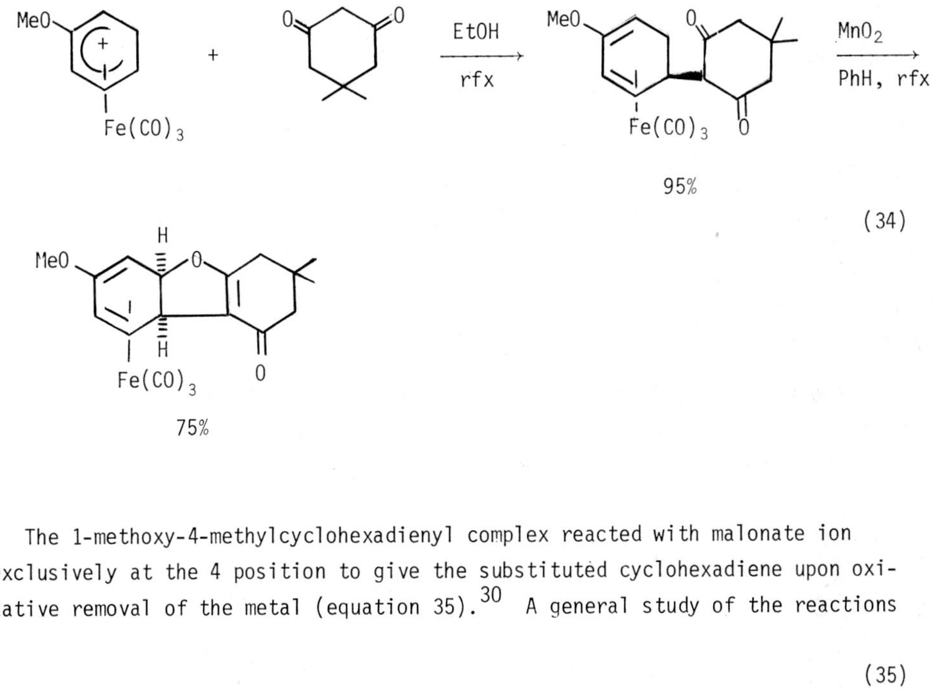

95%

(34)

75%

The 1-methoxy-4-methylcyclohexadienyl complex reacted with malonate ion exclusively at the 4 position to give the substituted cyclohexadiene upon oxidative removal of the metal (equation 35).[30] A general study of the reactions

(35)

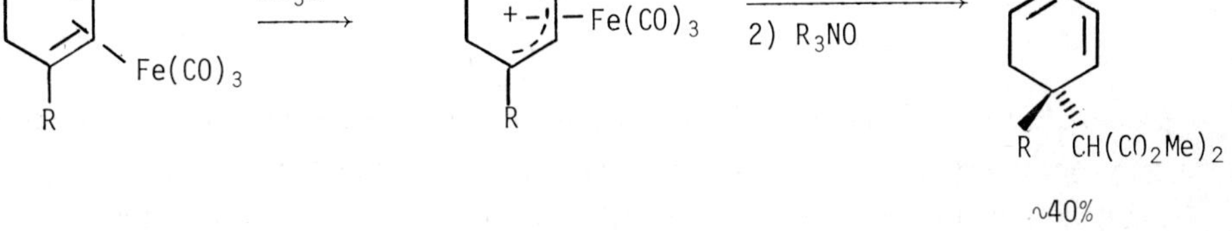

~40%

of 4-substituted-1-methoxycyclohexadienyl complexes of iron with stabilized carbanions showed that the regioselectivity depended on the steric bulk of both the
carbanion and the 4-substituent. Steric hindrance in either the nucleophile or
the 4-substituent of the diene increased the amount of alkylation at the 2-posi-
tion (equation 36).[31] This reaction has been effectively used for the introduction of stabilized anions into cyclohexyl ring systems thereby generating intermedia

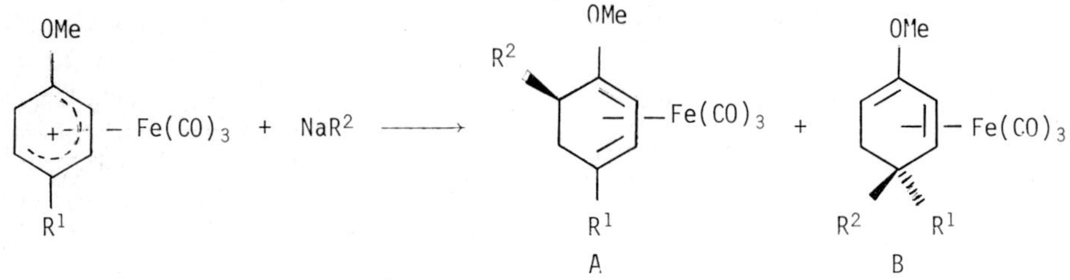

R^1	R^2	A:B
CH_2CO_2Me	$CH(CO_2Me)_2$	37:63
CH_2CH_2OAc	$CH(CN)_2$	64:36
CH_2CH_2OMe	$CH(CO_2Me)_2$	50:50
	$CH(CN)_2$	25:75
	$CH(COOMe)_2$	28:72
	$CH(CN)_2$	10:90

(36)

potentially useful in natural product synthesis. (See equations 37,[32] 38,[33]
and 39.)[34] (In equation 39, the quaternary carbon formed corresponds to C-20 in
the aspidosperma alkaloids.)

(37)

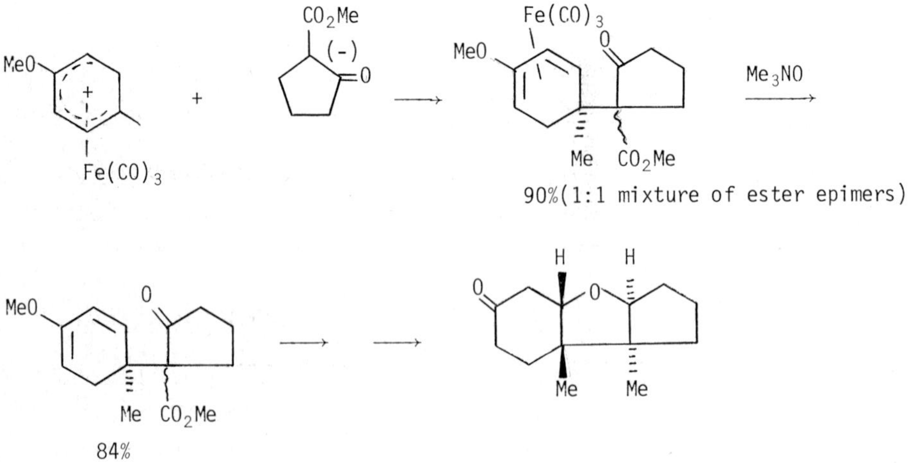

90%(1:1 mixture of ester epimers)

84%

(38)

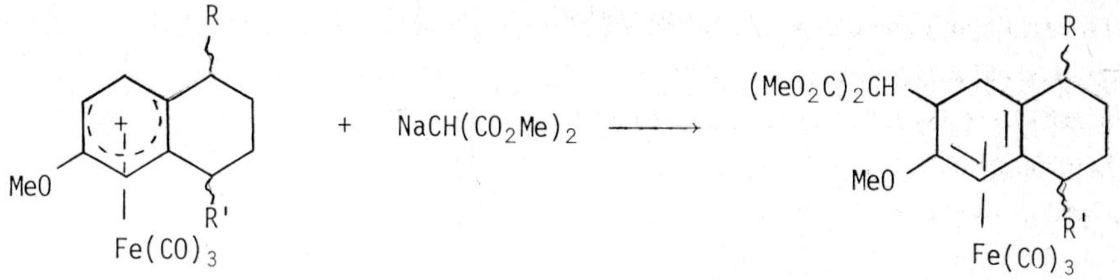

(39)

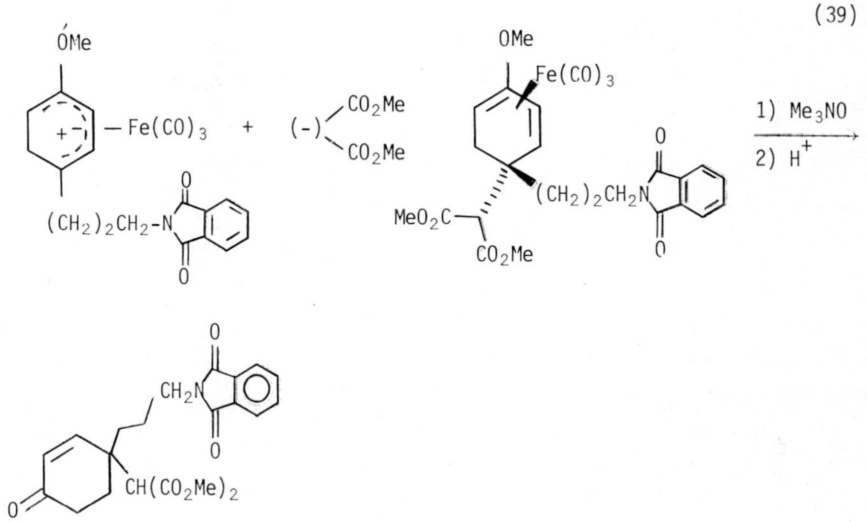

Intramolecular versions of this alkylation have also been successfully developed and have been used to generate both spiro-fused (equation 40)[33] and fused bicyclic ring systems (equation 41).[34] Only the spiro 6,6 ring system is accessible by this reaction. With side chains one and two carbons shorter, no cyclization was noted. This observation was rationalized by invoking "Baldwin's Rules" for cyclization which state that "6-endo trig" closures are favored.

(40)

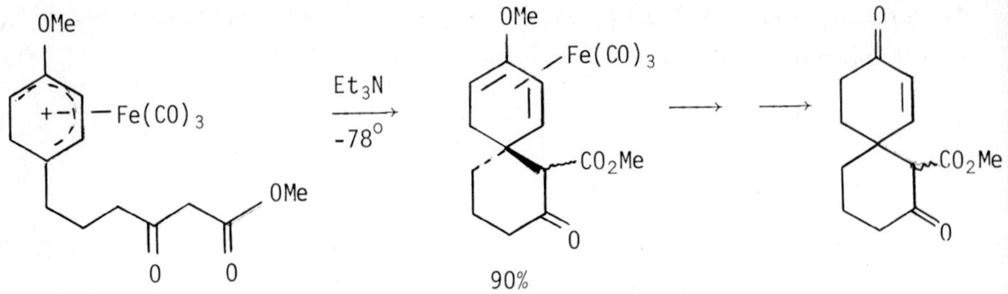
OMe
Fe(CO)3
OMe
O
O
Et3N
-78°
OMe
Fe(CO)3
CO2Me
O
90%
O
CO2Me
O

(41)

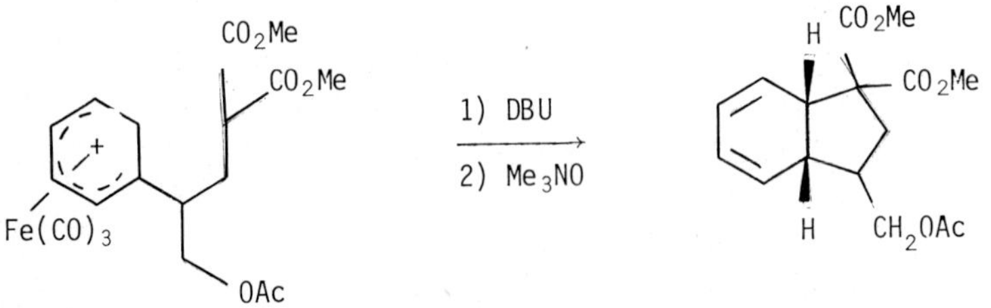
CO2Me
CO2Me
Fe(CO)3
OAc
1) DBU
2) Me3NO
H
CO2Me
CO2Me
H
CH2OAc

Trimethylsilyl enol ethers are sufficiently nucleophilic to alkylate cationic cyclohexadienyliron complexes. With trimethylsilyl enol ethers of simple aliphatic and alicyclic ketones, specifically substituted cyclohexadiene complexes or aromatic compounds were obtained (equation 42).[35, 36] Replacement of the cyclohexadienyl complexes by the corresponding cationic tertralin iron complexes leads to acylated naphthalene derivatives. By using bis-trimethylsilyl

(42)

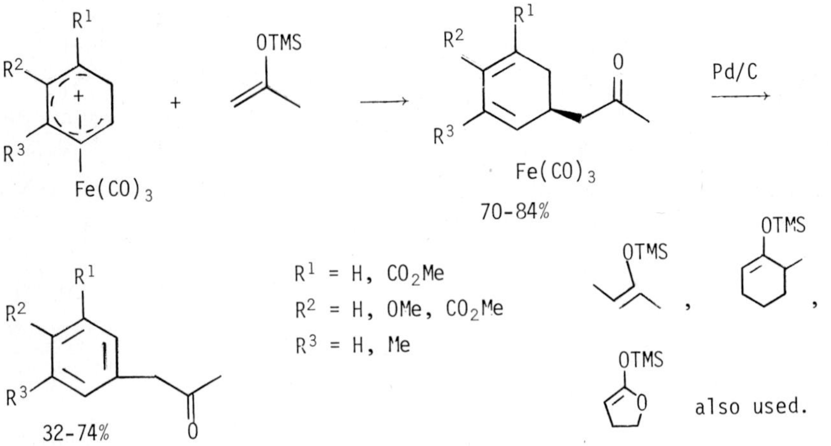
R1
R2
R3
Fe(CO)3
+
OTMS
R1
R2
R3
O
Fe(CO)3
70-84%
Pd/C
R1
R2
R3
O
32-74%
R1 = H, CO2Me
R2 = H, OMe, CO2Me
R3 = H, Me
OTMS
OTMS
OTMS
O
also used.

derivative of the enol of 2-hydroxycyclopentanone as the nucleophile in this type
of reaction, 2-arylcyclopentenones having specific aryl substitution patterns
were available in modest yield (equation 43).[37, 38]

(43)

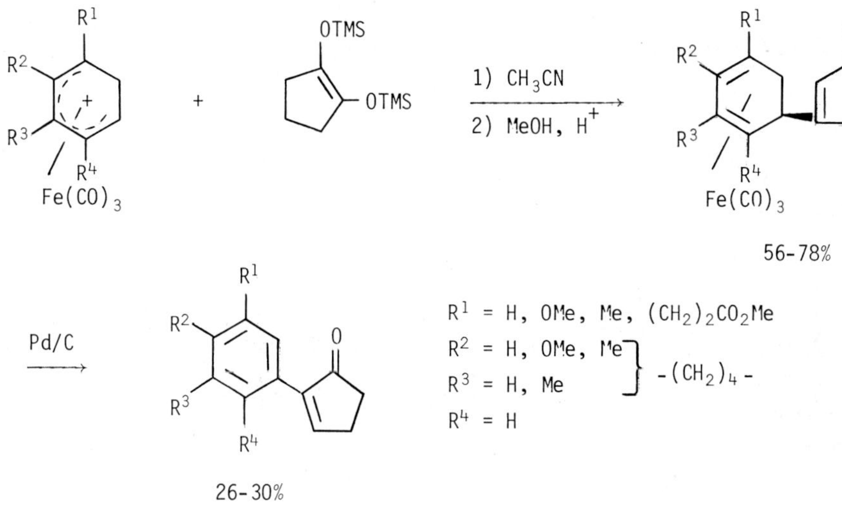

56-78%

Pd/C

R^1 = H, OMe, Me, $(CH_2)_2CO_2Me$
R^2 = H, OMe, Me
R^3 = H, Me $\left.\right\}$ -$(CH_2)_4$-
R^4 = H

26-30%

Enamines (equation 44) and ketone enolates (equation 45) are also reactive
toward cationic cyclohexadienyliron complexes and have been used to synthesize
substituted cyclohexenones and cyclohexadienes.[39]

(44)

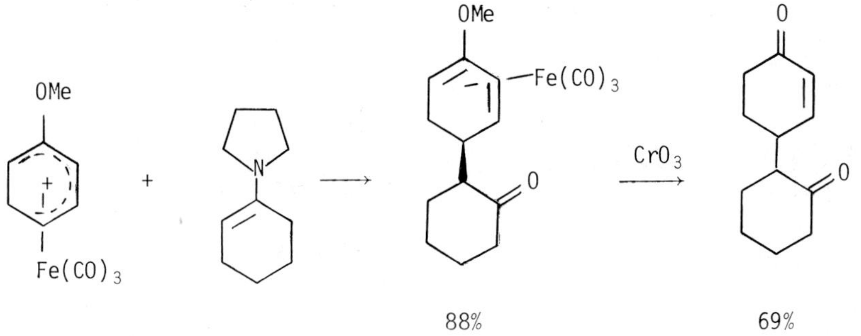

88%   69%

(45)

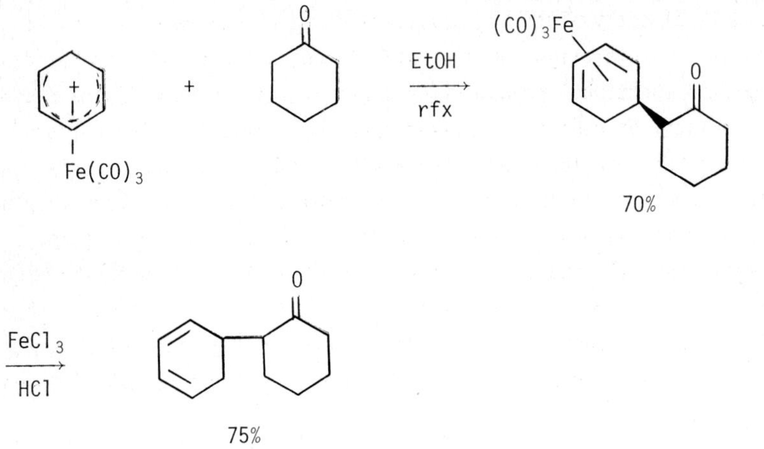

70%

75%

In summary, conjugated dienes complex strongly to iron carbonyl, and once complexed are quite inert to nucleophilic attack. However, removal of an allylic hydride generates a cationic dienyliron complex which is highly reactive towards nucleophiles. Nucleophilic attack regenerates the very stable irondiene complex (equation 46). The nucleophile always attacks from the face opposite the metal.

(46)

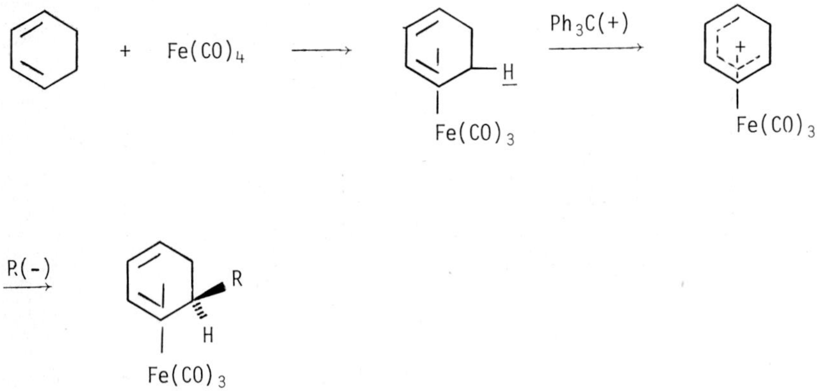

B. Metal-Stabilized Propargyl Cations

In the chemistry presented in section II, complexation to a transition metal was utilized to directly alter the chemistry of the complexed unsaturated organic group. However, complexation can also affect the chemistry of centers adjacent to the complexed functional group. An example of this is seen in the recently-developed chemistry of cobalt-stabilized propargyl cations. Alkynes react with dicobalt octacarbonyl to form very stable alkyne-cobalt complexes. When propargyl alcohols were complexed in this fashion, reaction with HBF_4 produced very stable complexed propargyl cations. These underwent alkylation with β-dicarbonyl compounds,[40] enol ethers,[41] and enol acetates (equation 47).[42] Since the cobalt could be removed by oxidation with ferric compounds, this represents a method for the alkylation of propargylic systems without the formation of allenic products which often plagues classical alkylations of propargyl halides, tosylates, or acetates.

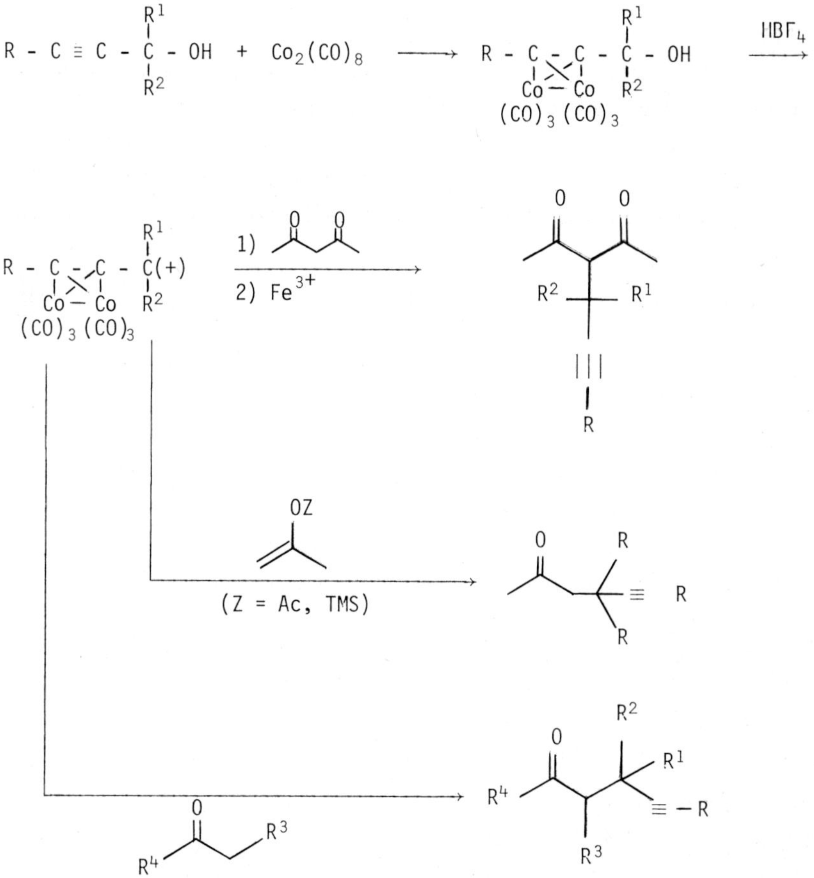

(47)

30

IV.   ALKYLATION OF π-ALLYLMETAL COMPLEXES

A.   Stoichiometric Alkylation of π-Allylpalladium Complexes

π-Allylpalladium halide complexes are air stable yellow crystalline solids which are generally prepared directly from olefins by treatment with palladium (II) chloride and sodium acetate in glacial acetic acid.[43] (Many other conditions also work; for example, the more electrophilic palladium (II) trifluoroacetate is considerably more reactive in this regard, and is useful in preparing π-allylpalladium complexes from less reactive olefins.[44]) The reaction involves ultimate abstraction of an allylic proton from a π-olefin palladium complex, a process likely to occur by insertion of the metal into a <u>syn</u> allylic C-H bond (equation 48). However, direct proton abstraction by a base cannot be ruled

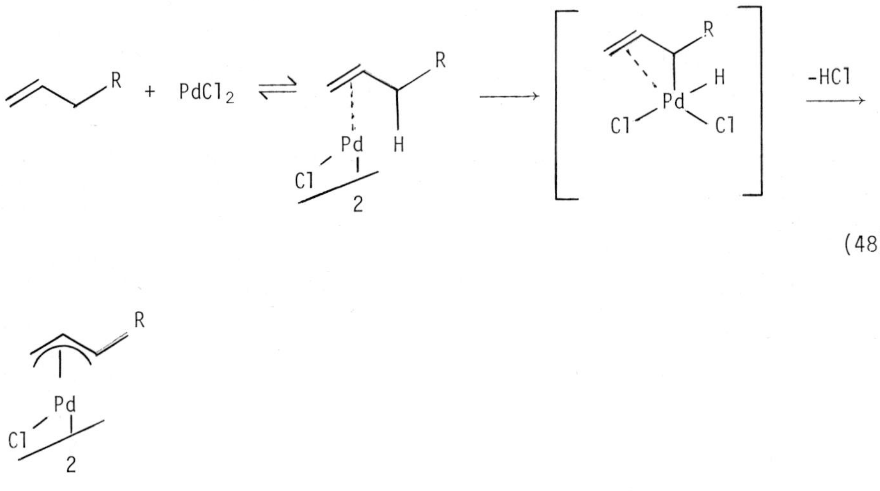

$$(48)$$

out. Nonconjugated olefins form π-allylpalladium chloride complexes more readily than conjugated enones. The latter will, however, react under more stringent conditions. With unsymmetrical internal olefins, mixtures of isomeric π-allylpalladium complexes are obtained, but with preferential loss of the hydrogen allylic to the <u>most</u> substituted end of the olefin. However, both steric and electronic factors influence the course of this reaction.

The synthetic value of these π-allylpalladium complexes lies in their facile reaction with stabilized carbanions in the presence of strongly coordinating ligands, particularly phosphines (equation 49). This reaction had been reported over fifteen years ago but was only recently developed into a useful synthetic procedure.[46, 47]

(49)

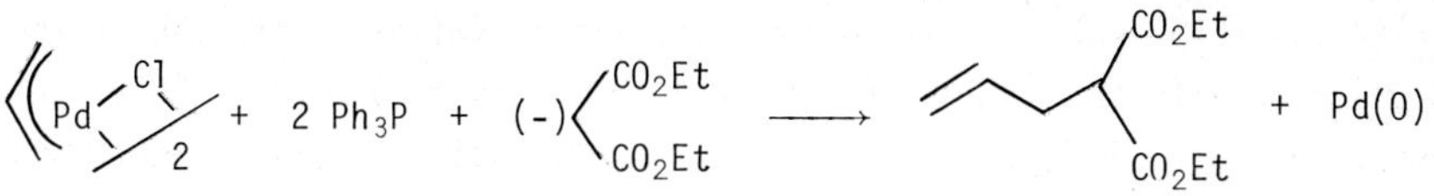

$$\left(Pd \underset{2}{\overset{Cl}{\diagdown}}\right) + 2\ Ph_3P + (-)\diagup\!\!\!\!\begin{array}{c} CO_2Et \\ CO_2Et \end{array} \longrightarrow \diagdown\!\!\!\diagup\!\!\!\overset{CO_2Et}{\underset{CO_2Et}{\diagup}} + Pd(0)$$

With unsymmetrical π-allylpalladium chloride complexes, the regioselectivity of attack depends both on the specific structure of the complex, and on the nature of the carbanion, although alkylation at the <u>less</u> substituted end of the π-allyl system usually predominates (equation 50). With cyclic π-allyl complexes, attack at the <u>exo</u> cyclic position predominates (equation 51). The carbanion was shown to attack from the face opposite the metal atom.[47]

(50)

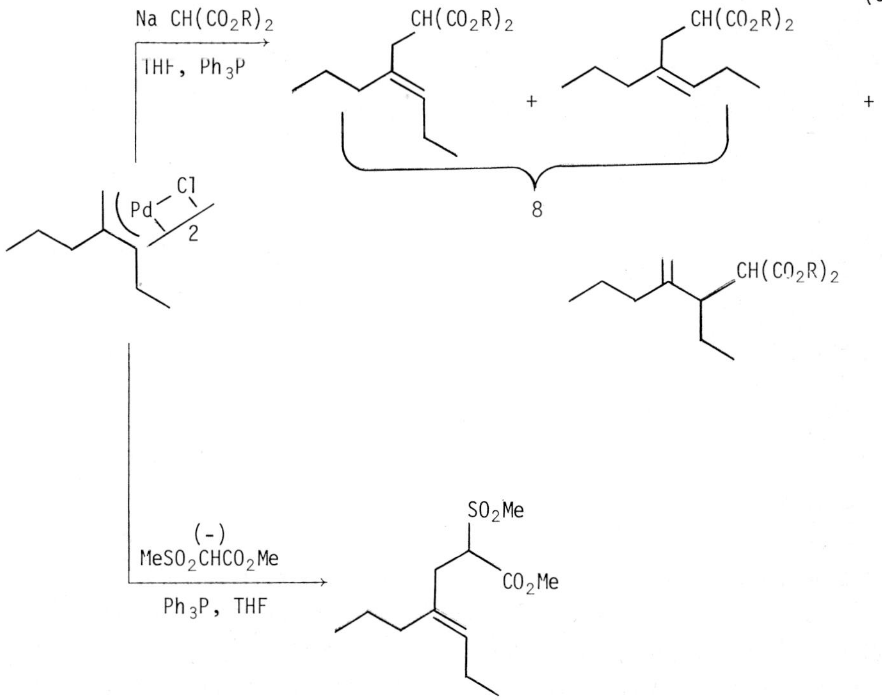

Na CH(CO_2R)_2
―――――→
THF, Ph_3P

(-)
MeSO_2CHCO_2Me
―――――→
Ph_3P, THF

CH(CO_2R)_2   +   CH(CO_2R)_2   +

8

CH(CO_2R)_2

SO_2Me
CO_2Me

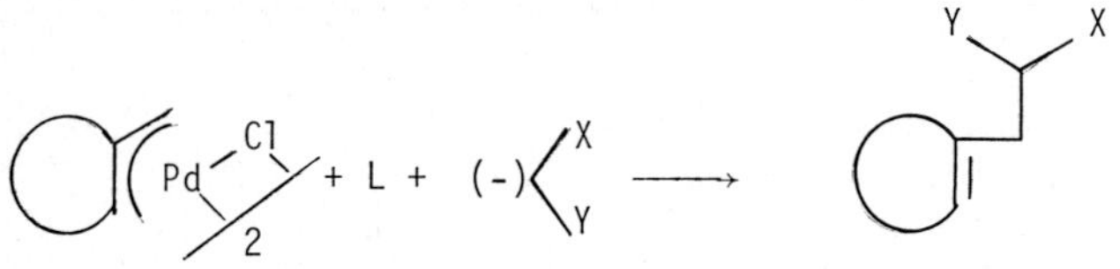

$$\text{structure} + L + (-)\overset{X}{\underset{Y}{\diagdown}} \longrightarrow \text{structure} \tag{51}$$

This chemistry has been used in the synthesis of a number of interesting organic compounds. The π-1-carbethoxyallylpalladium chloride complex reacted with stabilized carbanions to produce substituted acrylates (equation 52).[48, 49]

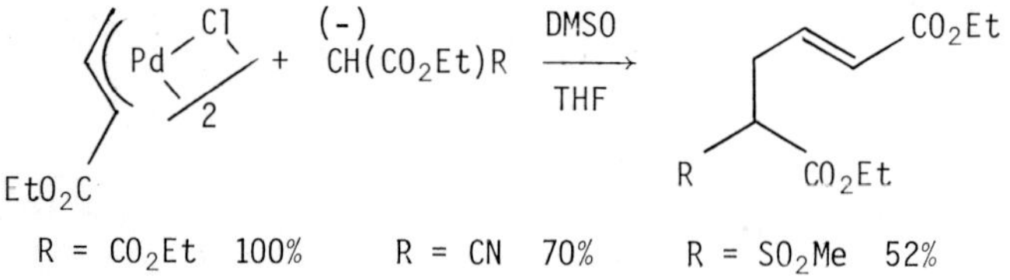

$$\left(\text{Pd}\diagup\diagdown_{Cl}\right)_2 + \overset{(-)}{CH(CO_2Et)R} \xrightarrow[\text{THF}]{\text{DMSO}} \text{product} \tag{52}$$

EtO$_2$C

R = CO$_2$Et 100%   R = CN 70%   R = SO$_2$Me 52%

Similar chemistry was used to alkylate cholestanone and testosterone at position 6 utilizing dimethyl malonate as base. The resulting product was decarbomethoxylated using LiI/DMF to afford the product shown in equation 53.[50] Steroids

$$\tag{53}$$

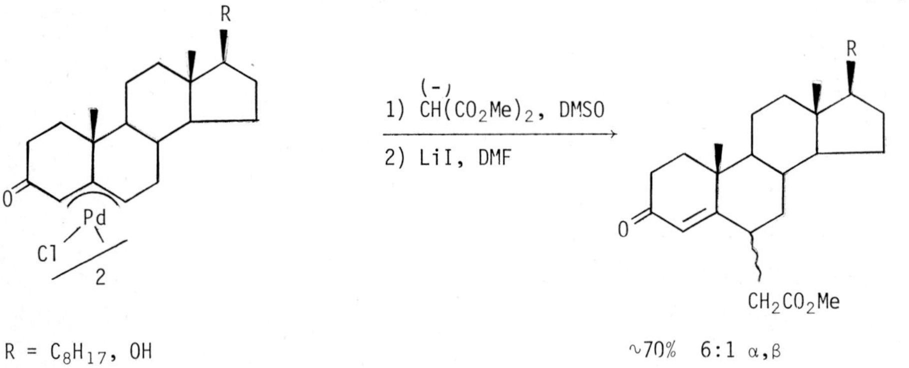

1) $\overset{(-)}{CH(CO_2Me)_2}$, DMSO

2) LiI, DMF

R = C$_8$H$_{17}$, OH

~70% 6:1 α,β

CH$_2$CO$_2$Me

possessing abnormal stereochemistry at C-20 have also been synthesized in this fashion (equation 54).[51] A new "prenylation" procedure using sulfone stabilized allylic carbanions was used to synthesize geranylgeraniol from methyl farnesoate (equation 55).[52] Vitamin A and related compounds were synthesized using analogous chemistry (equation 56).[53] Finally, π-allylpalladium chloride complexes reacted with sodium acetylacetonate in DMSO to produce diallylated products.[54]

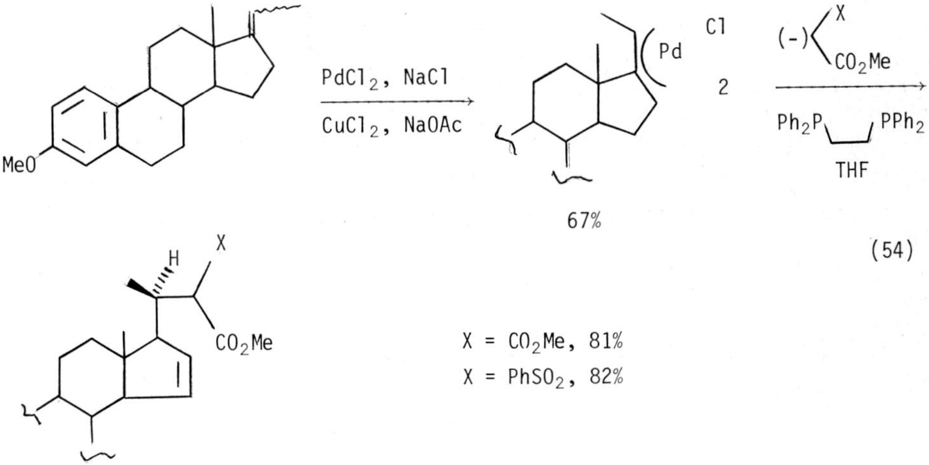
MeO
PdCl2, NaCl
CuCl2, NaOAc
Pd
Cl
2
67%
(-)
X
CO2Me
Ph2P
PPh2
THF
H
X
CO2Me
X = CO2Me, 81%
X = PhSO2, 82%

(54)

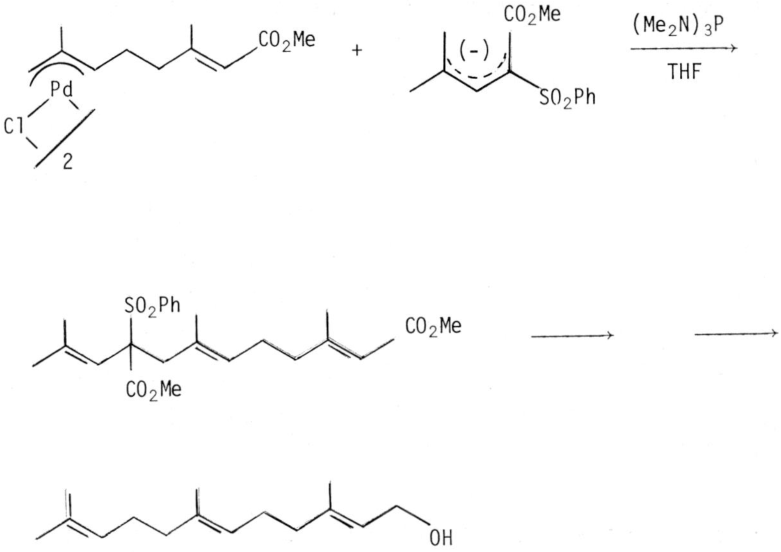
CO2Me
Pd
Cl
2
+
CO2Me
(-)
SO2Ph
(Me2N)3P
THF
SO2Ph
CO2Me
CO2Me
OH

(55)

(56)

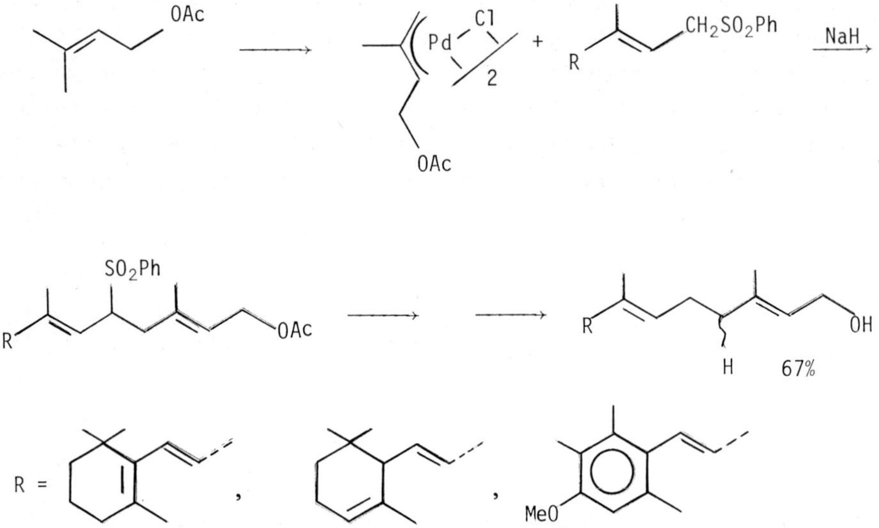

All of the above reactions involve alkylation at the 1- or 3- positions of the π-allylpalladium complex. Under different conditions (in the presence of triethylamine and HMPA), α-branched ester enolates reacted at the <u>central</u> carbon of the π-allyl complex, producing cyclopropanes (equation 56a).[54a] This reaction was restricted to α-branched enolates and to π-allylpalladium complexes lacking substituents on the central carbon. This is an unusual reaction, and has not been studied mechanistically. A palladiacyclobutane was proposed as an intermediate.

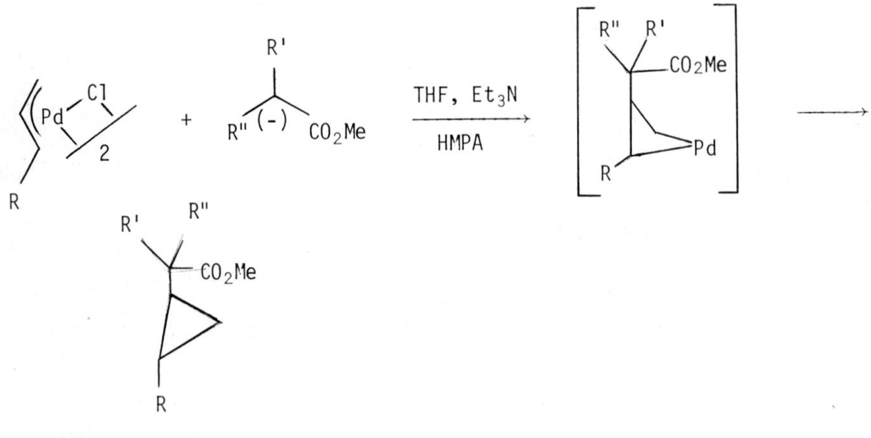

(56a)

B. Palladium Catalyzed Alkylation of Allyl Acetates

Although the chemistry presented above is quite useful, it suffers from the necessity of using stoichiometric amounts of expensive palladium salts. Although no solution to this problem has been found for the allylic alkylation of olefins, an extremely useful catalytic allylic alkylation of allylic acetates has been developed (equation 57).[55] This process relies on the facile oxidative addition of palladium (0) phosphine complexes to allylic acetates thereby yielding π-allylpalladium complexes. Alkylation of these complexes then regenerates the

(57)

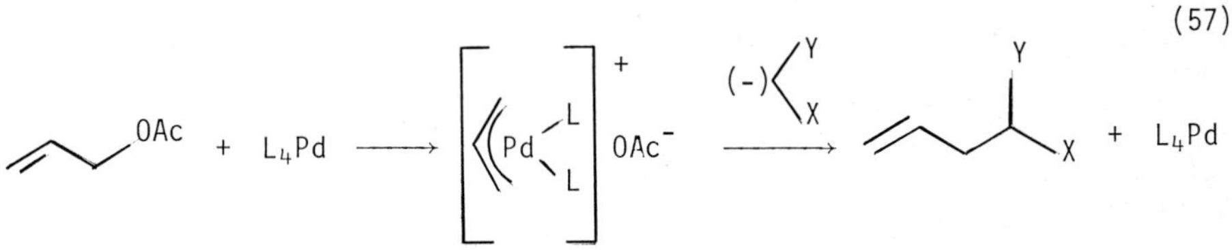

palladium (0) complex. This process was shown to proceed with net retention of configuration at the acetate-bearing carbon in a study of the catalytic alkylation at C-20 of a steroid (equation 58).[51] The catalytic alkylation gave

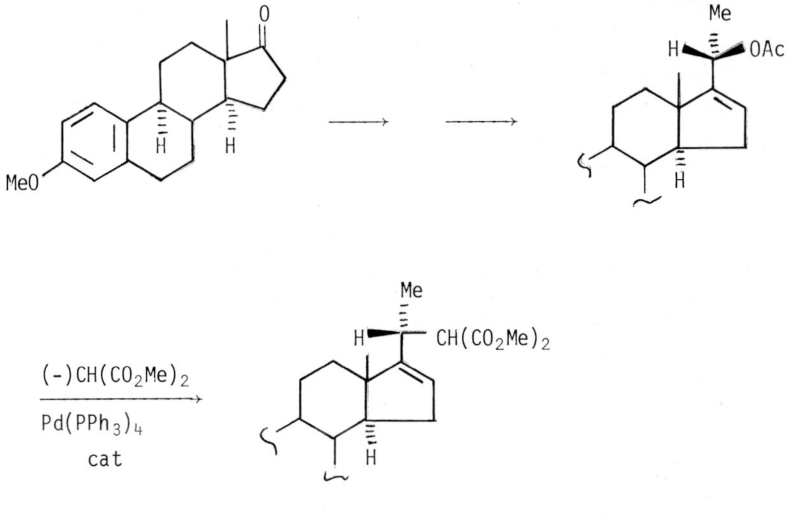

(58)

products having the naturally occurring configuration at C-20, since both π-allyl complex formation [oxidative addition to Pd(0)] and the alkylation occurred with inversion, and two inversions result in net retention. Similar chemistry was used to elaborate the side chain of the steroid ecdysone.[56]

Since the carbon-carbon bond forming step results from the reaction of a carbanion with a π-allylpalladium complex, the catalytic process shares many of the features observed in the stoichiometric reactions presented above. These include a preference for attack at the _exo_ terminus of cyclic allyl acetates (equation 59), and a dependence of regioselectivity on the nature of the carbanion as well as the substrate when unsymmetrical acyclic allyl acetates are involved.[55] This catalytic process is quite specific for allylic acetates, and

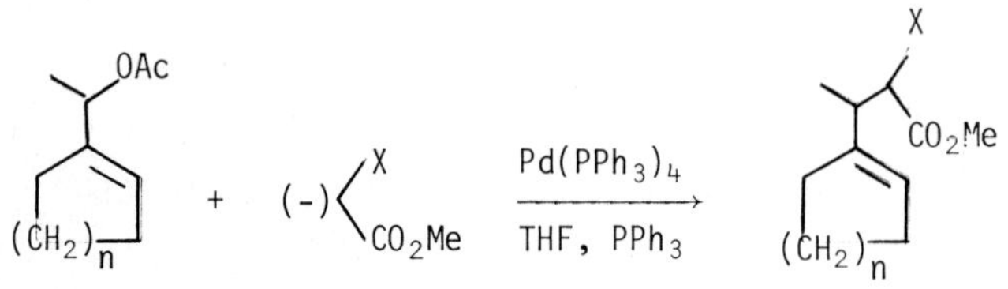

$$(59)$$

$$n = 1, 3 \qquad\qquad X = CO_2Me, SO_2Ph$$

can readily tolerate the presence of a primary bromide (equation 60).[55] Allylic acetates of enol ethers undergo clean alkylation by stabilized carbanions without loss of, or reaction at, the sensitive enol ether group (equation 61).[57] The mixed allyl acetate alcohol in equation 62 reacted cleanly at the acetate-bearing position to produce a chrysanthemic acid precursor.[58]

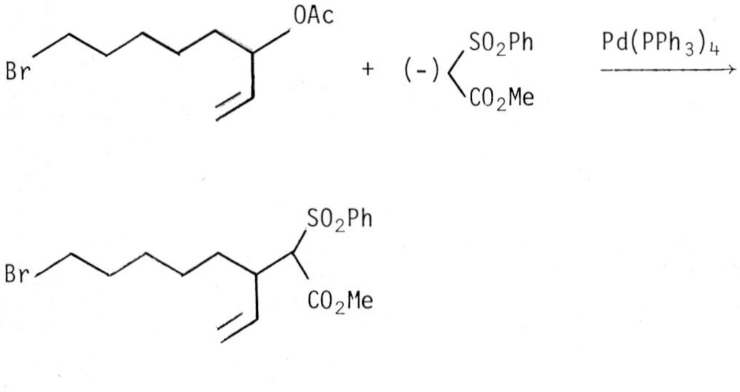

$$(60)$$

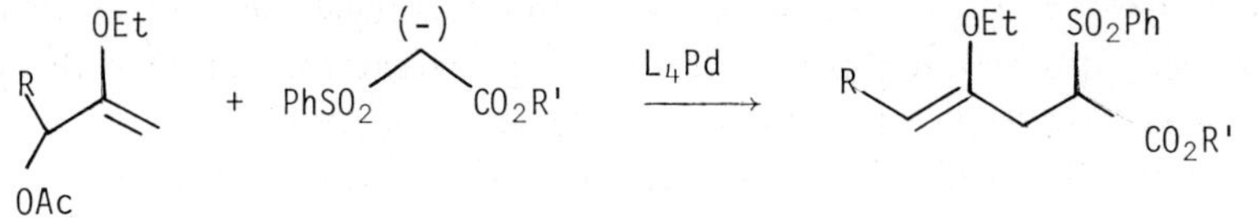

(61)

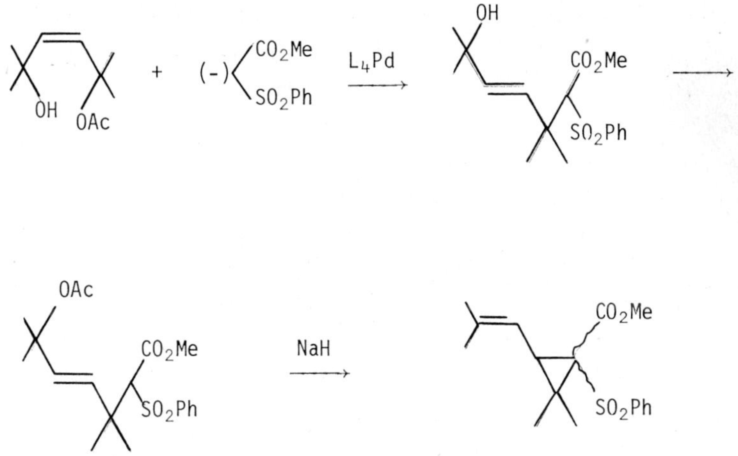

(62)

While the anions of allylic sulfones were acceptable nucleophiles in palladium catalyzed reactions of allyl acetates (equation 63),[46b, 59] allyl sulfones themselves were reactive towards stabilized carbanions in the presence
of palladium (0) catalysts (equation 64).[59]

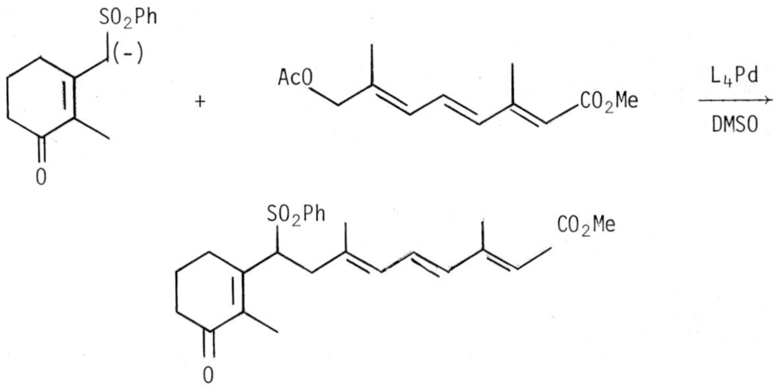

(63)

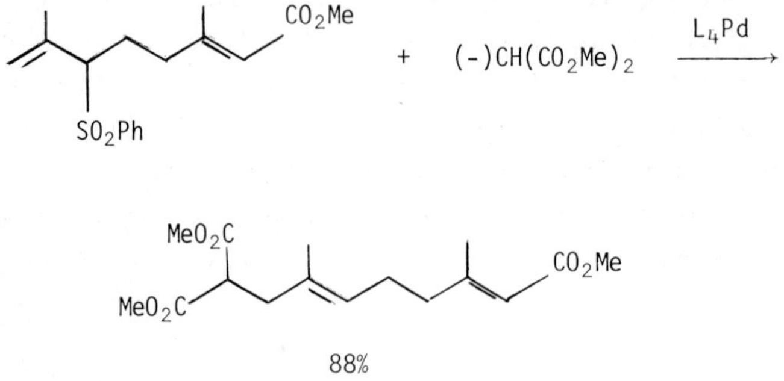

38

(64)

88%

Control of stereochemistry in acyclic systems is among the most difficult of the challenges faced in organic synthesis. The stereospecific nature of both the formation of π-allylpalladium halide complexes, and their reactions with carbanions has been used to relay the stereochemistry of one center to a remote position in conformationally nonrigid systems.[60] The system studied used organopalladium chemistry to transfer the chirality found in a vinyl lactone to produce an intermediate having two chiral centers in a 1,5 relationship useful for ultimate conversion to the vitamin E sidechain (equation 64a). For this

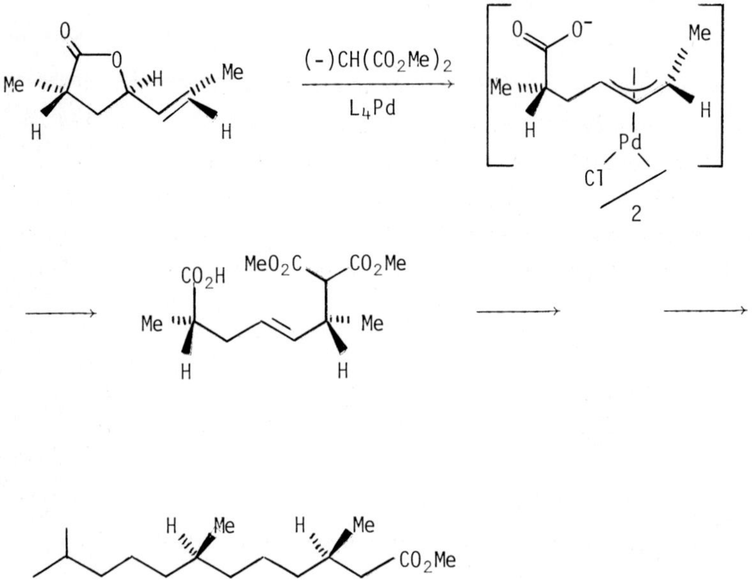

(64a)

process to be successful, formation of the π-allylmetal complex must proceed only from the conformation shown, the π-allylmetal complex must maintain its stereochemistry, and the nucleophile must attack regiospecifically at the terminal carbon of the π-allyl system. All of these criteria were met, and the reaction went in greater than 90% yield and greater than 95% stereospecificity.

The bifunctional allylic acetate in equation 65 reacted with stabilized carbanion precursors to produce either normal alkylation products, or desilylated alkylation products depending on the carbanion structure. The desilylated products were claimed to arise from a trimethylenemethane palladium complex (equation 65).[61]

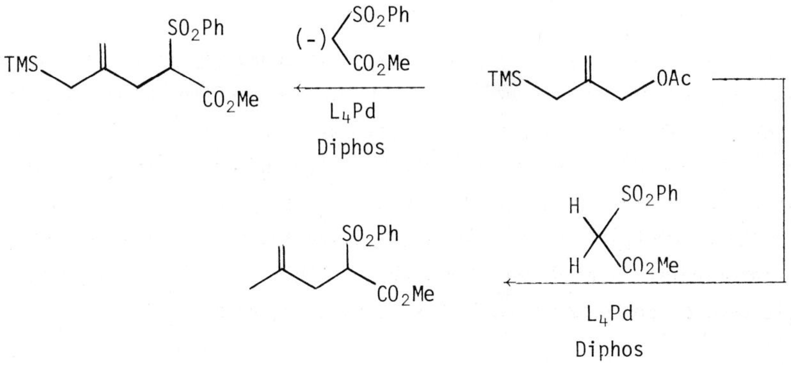

(65)

Palladium catalyzed allylic alkylation has found extensive use in intramolecular processes, to form cyclic compounds, and the results of a detailed study of this process have recently been reported.[62] Virtually any ring size is potentially accessible by this route. Both 6,4 fused ring systems and bicyclo-[2.2.2]octane ring systems resulted from intramolecular attack by an α-sulfonylacetate carbanion on both positions of a cyclohexenylacetate (equation 66).

(66)

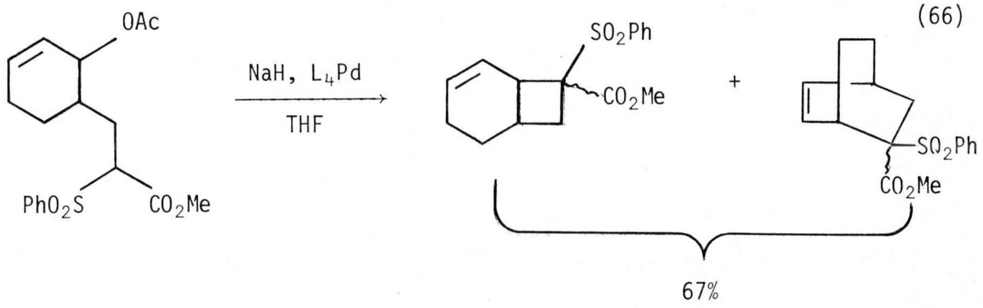

67%

The one-carbon homolog closed cleanly to yield the 6,5 fused ring systems (equation 67). Macrolide esters are also available by this procedure, as evidenced by the synthesis of fourteen and sixteen member macrocycles including exaltolide (equation 68), and the twelve member lactone recifeiolide (equation 69). Medium size rings have also been synthesized using this procedure (equations 70-72). In these cases, the larger of the two possible ring sizes

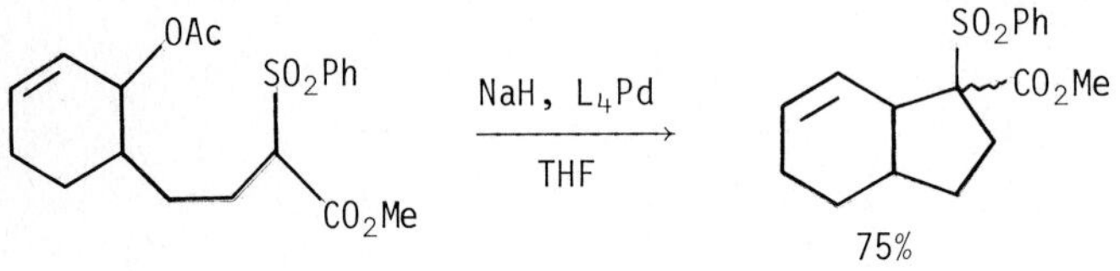

$$(67)$$

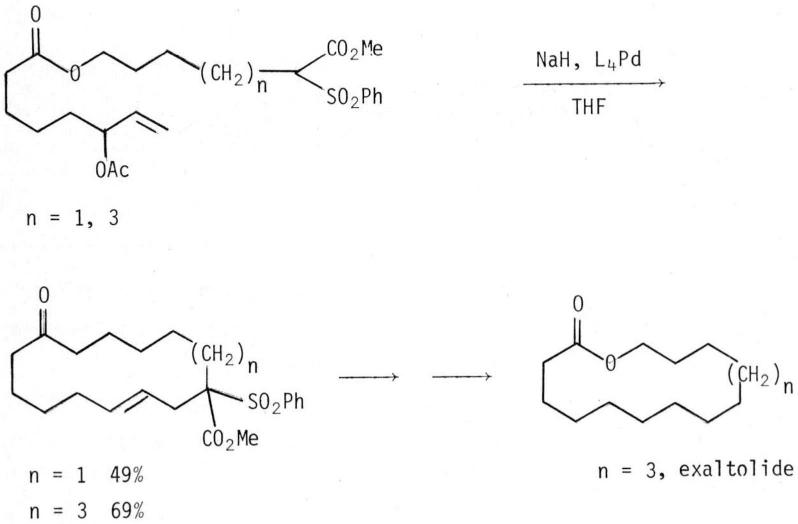

75%

n = 1, 3

$$(68)$$

n = 1   49%
n = 3   69%

n = 3, exaltolide

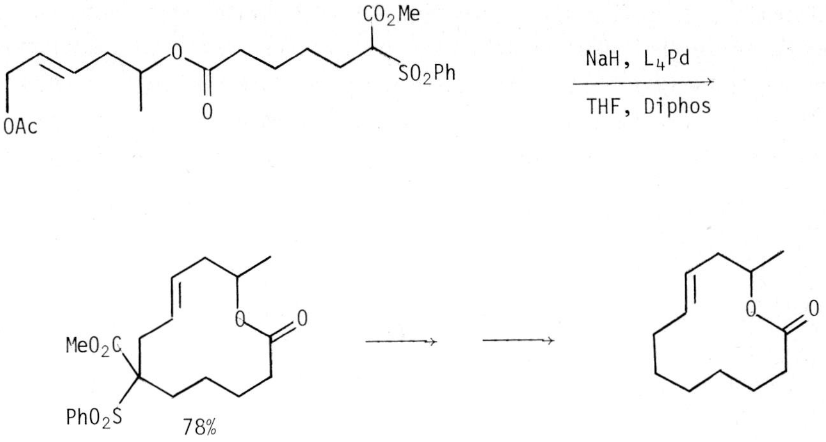

CO$_2$Me

SO$_2$Ph

OAc

NaH, L$_4$Pd

THF, Diphos

(69)

MeO$_2$C

PhO$_2$S   78%

$\longrightarrow$ $\longrightarrow$

PhSO$_2$

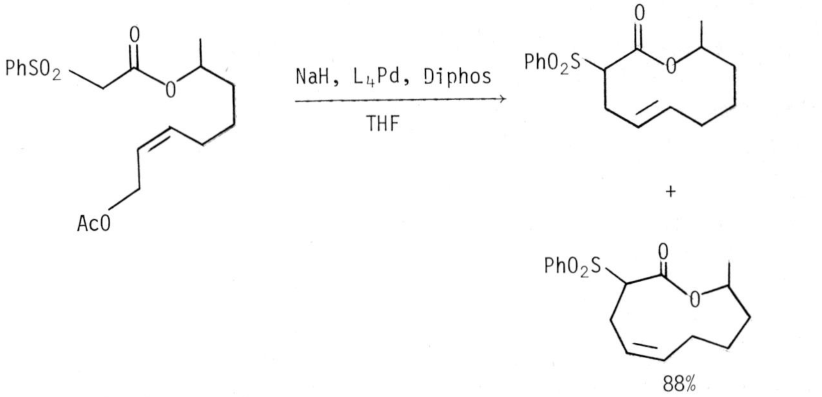

O

AcO

NaH, L$_4$Pd, Diphos

THF

PhO$_2$S

+   (70)

PhO$_2$S

88%

(71)

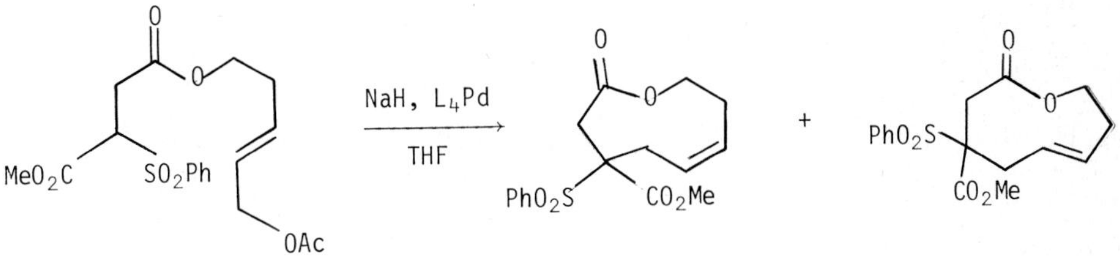

MeO$_2$C   SO$_2$Ph

OAc

NaH, L$_4$Pd

THF

PhO$_2$S   CO$_2$Me

+

PhO$_2$S

CO$_2$Me

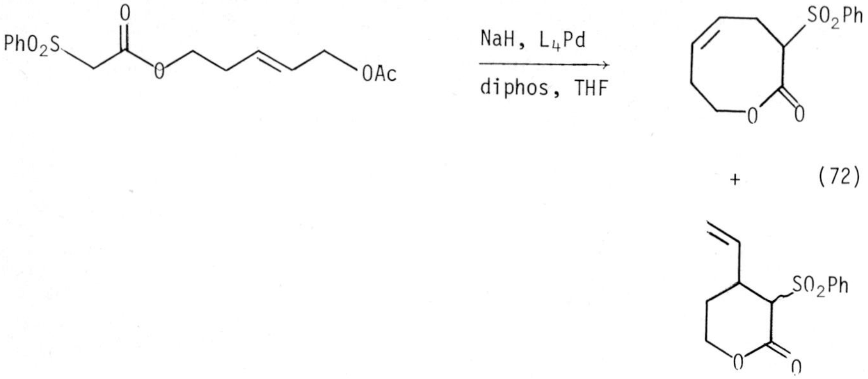

(72)

was always favored (10 vs. 8, 9 vs. 7, and 8 vs. 6). This also corresponded, in all cases, to alkylation at the least substituted (most favored) terminus of the π-allyl ligand. The macrocyclic terpenoid humulene was synthesized using this approach (equation 73).[63] Allyl phenyl ethers also underwent a related palladium catalyzed cyclization, but with these substrates, formation of small rings by nucleophilic attack at the most substituted π-allyl terminus predominated. Thus, five member ring formation was favored over seven and six member ring formation, over eight (equations 74 and 75).[64] This process has been used to synthesize a number of highly substituted 5- and 6-membered rings as illustrated in equations 76 and 77).[65]

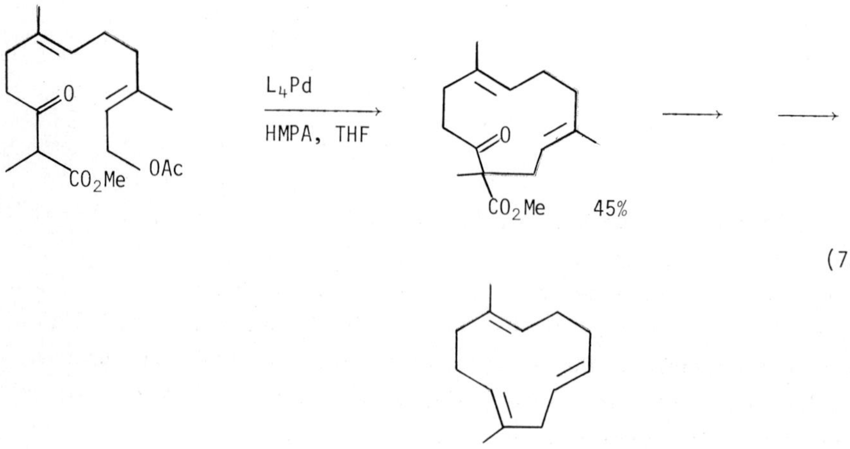

(73)

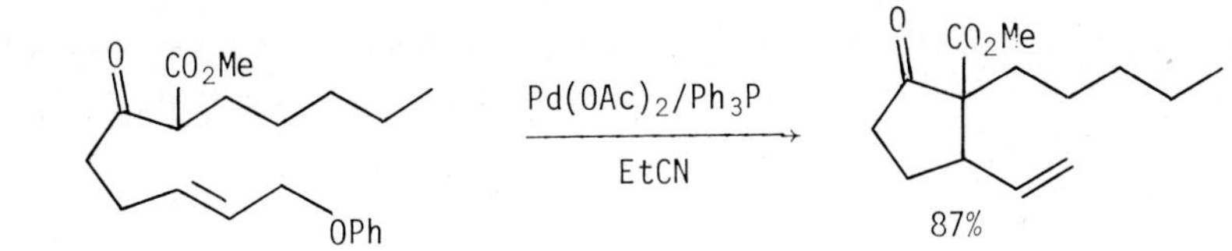

Pd(OAc)$_2$/Ph$_3$P

EtCN

87%

(74)

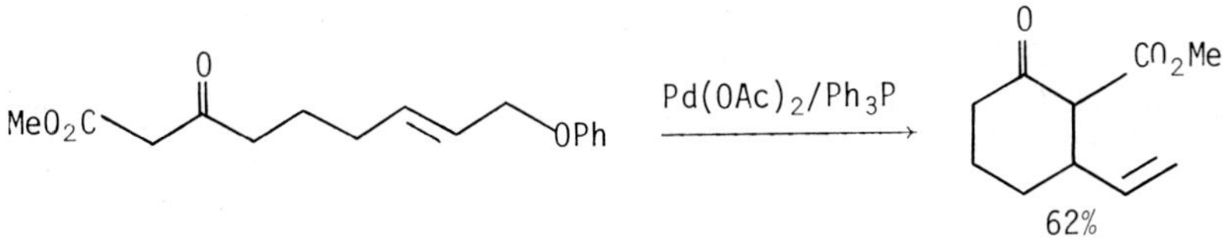

MeO$_2$C

Pd(OAc)$_2$/Ph$_3$P

OPh

62%

(75)

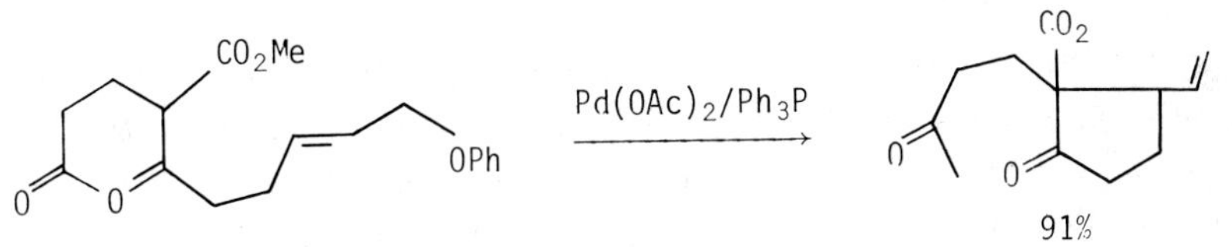

CO$_2$Me

Pd(OAc)$_2$/Ph$_3$P

OPh

91%

(76)

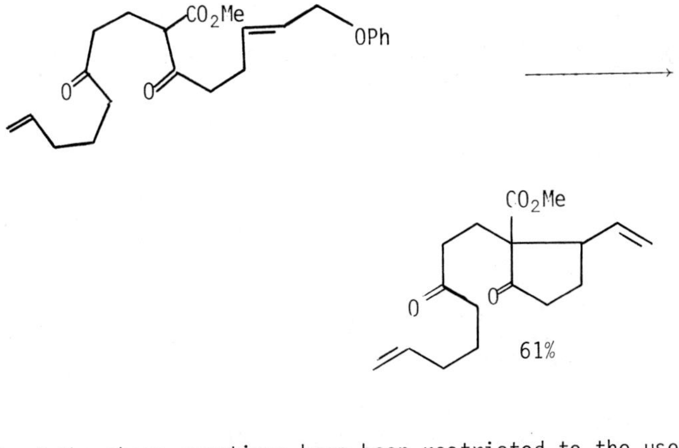

CO$_2$Me

OPh

(77)

CO$_2$Me

61%

All of the above reactions have been restricted to the use of stabilized carbanions having pK$_a$'s ranging from ~10 to 20. More reactive carbanions did not
react productively in these systems. It was recently observed that tin enolates
of simple ketones reacted with allyl acetate in the presence of palladium (0)
catalysts, ostensibly by the mechanism of the reactions discussed above
(equation 78).[66] However, another research group studying the chemistry of
trimethylsilylenol ethers with palladium (II) complexes observed formation of

oxallylpalladium (II) complexes, which then inserted unsaturated substrates
to form rings (equation 79).[67-69] This process is fundamentally different from
the external nucleophilic attack on cationic π-allylpalladium complexes proposed above, and it indicates the variety of reaction pathways available to
organometallic species.

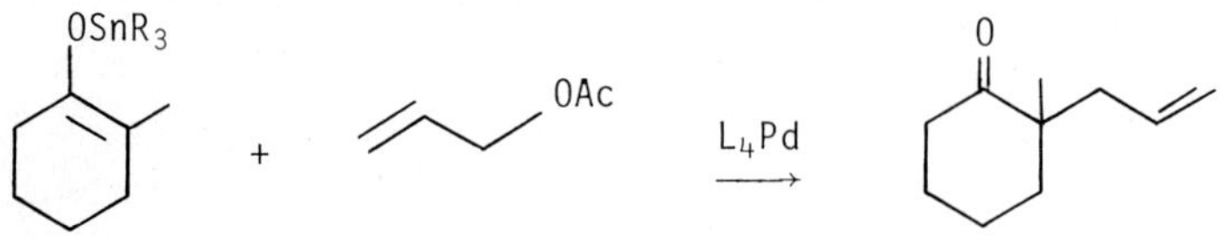

(78)

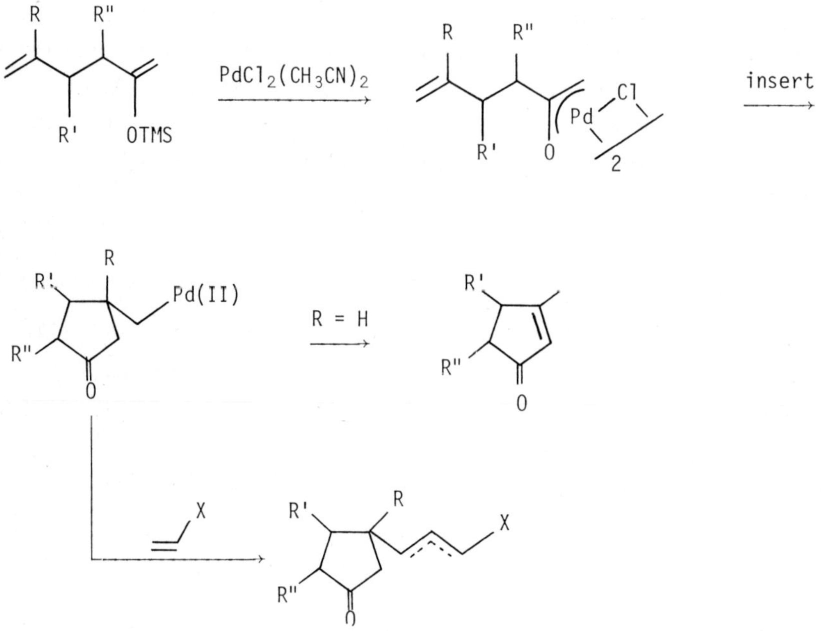

(79)

C. Palladium Catalyzed Alkylation-Dimerization Reactions of Dienes

Butadiene reacts with a variety of nucleophiles in the presence of palladium
catalysts to produce substituted octadienes resulting from a combined
dimerization-nucleophilic attack process. With a carbanion as the nucleophile
two carbon-carbon bonds are formed. This is illustrated in equation 80 in

(80)

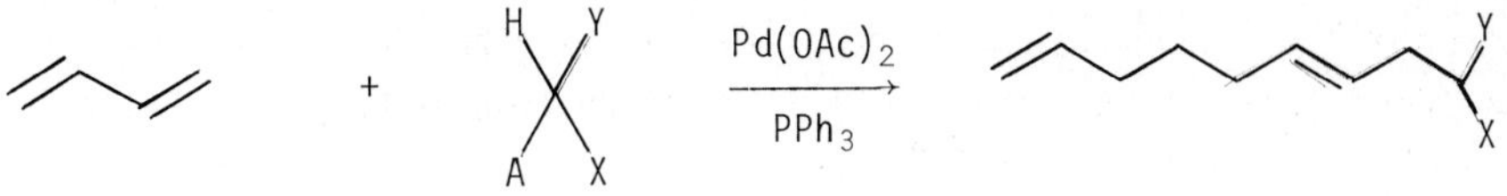

$$X = Y = CO_2Et, \quad A = H$$
$$X = CO_2Et, \quad COMe, \quad A = H$$
$$X = SO_2Ph, \quad Y = CO_2Me, \quad A = H$$
$$X = Y = CO_2Et, \quad A = NHAc$$
$$X = COMe, \quad Y = NHAc, \quad A = R$$
$$X = COR, \quad Y = OH, \quad A = R$$
$$X = NO_2, \quad Y = CH_3, \quad A = H$$

which diethyl malonate serves as the anion.[70, 71] Exactly the same process occurred with methyl acetoacetate[72] (used for the synthesis of methyl dihydro-jasmonate), α-sulfonyl ester stabilized carbanions,[73] diethyl acetamidomalo-nates,[74, 75] α-acetamidoketones,[75] α-hydroxyketones,[76] and nitroethane.[77] In contrast, nitromethane underwent dialkylation to produce a seventeen carbon tetraene used to synthesize muscone (equation 81).[78] The mechanism of this reaction is virtually unstudied.

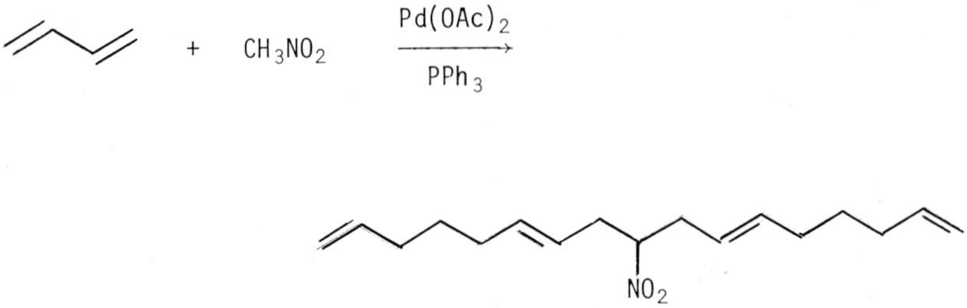

(81)

V. ALKYLATION OF π-ARENE TRANSITION METAL COMPLEXES

Although π-arenechromium tricarbonyl complexes have been known for a number of years, it is only recently that they have been utilized in synthetic organic chemistry. Coordination of arenes to the $Cr(CO)_3$ moiety significantly alters the chemistry of the arene, both at the ring positions and at the α and β positions of alkyl side chains. These effects are summarized in Figure 2.[79] The $Cr(CO)_3$ group is an electron withdrawing group and upon complexation, activates the arene ring to undergo nucleophilic aromatic substitution reactions.

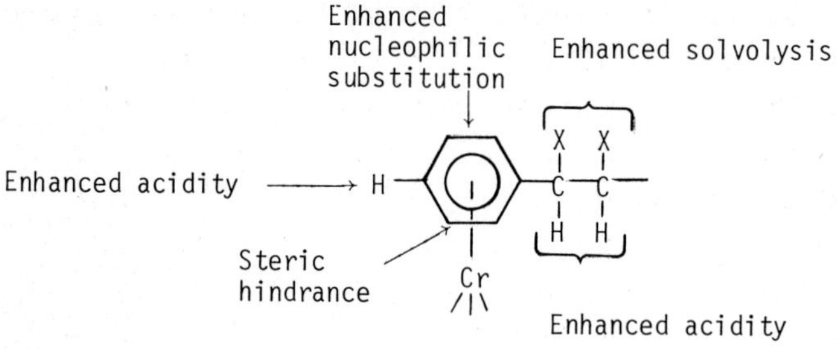

Figure 2. Effects of Complexation of Arenes to $Cr(CO)_3$

Among the nucleophiles found to react with $Cr(CO)_3$ complexed arenes were a number of carbanions. The first complexes studied were those of aryl halides, particularly chloro- and fluorobenzenes. Both stabilized carbanions, such as dimethyl malonate anion, and much less stabilized carbanions such as those from isobutyronitrile, ethyl isobutyrate, the dianion of isobutyric acid, and protected cyanohydrin anions cleanly alkylated the π-haloarene chromium tricarbonyl complexes to produce substituted arene complexes in excellent yield (equation 82).[80] The arene was freed from chromium by oxidation with iodine.

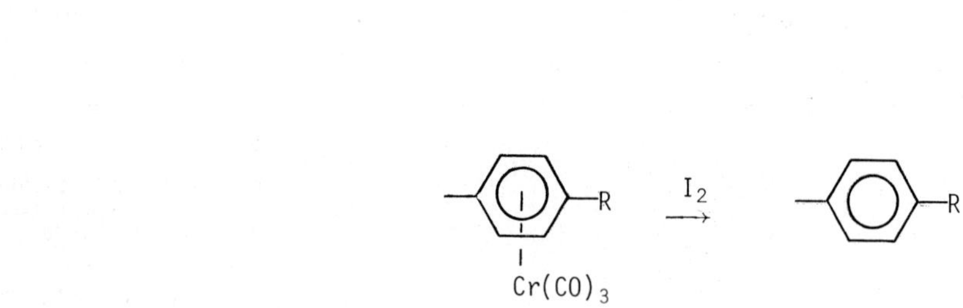

$$(82)$$

R = (-)CH(CO_2Me)_2, (-)C(Me)_2CO_2Et, (-)C(Me)_2CN, (-)C(Me)_2COO(-),

The reaction has been shown to proceed by an addition-elimination (nucleophilic aromatic substitution) reaction rather than by a benzyne reaction using p-methyl-chlorobenzene complex, from which only para-disubstituted products were obtained. (A benzyne mediated reaction would be expected to produce both p and m substitution products.) This reaction was somewhat limited in the range of carbanions which reacted cleanly. Methyl, allyl, phenyl, and t-butylmagnesium halides, as well as lithium dimethylcuprate failed to react at low temperatures, and led to intractable materials at higher temperatures. In contrast, $LiCH_2CO_2Me$, $LiC(Me)_2CO_2Me$, $LiCH_2CO_2(t\text{-}Bu)$, $LiCH_2CN$, $LiC{\equiv}CH$, and 2-lithio-1,3-dithiane reacted even at $-78°$, but instead of displacing chloride, alkylated the ring both o and m to the halogen, producing stable anionic cyclohexadienyl complexes. Oxidation of these led to o- and m-alkylated chlorobenzenes (equation 83).[80] The tertiary

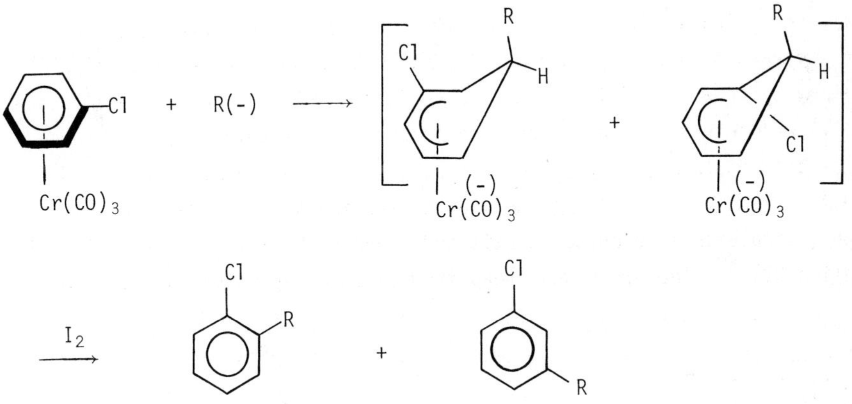

$$(83)$$

anions that displaced the halogen in these substitution reactions appeared to also react initially o and m to the halogen, but subsequently were able to migrate to the halogen-bearing carbon resulting in formal substitution products.

These results indicated that direct alkylation by nucleophilic aromatic substitution reactions of unsubstituted arenes was possible. Hence, treatment of π-benzenechromium tricarbonyl itself with a wide range of carbanions produced stable anionic cyclohexadienyl complexes which were characterized by X-ray crystallographic structure determinations. Oxidation by iodine produced excellent yields of the alkylated arene, but reaction with electrophiles which were potential hydride acceptors (CH_3I, benzophenone, Ph_3C^+, Et_3B) regenerated the starting arene complex (equation 84).[81, 82]

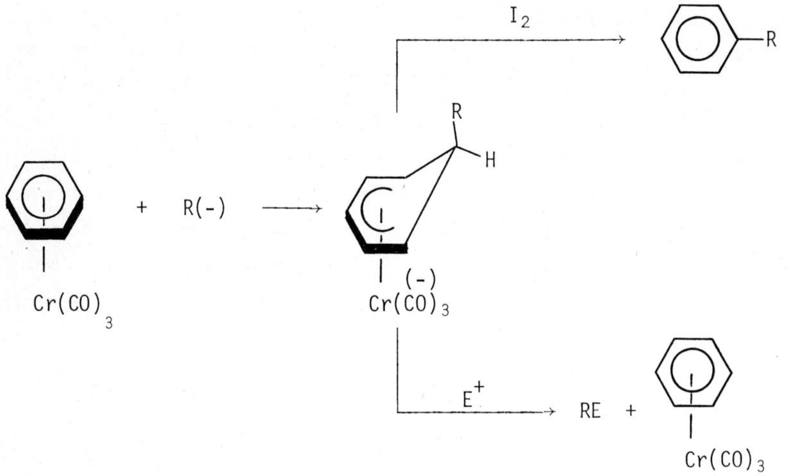

$$(84)$$

Anions more stable than ester enolates, and also Grignard and dialkylcopper reagents, lacked sufficient reactivity to attack the complexed arene, and with these species, no reaction was observed. Anions which reacted in high yield were $LiC(Me)_2CO_2R$, $LiCH(Me)CO_2R$, $LiCH_2CO_2R$, p-tolyllithium, $LiC(Me)_3$, $LiCH(SPh)_2$, $LiC(CN)(OR)R'$, $LiCH[S(CH_2)_3S]$, $LiCH_2CN$, and $LiC(Me)_2CN$. Strongly basic anions such as methyl or n-butyllithium abstracted an aryl proton to produce complexed aryllithiums, reactions of which will be discussed below.

In general, substituted arenechromium tricarbonyl complexes displayed both the reactivity and the regioselectivity expected for a nucleophilic aromatic substitution reaction.[83, 84] Thus, the order of reactivity towards carbanions was $PhCl>PhCH_3>PhOCH_3$. Methoxy groups were powerful meta directors, and very little ortho (<5%), and no para substitution was observed in the alkylation of chromium complexed anisole. Methyl groups were also meta directing, but less strongly so (from 43 to 92% meta substitution was observed, depending on the carbanion). Surprisingly, even chlorine favored meta and ortho alkylation over para alkylation. In contrast, trimethylsilyl and trifluoromethyl groups were para directors.

This chemistry has been used for the direct alkylation of the B ring of indoles (equation 85).[85] Similarly, dihydropyridines were stabilized by complexation to the chromium tricarbonyl group,[86] then alkylated by cyano-stabilized carbanions (equations 86, 87).[87]

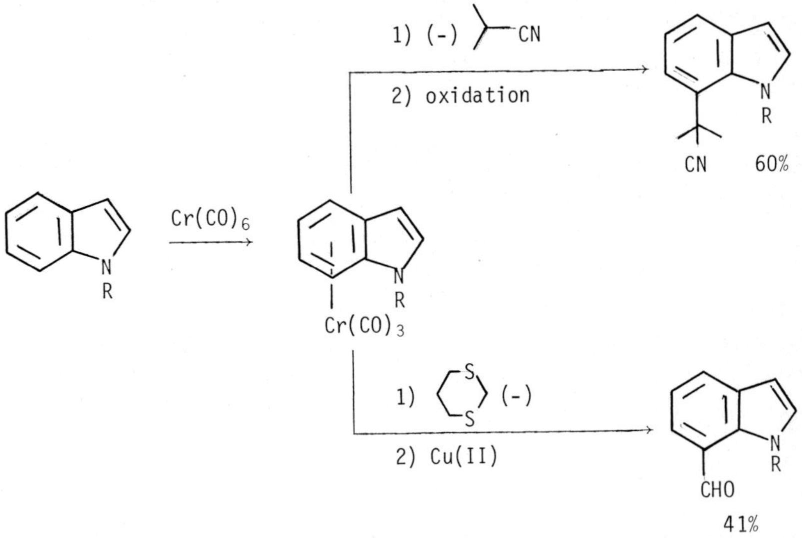

1) (-) ⟩—CN

2) oxidation

(85)

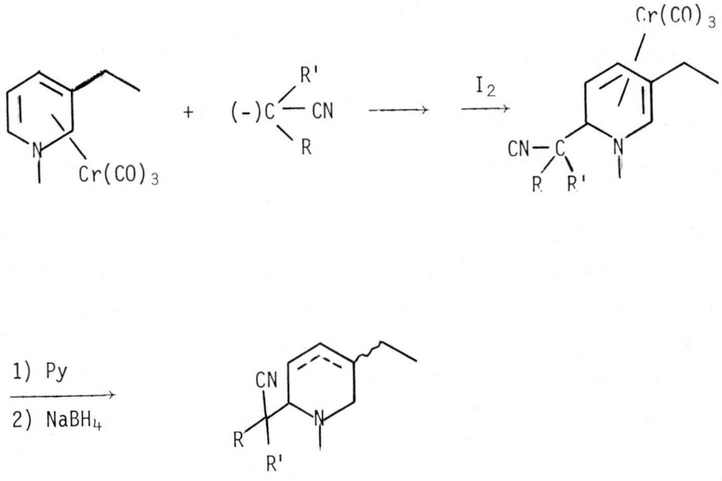

$Cr(CO)_6$

$Cr(CO)_3$

CN   60%

1) ⟨S S⟩ (-)

2) Cu(II)

CHO

41%

+ (-)C⟨R' CN / R

I_2

$Cr(CO)_3$

$Cr(CO)_3$

CN—C⟨R R'⟩

(86)

1) Py

2) $NaBH_4$

CN

R R'

R = R' = CH_3

R = H, R' = CH_3

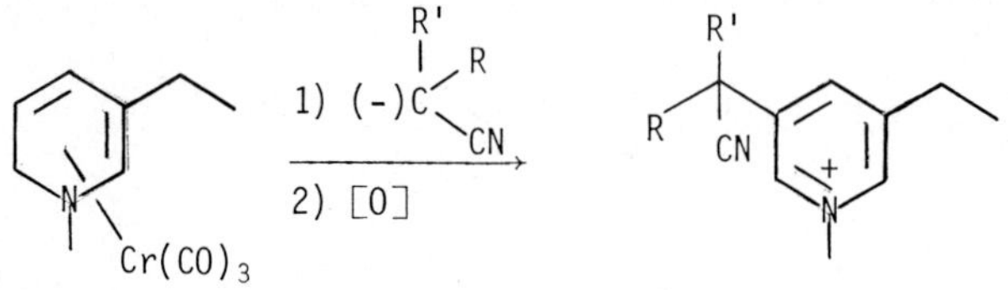

$$(87)$$

The chemistry presented above relies upon an oxidative removal of the chromium from a hexadienyl chromium tricarbonyl species, a process which generated a substituted arene. In contrast, acid cleavage of these π-cyclohexadienyl chromium complexes generates free cyclohexadienes. This is a general process (equation 88). With π-anisole chromium tricarbonyl as substrate a route to 3-substituted cyclohexenones has been developed (equation 89).[88]

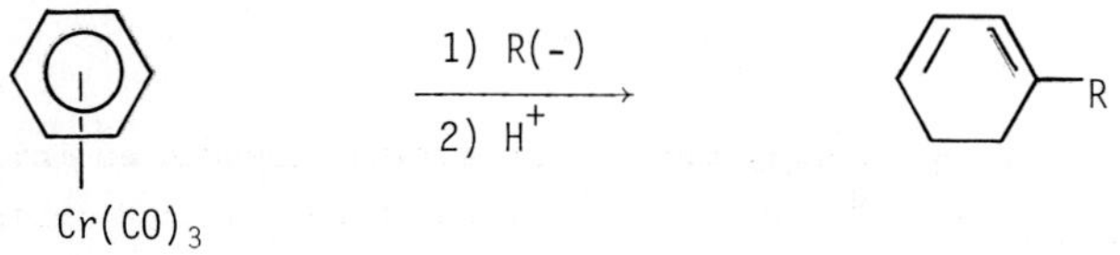

$$(88)$$

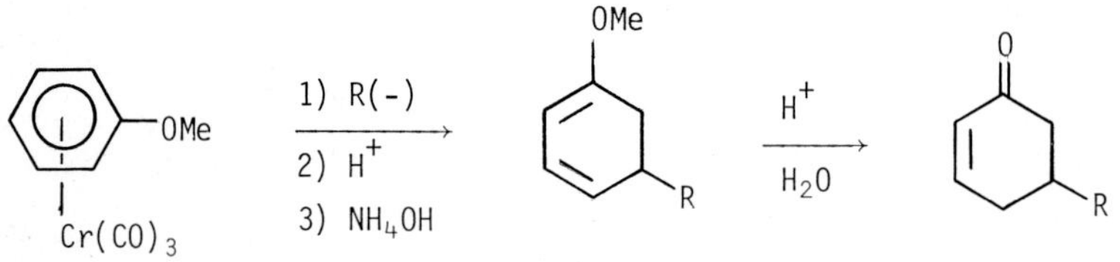

$$(89)$$

Particularly useful in organic synthesis have been intramolecular alkylation reactions of π-arenechromium tricarbonyl complexes by cyano-stabilized carbanions. The product formed depended both on the chain length and the substituents on the aromatic ring. The nitrile with a four carbon chain cyclized cleanly to the corresponding tetralin derivative (equation 90). The next lower homolog dimerized to the metacyclophane (equation 91) rather than forming the indane ring systems. In contrast, the next higher homolog cyclized to a

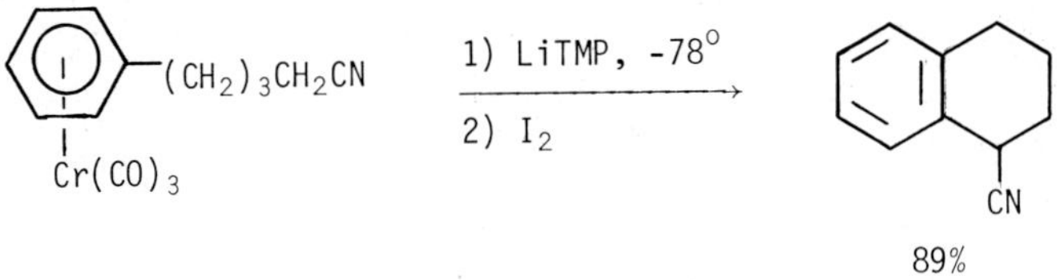

$$\text{(90)}$$

89%

(LiTMP = lithium tetramethylpiperidide)

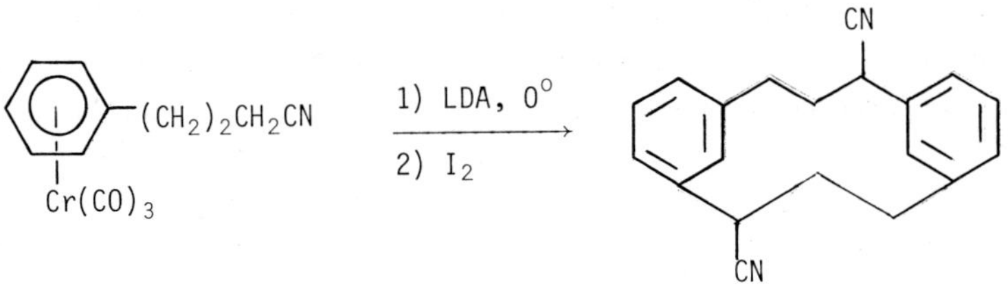

$$\text{(91)}$$

mixture of fused and spiro fused products whose composition depended on the reaction conditions (equation 92).[89] This chemistry has been combined in an elegant synthesis of Acorenone B in which two key steps utilized the unique reactivity of the π-arene chromium tricarbonyl system (equation 93).[90] In this case, the powerful _meta_ directing influence of the methoxy group was clearly responsible for exclusive spiro ring formation.

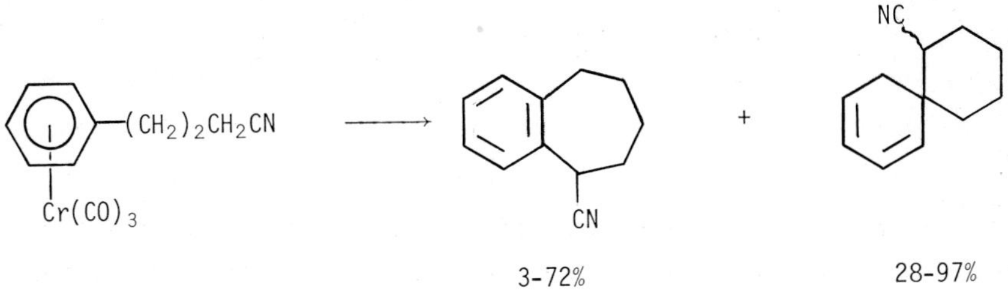

$$\text{(92)}$$

3-72%   28-97%

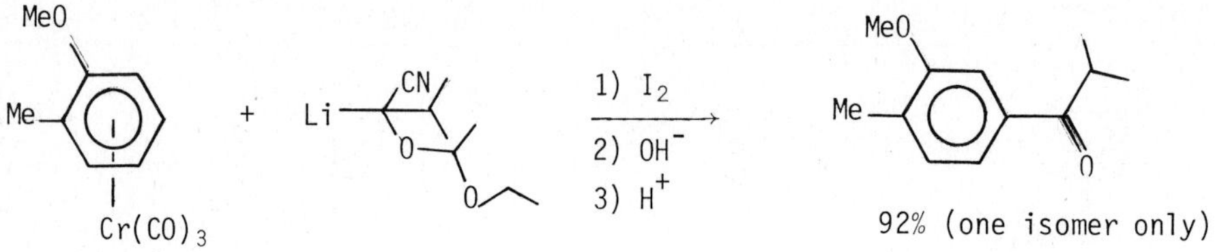
MeO
Me
Cr(CO)3
+
Li
CN
O
O
1) I2
2) OH⁻
3) H⁺
MeO
Me
O
92% (one isomer only)

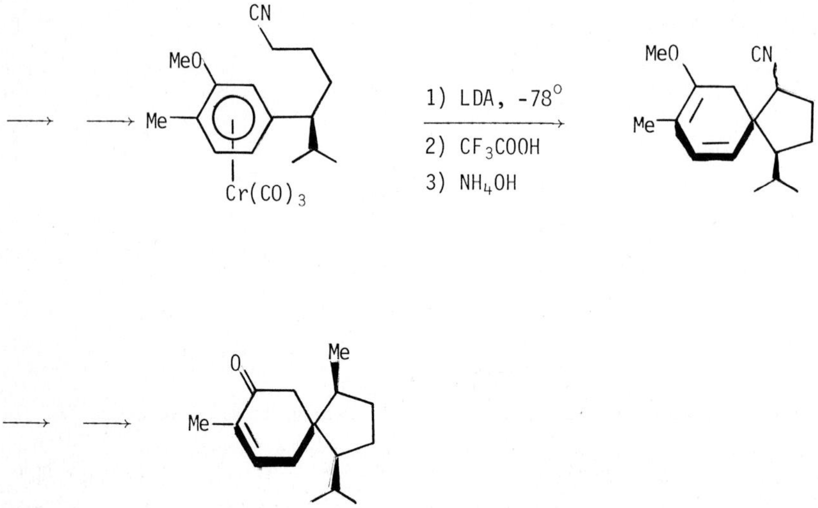
CN
MeO
Me
Cr(CO)3
1) LDA, -78°
2) CF3COOH
3) NH4OH
MeO
CN
Me
O
Me
Me

(93)

As presented above, n-butyllithium did not alkylate π-arenechromium tricarbonyl complexes, but rather abstracted a phenyl proton to give a chromium stabilized phenyllithium. This reagent was quite reactive towards electrophiles and permitted functionalization of the aromatic ring in a variety of useful ways. With substituted arene complexes, lithiation occurred exclusively <u>ortho</u> to methoxy, fluoro, and chloro derivatives, and <u>meta</u> to methyl derivatives. This chemistry is summarized in equation 94.[91-93] Bipyridines were obtained when dihydropyridine $Cr(CO)_3$ complexes were lithiated (equation 95).[94]

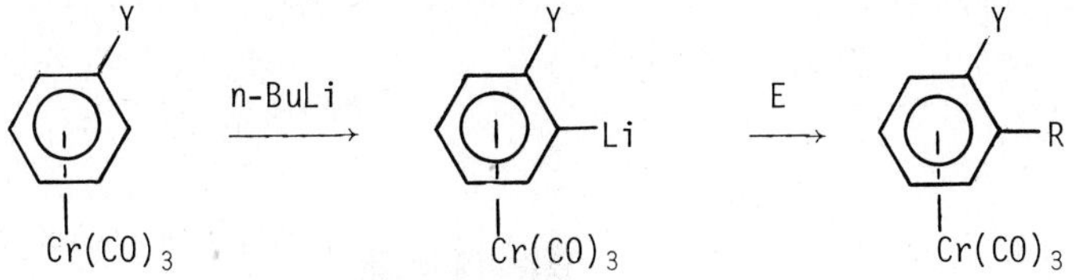

$$(94)$$

Y = H, OMe, F, Cl

E = CO_2, MeI, PhCHO, TMSCl, MeCOMe

R = CO_2Me, Me, PhCHOH, TMS, Me_2COH

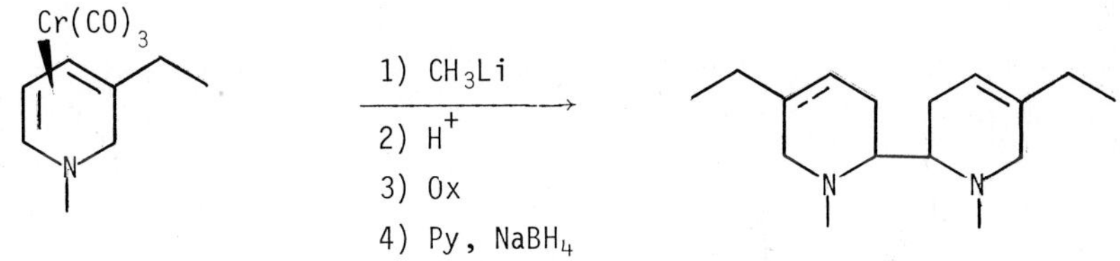

1) CH_3Li

2) H^+

3) Ox

4) Py, $NaBH_4$

$$(95)$$

The influence of the chromium in π-arenechromium carbonyl complexes extends to substituents on the aromatic ring. For example, complexed styrenes underwent β-alkylation with carbanions to give chromium stabilized benzyl anions. These were trapped with a number of electrophiles, resulting in difunctionalization of the double bond (equation 96).[95]

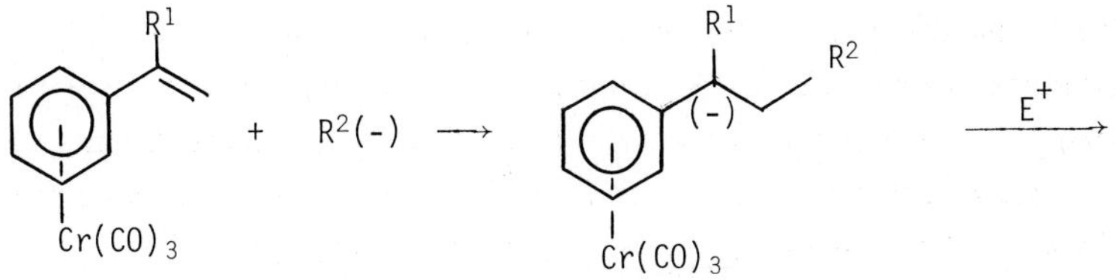

$$(96)$$

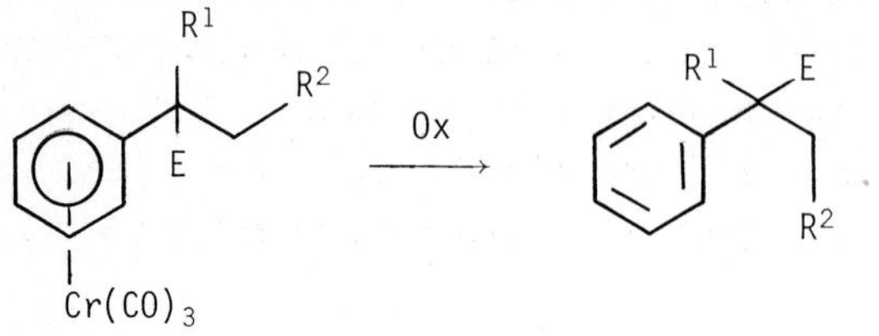

$$(96)$$

R^1 = H, Me, SEt

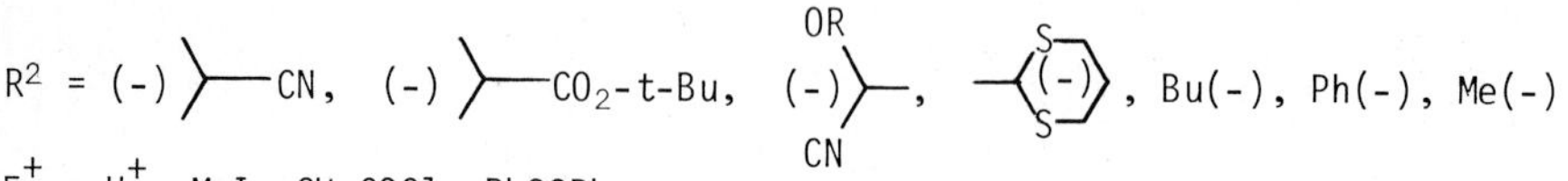

R^2 = (-)$\langle$>—CN, (-)$\langle$>—CO$_2$-t-Bu, (-)$\langle$>—$\overset{OR}{\underset{CN}{}}$, —$\langle$(-)$\rangle\overset{S}{\underset{S}{}}$, Bu(-), Ph(-), Me(-)

E^+ = H^+, MeI, CH$_3$COCl, PhSSPh

This ability of chromium to stabilize anions at the benzylic position has found use in the facile dialkylation of complexed methyl phenylacetate under conditions where the free ligand did not react (equation 97).[96] This benzylic

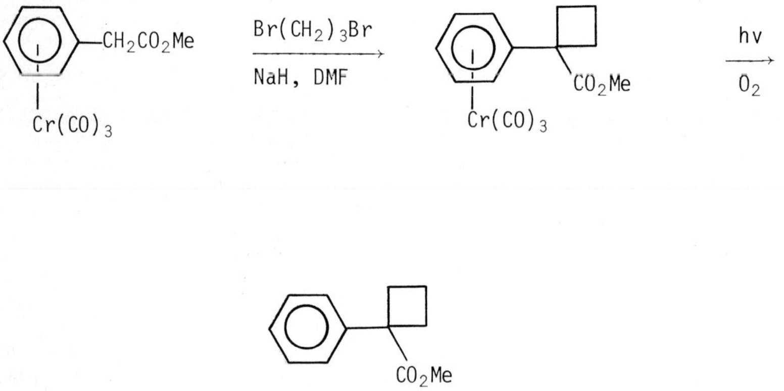

$$(97)$$

activation was even more strikingly illustrated by the reaction of π-toluene-chromium tricarbonyl with potassium t-butoxide and methyl iodide to give a mix-ture of ethylbenzene (34%) and isopropylbenzene (49%) after decomplexation.[97] The π-ethylbenzene chromium complex was converted to isopropylbenzene (71%) and t-butylbenzene (7%) under similar conditions. In contrast, π-arenechromium tricarbonyl complexes of methyl phenylacetate and methyl 3-phenylpropionate alkylated exclusively α to the carbonyl group, in preference to alkylation at the benzylic position.[96, 97]

A final consequence of complexation of arenes to chromium tricarbonyl moieties
is the steric hindrance introduced to the face of the molecule occupied by the
chromium. Reaction of the complex of o-methoxyacetophenone with Grignard
reagents resulted in predominant alkylation from the face opposite the chromium
group (equation 98).[98] The same effect was used for the stereospecific alkylation

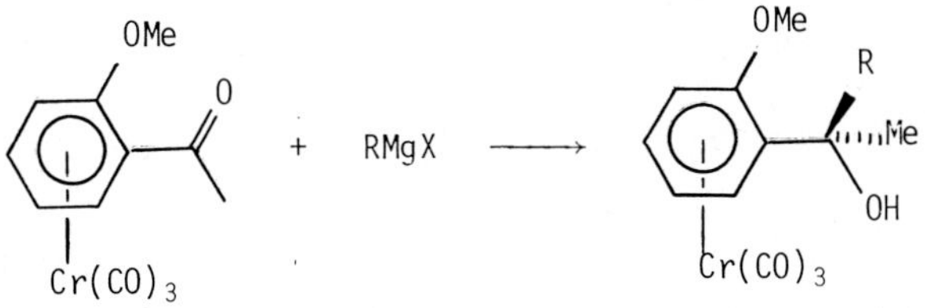

$$\text{(98)}$$

of indanones. Racemic 1-indanol was complexed to chromium in a reaction that
placed the chromium exclusively on the same face as the hydroxyl group. Resolution and oxidation of these complexes was followed by alkylation of one enantiomer
with a Grignard reagent and decomplexation to give the optically active indanol with
100% optical purity (equation 99).[99] Similar chemistry has been carried out

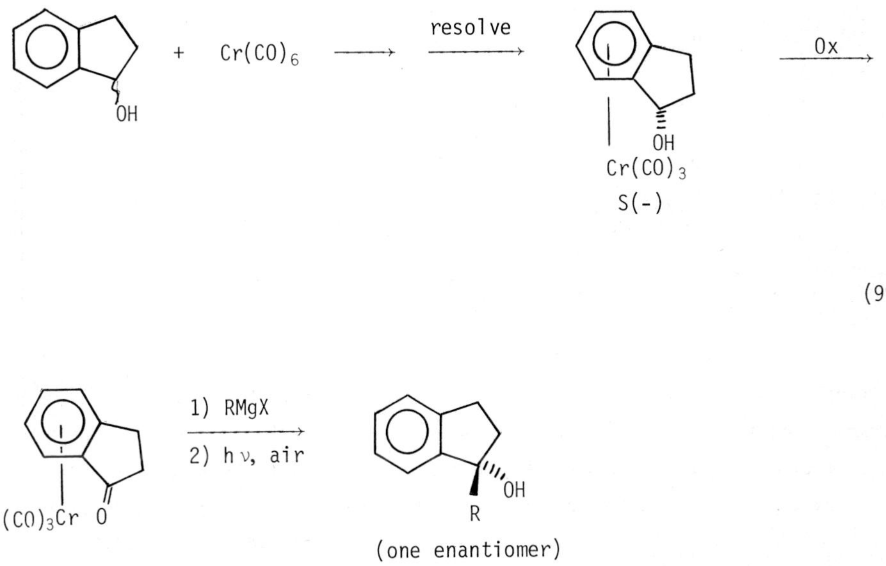

$$\text{(99)}$$

with tetralones.[100] Indane esters were alkylated stereospecifically from the face opposite the chromium substituent (equation 100).[101, 102]

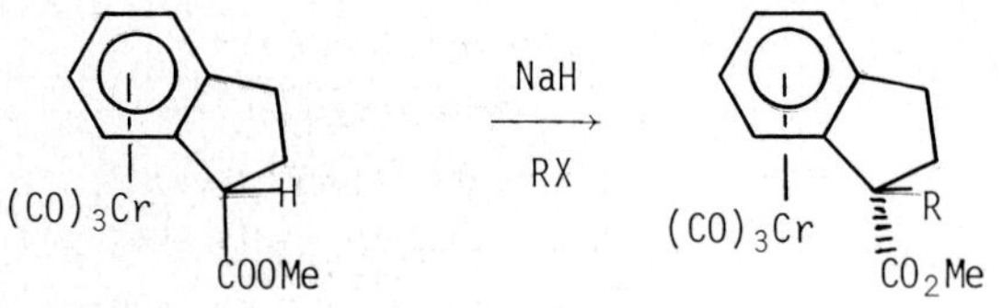

$$(100)$$

A stereospecific synthesis of optically active benzobicyclic ring systems which uses both the steric directing effects and the benzylic position activation afforded by complexation to chromium has been developed (equation 101).[103]

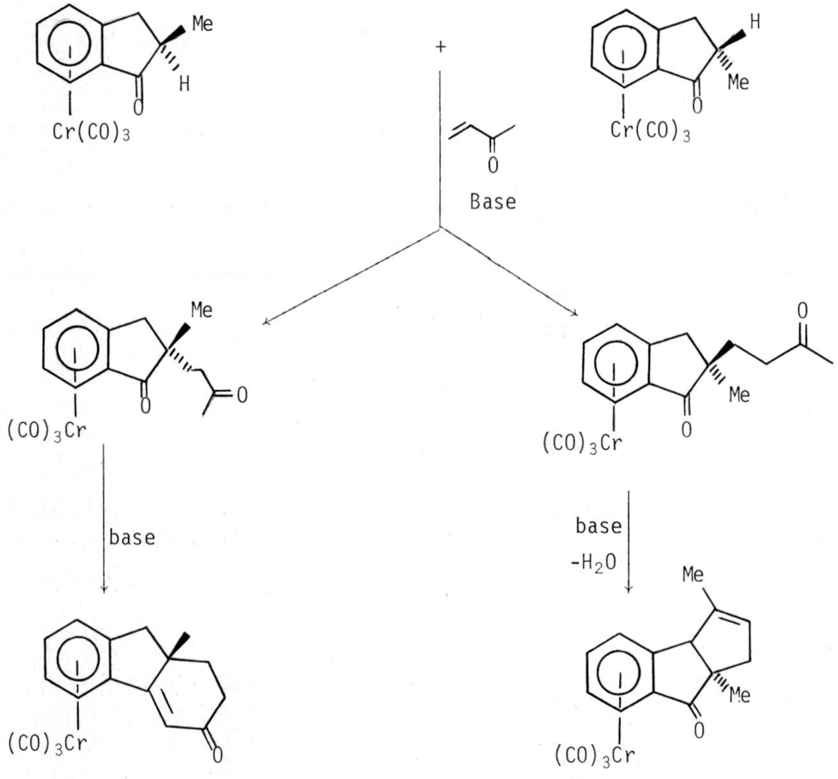

$$(101)$$

Michael addition of the carbanion derived from a mixture of diastereoisomers of the chromium complex of 2-methyltetralone to methylvinyl ketone produce two adducts. Annelation by a base-catalyzed aldol condensation converted the _endo_ adduct exclusively to the classical Robinson annelation product. In contast, the _exo_ adduct underwent cyclization by attack of the (Cr activated) benzylic carbon on the methyl ketone. The same chemistry resulted when the chromium complexes of 2-methyltetralone were subjected to the series of reactions. Since the chiral chromium complexes involved in these reactions are easily resolved, the annelation is stereospecific, and the chromium tricarbonyl group is easily removed, this sequence represents a useful method for the synthesis of optically active indanone and tetralone derivatives.

A number of other transition metals complex arenes, and the resulting complexes are reactive toward carbanions. These include the _bis_-arene complexes of iron, and the arenetricarbonyl manganese complexes. However, they have been little studied to date, and have not been used in organic synthesis. See references 79 and 102 for more information on these systems.

VI. REACTIONS OF CARBANIONS WITH METAL CARBONYLS

When coordinated to transition metals, carbon monoxide is frequently reactive towards carbanions forming acylmetal carbonyl complexes. The properties of these acylmetal carbonyl complexes depend strongly on the specific transition metal involved. For example, group VI metal carbonyls (Cr, Mo, W) react readily with organolithium complexes to produce acylmetal carbonylates in which much of the charge resides on oxygen. Reaction with an alkylation agent produces a relatively stable metal "carbene" complex (equation 102). These are potentially useful complexes but to date relatively little has been done with them

$$Cr(CO)_6 \quad + \quad RLi \quad \longrightarrow \quad [(CO)_5Cr\overset{\overset{(-)}{O}}{\underset{\parallel}{-}}C-R \quad \longleftrightarrow \quad (CO)_5Cr=\overset{\overset{O(-)}{|}}{C}-R]$$

$$(102)$$

$$\xrightarrow{Me_3OBF_4} \quad (CO)_5Cr{=}\begin{array}{c} {\diagup} OMe \\ {\diagdown} R \end{array}$$

synthetically.[104] Complexes of this type have been implicated in the olefin metathesis reaction, a topic outside the scope of this review.

The considerably more reactive acylmetal carbonylates of nickel have found more applications in synthetic organic chemistry. Reaction of nickel carbonyl with organolithium reagents gave unstable species whose assumed formula $RCNi(CO)_3^-$ was consistent with their chemistry. For example, allylic halides reacted with these acylnickel carbonylates to produce high yields of β,γ-unsaturated ketones. The reaction proceeded without allylic transposition (equation 103).[105] Only

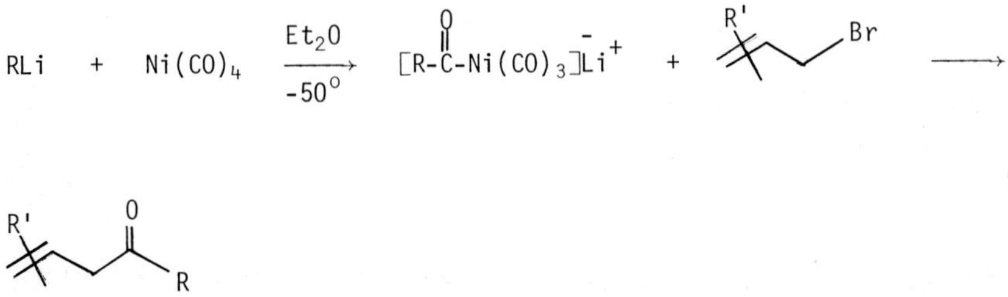

$$RLi \quad + \quad Ni(CO)_4 \quad \xrightarrow[-50°]{Et_2O} \quad [R\overset{O}{\overset{\|}{C}}-Ni(CO)_3]^-Li^+ \quad + \quad \text{(allyl-Br)} \quad \longrightarrow$$

(103)

(product structure)

methyl- and n-butyllithium were studied. The allylic substrates included 3-bromocyclooctene, trans-geranyl bromide, trans-cinnamyl bromide, and trans, trans-1',12-dibromo-2,10-dodecadiene. Aryl iodides, benzoyl chloride and aliphatic halides were unreactive toward these acylnickel complexes even under conditions sufficiently severe to lead to decomposition.

In contrast, vinyl halides reacted readily, but underwent diacylation to produce 1,4-diones (equation 104).[106] This was shown to result from acylation

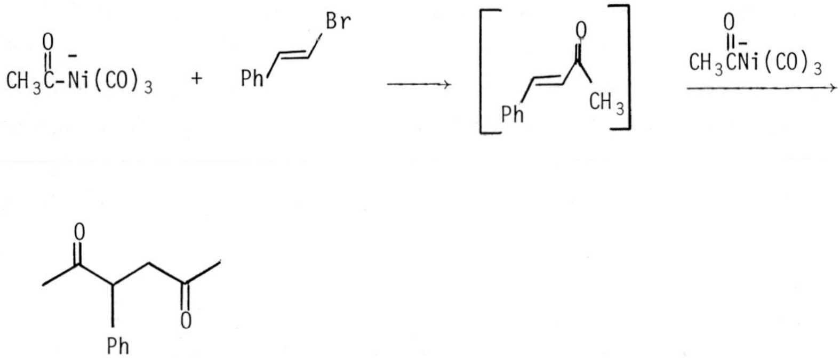

$$CH_3\overset{O}{\overset{\|}{C}}-Ni(CO)_3^- \quad + \quad Ph\text{(vinyl-Br)} \quad \longrightarrow \quad \left[\underset{Ph}{\overset{O}{\text{(enone)}}}CH_3 \right] \quad \xrightarrow{CH_3\overset{O}{\overset{\|}{C}}Ni(CO)_3^-}$$

(104)

(1,4-dione structure with Ph)

of the vinyl halide to produce a conjugated enone, followed by conjugate acylation of the enone. This 1,4-acylation of conjugated enones is a general process (equation 105). Monosubstituted alkynes underwent a similar diacylation to produce 1,4-diones.[107]

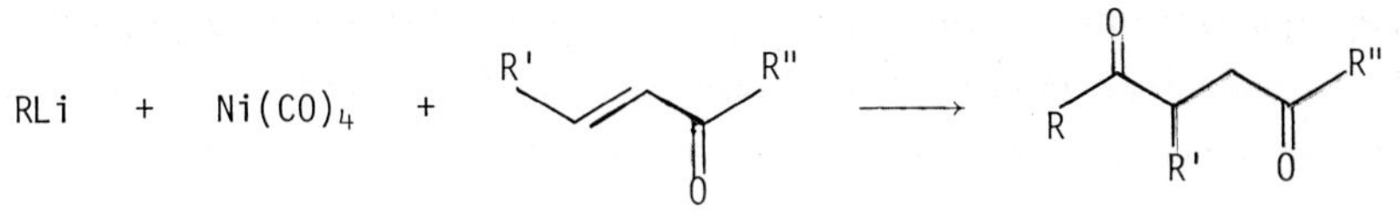

$$\text{RLi} \quad + \quad \text{Ni(CO)}_4 \quad + \quad \underset{\overset{|}{O}}{R'} \diagdown = \diagup R'' \quad \longrightarrow \quad R \overset{O}{\underset{R'}{\diagdown}} \diagup \overset{R''}{\underset{O}{\diagup}}$$

(105)

R = Me, n-Bu, Ph

R' = Ph, (Me)$_2$, Me, t-Bu $\left.\right\}$ -(CH$_2$)$_3$ -

R" = Me, OMe

Phenacylnickel carbonylates produced from aryllithium and nickel carbonyl produced α-diketones when treated with ethanolic HCl, whereas those from alkyllithium reagents gave symmetrical ketones.[108]

VII. CONCLUSION

The ability of transition metals to activate unsaturated organic compounds toward reaction with carbanions is becoming increasingly important in synthetic organic chemistry. These new methods of carbon-carbon bond formation permit the direct alkylation of unusual functional groups such as simple olefins, arenes, and carbon monoxide, and greatly enhances the synthetic chemist's ability to synthesize complex molecules.

REFERENCES

(1) B. F. G. Johnson, J. Lewis, and M. S. Subramanian, Chem. Commun., (1966) 117.

(2) B. F. G. Johnson, J. Lewis, and M. S. Subramanian, J. Chem. Soc. A, (1968) 1993.

(3) J. Tsuji and H. Takahashi, J. Am. Chem. Soc., 87 (1965) 3275.

(4) J. Tsuji and H. Takahashi, J. Am. Chem. Soc., 90 (1968) 2387.

(5) J. K. Stille and D. B. Fox, J. Am. Chem. Soc., 92 (1980) 1274.

(6) C. R. Kistner, J. H. Hutchinson, J. R. Doyle, and J. C. Storlie, Inorg. Chem., 2 (1963) 1255.

(7) R. A. Holton and R. A. Kjonaas, J. Am. Chem. Soc., 99 (1977) 4177.

(8) R. A. Holton, J. Am. Chem. Soc., 99 (1977) 8083.

(9) R. A. Holton and R. A. Kjonaas, J. Organomet. Chem., 142 (1977) C15.

(10) H. Hirai, N. Ishii, H. Suzuki, Y. Moro-Oka, and T. Ikawa, Chem. Lett., (1979) 1113.

(11) S-i Murahashi, M. Yamamura, and N. Mita, J. Org. Chem., 42 (1977) 2870.

(12) L. S. Hegedus, R. E. Williams, M. A. McGuire, and T. Hayashi, J. Am. Chem. Soc., 102 (1980) 4973.

(13) J. P. Haudegond, Y. Chauvin, and D. Commereuc, J. Org. Chem., 44 (1979) 3063.

(14) L. S. Hegedus and W. H. Darlington, J. Am. Chem. Soc., 102 (1980) 4980.

(15) B. W. Roberts and J. Wong, J. C. S. Chem. Comm., (1977) 20.

(16) M. A. Baar and B. W. Roberts, J. C. S. Chem. Comm., (1979) 1129.

(17) B. W. Roberts, M. Ross, and J. Wong, J. C. S. Chem. Comm., (1980) 428.

(18) A. Rosan and M. Rosenblum, J. Org. Chem., 40 (1975) 3622.

(19) P. Lennon, A. M. Rosan, and M. Rosenblum, J. Am. Chem. Soc., 99 (1977) 8426.

(20) T. C. T. Chang, M. Rosenblum, and S. B. Samuels, J. Am. Chem. Soc., 102 (1980) 5930.

(21) H. Kurosawa and N. Asada, Tetrahedron Lett., (1979) 255.

(22) H. Kurosawa, T. Majima, and N. Asada, J. Am. Chem. Soc., 102 (1980) 6996.

(23) For a review covering the literature through 1975 see A. J. Birch and I. D. Jenkins in H. Alper, ed., Transition Metal Organometallics in Organic Synthesis, (Academic Press, 1976) Vol 1, pp 1-75.

(24) A. J. Birch and G. S. R. Subba Rao, Tetrahedron Lett., (1968) 3797.

(25) A. J. Pearson, J. C. S. Chem. Comm., (1977) 339.

(26) A. J. Birch and A. J. Pearson, Tetrahedron Lett., (1975) 2379.

(27) A. J. Pearson, Aust. J. Chem., 29 (1976) 1101.

(28) A. J. Pearson, Aust. J. Chem., 30 (1977) 345.

(29) A. J. Birch and D. H. Williamson, J. Chem. Soc., Perkin I, (1973) 1892.

(30) A. J. Pearson, J. C. S. Perkin I, (1979) 1255.

(31) A. J. Pearson and M. Chandler, J. C. S. Perkin I, (1980) 2238.

(32) A. J. Pearson and P. R. Rathby, J. C. S. Perkin I, (1980) 395.

(33) A. J. Pearson, J. C. S. Perkin I, (1980) 400.

(34) A. J. Pearson and M. Chandler, Tetrahedron Lett., 21, 3933 (1980).

(35) L. F. Kelley, A. S. Narula, and A. J. Birch, Tetrahedron Lett., (1979) 4107.

(36) L. F. Kelley, A. S. Narula, and A. J. Birch, Tetrahedron Lett., 21 (1980) 2451.

(37) A. J. Birch, A. S. Narula, P. Dahler, G. R. Stephenson, and L. F. Kelley, Tetrahedron Lett., 21 (1980) 979.

(38) A. J. Birch, P. Dahler, A. S. Narula, and G. R. Stephenson, Tetrahedron Lett., 21 (1980) 3817.

(39) A. J. Birch and K. B. Chamberlain, Org. Synth., 53 (1973) 1859.

(40) H. D. Hodes and K. M. Nicholas, Tetrahedron Lett., (1978) 4349.

(41) K. M. Nicholas, M. Mulvaney, and M. Bayer, J. Am. Chem. Soc., 102 (1980) 2508.

(42) S. Padmanabhan and K. M. Nicholas, Synth. Comm., 10 (1980) 503.

(43) B. M. Trost, P. E. Strege, L. Weber, T. J. Fullerton, and T. J. Dietsch,
      J. Am. Chem. Soc., 100 (1978) 3407.

(44) B. M. Trost and P. Metzner, J. Am. Chem. Soc., 102 (1980) 3572.

(45) J. Tsuji, H. Takahashi, and M. Morkawa, Tetrahedron Lett., (1965) 4387.

(46) For reviews on the subject of allylic alkylation via palladium complexes
      see: (a) B. M. Trost, Tetrahedron, 33 (1977) 2615; (b) B. M. Trost
      Accts. Chem. Res., 13 (1980) 385.

(47) B. M. Trost, L. Weber, P. E. Strege, T. J. Fullerton, and T. J. Dietsche,
      J. Am. Chem. Soc., 100 (1978) 3416.

(48) W. R. Jackson and J. U. G. Strauss, Tetrahedron Lett., (1975) 2591.

(49) W. R. Jackson and J. U. G. Strauss, Aust. J. Chem., 30 (1977) 553.

(50) D. J. Collins, W. R. Jackson, and R. N. Timms, Tetrahedron Lett., (1976) 495.

(51) B. M. Trost and T. R. Verhoeven, J. Am. Chem. Soc., 100 (1978) 3435.

(52) B. M. Trost, L. Weber, P. Strege, T. J. Fullerton, and T. J. Dietsche,
      J. Am. Chem. Soc., 100 (1978) 3426.

(53) P. S. Marchand, H. S. Wong, and J. F. Blount, J. Org. Chem., 43 (1978)
      4769.

(54) W. R. Jackson and J. U. Strauss, Aust. J. Chem., 31 (1978) 1073.

(54a) L. S. Hegedus, C. E. Russell, and W. H. Darlington, J. Org. Chem., 45
      (1980) 5193.

(55) B. M. Trost and T. R. Verhoeven, J. Am. Chem. Soc., 102 (1980) 4730 and
      references cited therein.

(56) B. M. Trost and Y. Matsumura, J. Org. Chem., 42 (1977) 2036.

(57) B. M. Trost and F. W. Gowland, J. Org. Chem., 44 (1979) 3448.

(58) J. P. Genet, F. Piau, and J. Ficini, Tetrahedron Lett., 21 (1980) 3183.

(59) B. M. Trost, N. R. Schmuff, and M. J. Miller, J. Am. Chem. Soc., 102
      (1980) 5979.

(60) B. M. Trost and T. P. Klun, J. Am. Chem. Soc., 101 (1979) 6756.

(61) B. M. Trost and D. M. T. Chan, J. Am. Chem. Soc., 102 (1980) 6361.

(62) B. M. Trost and T. R. Verhoeven, J. Am. Chem. Soc., 102 (1980) 4743.

(63) Y. Kitagawa, A. Itoh, S. Hashimoto, H. Yamamoto, and H. Nozaki, J. Am. Chem.
      Soc., 99 (1977) 3864.

(64) J. Tsuji, Y. Kobayashi, H. Katoaka, and T. Takahashi, Tetrahedron Lett.,
      21 (1980) 1475.

(65) J. Tsuji, Y. Kobayashi, H. Kotoaka, and T. Takahashi, Tetrahedron Lett.,
      21 (1980) 3393.

(66) B. M. Trost and E. Keinan, Tetrahedron Lett., 21, (1980) 2591.

(67) Y. Ito, A. Aoyama, T. Hirao, A. Mochizuki, and T. Saegusa, J. Am. Chem.
      Soc., 101 (1979) 494.

62

(68) Y. Ito, A. Aoyama, and T. Saegusa, J. Am. Chem. Soc., 102 (1980) 4519.

(69) Y. Ito, M. Nakatsuka, N, Kise, and T. Saegusa, Tetrahedron Lett., 21 (1980) 2873.

(70) J. Tsuji, K. Masoaka, and T. Takahashi, Tetrahedron Lett., (1977) 2267.

(71) R. Baker and P. M. Winton, Tetrahedron Lett., 21 (1980) 1175.

(72) J. Tsuji, K. Kasuga, and T. Takahashi, Bull. Chem. Soc. Japan, 52 (1979) 216.

(73) A. I. Lazutkin, A. M. Lazutkin, and Yu. I. Yermakov, React. Kinet. Katal. Lett., 8 (1978) 353.

(74) J. P. Haudegond, Y. Chauvin, and D. Commereuc, J. Org. Chem., 44 (1979) 3063.

(75) Y. Tamaru, R. Suzuki, M. Kagotani, and Z. Yoshida, Tetrahedron Lett., 21 (1980) 3791.

(76) Y. Tamaru, R. Suzuki, M. Kagotani, and Z. Yoshida, Tetrahedron Lett., 21 (1980) 3787.

(77) J. Tsuji, T. Yamakawa, and T. Mandai, Tetrahedron Lett., (1978) 565.

(78) J. Tsuji and T. Mandai, Tetrahedron Lett., 21 (1980) 3285.

(79) This figure is adapted from an excellent review on the chemistry of arenemetal tricarbonyl complexes by M. F. Semmelhack, J. Organomet. Chem. Libr., 1 (1976) 361.

(80) M. F. Semmelhack and H. T. Hall, Jr., J. Am. Chem. Soc., 96 (1974) 7091; 7092.

(81) M. F. Semmelhack, H. T. Hall, Jr., R. Farina, M. Yoshifuji, G. Clark, T. Barger, K. Hirotsu, and J. Clardy, J. Am. Chem. Soc., 101 (1979) 3535.

(82) M. F. Semmelhack, H. T. Hall, Jr., M. Yoshifuji, and G. Clark, J. Am. Chem. Soc., 97 (1975) 1247.

(83) M. F. Semmelhack, G. R. Clark, R. Farina, and M. Saeman, J. Am. Chem. Soc., 101 (1979) 217.

(84) M. F. Semmelhack and G. Clark, J. Am. Chem. Soc., 99 (1977) 1675.

(85) A. P. Kozikowski and K. Isobe, J. C. S. Chem. Comm., (1978) 1076.

(86) J. P. Kutney, R. A. Badger, W. R. Cullen, R. Greenhouse, M. Noda, V. E. Ridaura-Sanz, Y. H. So, A. Zanarotti, and B. R. Worth, Can. J. Chem., 57 (1979) 300.

(87) J. P. Kutney, M. Noda, and B. R. Worth, Heterocycles, 12 (1979) 1269.

(88) M. F. Semmelhack, J. J. Harrison, and Y. Thebtaranoth, J. Org. Chem., 44 (1979) 3275.

(89) M. F. Semmelhack, V. Thebtaranoth, and L. Keller, J. Am. Chem. Soc., 99 (1977) 959.

(90) M. F. Semmelhack and A. Yamashita, J. Am. Chem. Soc., 102 (1980) 5926.

(91) M. F. Semmelhack, J. Bisaha, and M. Czarny, J. Am. Chem. Soc., 101 (1979) 76

(92) R. J. Card and W. S. Trahanovsky, <u>J. Org. Chem.</u>, 45 (1980) 2555; 2560.

(93) M. Uemura, N. Nishiwaka, and Y. Hayashi, <u>Tetrahedron Lett.</u>, (1980) 2069.

(94) J. P. Kutney, T. C. W. Mak, D. Mostowicz, J. Trotter, and B. R. Worth, <u>Heterocycles</u>, 12 (1979) 1517.

(95) M. F. Semmelhack, W. Seufert, and L. Keller, <u>J. Am. Chem. Soc.</u>, 102 (1980) 6586.

(96) G. Jaouen, A. Meyer, and G. Simonneaux, <u>J. C. S. Chem. Comm.</u>, (1975) 813.

(97) G. Jaouen and G. Simonneaux, <u>Tetrahedron</u>, 35 (1979) 2249.

(98) J. Besancon, J. Tirouflet, A. Card, and Y. Dusausoy, <u>J. Organomet. Chem.</u>, 59 (1973) 267.

(99) A. Meyer and G. Jaouen, <u>J. C. S. Chem. Comm.</u>, (1974) 787.

(100) G. Jaouen and A. Meyer, <u>J. Am. Chem. Soc.</u>, 97 (1975) 4667.

(101) H. des Abbayes and M. A. Boudeville, <u>J. Org. Chem.</u>, 42 (1977) 4104.

(102) For a review on this subject see G. Jaouen, "Arene Complexes in Organic Synthesis," in <u>Transition Metal Organometallics in Organic Synthesis</u>, 33 II (1978) 65, H. Alper, Ed.

(103) A. Meyer and O. Hofer, <u>J. Am. Chem. Soc.</u>, 102 (1980) 4410, and references therein.

(104) C. P. Casey, "Metal-Carbene Complexes in Organic Synthesis," in <u>Transition Metal Organometallics in Organic Synthesis</u>, I (1976) 190.

(105) L. S. Hegedus, PhD Thesis, Harvard University, Cambridge, Mass., (1970) pp 85-87.

(106) E. J. Corey and L. S. Hegedus, <u>J. Am. Chem. Soc.</u>, 91 (1969) 4926.

(107) Y. Sawa, I. Hashimoto, M. Ryang, and S. Tsutsumi, <u>J. Org. Chem.</u>, 33 (1968) 2159.

(108) S. K. Myeong, Y. Sawa, M. Ryang, and S. Tsutsumi, <u>Bull. Chem. Soc. Japan</u>, 38 (1965) 330.

CHAPTER 2

THE SYN EFFECT AND THE USE OF ENOLATE EQUIVALENTS IN SYNTHESIS

ROBERT R. FRASER

Ottawa-Carleton Institute for Research and Graduate Studies in Chemistry,
University of Ottawa, Ottawa K1N 9B4 Ottawa

CONTENTS

INTRODUCTION

   The use of organometallic reagents in organic synthesis has greatly expanded
both in scope and magnitude in recent years. One small segment of this expansion
which has attracted a great deal of interest is the chemistry of alpha-lithiated
carbonyl derivatives. This group, comprised of three main classes - hydrazones,
oximes and imines, serves as "enolate equivalents" or "masked enolates" in fulfilling
the role depicted in route a in [1], i.e., formation of the carbonyl derivative,

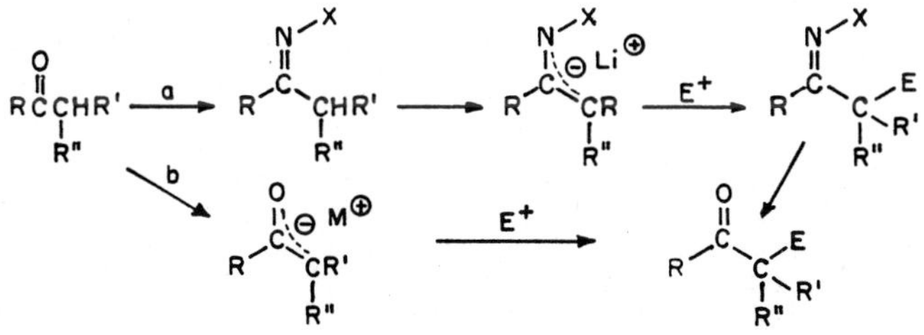

metallation, alkylation and finally hydrolysis to the new carbonyl compound. The
same transformation can be achieved more directly via the enolate (route b), yet the
more circuitous route (a) frequently proves to be more advantageous, since the
particular stereochemical properties of these intermediates and their consequent
manifestations in reactivity confer upon them certain unique chemical capabilities.

Specific lithiated derivatives of aldehydes and ketones to be considered are
hydrazones and their derivatives, 1, oximes and oxime ethers, 2 (including isoxazolines) and imines, 3. The reactions of these intermediates provide a useful

$\underline{1}$ X = NH$_2$, N(CH$_3$)$_2$, $\underline{2}$ R = Li, CH$_3$, $\underline{3}$ R = alkyl or aryl

NHOTs, or $\overset{\oplus}{N}$(CH$_3$)$_3$ $-CH_2CH_2C\alpha$

(R' = H or alkyl, R" = alkyl)

complement to the role of metal enolates and other enolate equivalents in synthesis.
In another chapter of this book, Heathcock[1] describes the use of metal enolates in
stereoselective aldol condensations. Additional coverages of enolate chemistry
include several recent monographs[2,3,4] and review articles[5,6,7,8].

II. THE SYN EFFECT

(A) General Aspects

In contrast to the chemical properties of enolates, the "masked enolates", 1, 2,
and 3 all exhibit a marked thermodynamic preference for the syn configuration at the
C-N partial double bond.[†] The large magnitude of this preference has generally pre-
cluded its determination' in only one of the rare cases[9] in which anti species has
been observed, could the position of the equilibrium be determined. This thermodynamic preference for the more sterically congested diastereomer confers two
additional features upon the reactions of these anions, high regioselectivity and
high stereoselectivity. The lithiation and electrophilic substitution of all ketone
derivatives has consistently been strongly regioselective, the site of attack being
the least substituted carbon atom. The reactions with electrophiles are in general
highly diastereoselective as a result of perpendicular or axial attack. Consequent

---

[†] At this point it is appropriate to justify our adoption of syn-anti nomenclature
to denote the stereochemistry at the C$_2$-N bond of the LiC$_1$-C$_2$-N$_3$-X$_4$ fragment, in
spite of its supercession by the E-Z system of IUPAC. We retain it for purposes
of clarity in designating the C$_2$-N bond as syn (Z) or anti (E) while using the E-
nomenclature to clearly indicate the stereochemistry of the C$_1$-C$_2$ bond.

the reaction sequence [1] in which X possesses chirality proves capable of forming carbonyl derivatives having high optical purities. Such applications to asymmetric synthesis will be discussed in detail at the end of this chapter.

The first examples of allyl anions which exhibited an unexpected preference for the cis or syn-like geometry were described over twenty years ago in studies of metallated allyl ethers[12]. Subsequently, metallation of allyl amines[13], thiols[14], and cis 2-butenes[15] were all found to produce anions having greater stability in the cis configuration. The first report of a syn preference was for a 2,3-diazaallyl ion as observed in metallated nitrosamines 4 whose contributing structure 4(b) is iso-electronic with the general formula for the masked enolates, 1, 2 and 3[16].

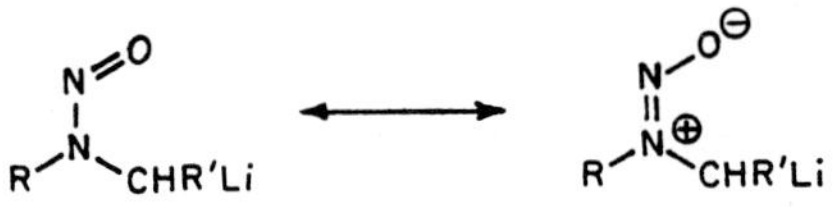

(R = alkyl, R'= H or alkyl)

Studies in our laboratory on the stereochemistry of the H-D exchange in a rigid nitrosamine and later on the metallation and alkylation of other conformationally biased nitrosopiperidines, established the same stereochemical characteristics in all metallated intermediates, which were subsequently duplicated in all respects in studies of analogues in the categories of 1, 2 and 3. For this reason we choose to describe, in detail, studies on the stereochemistry of metallated nitrosamines as representative examples of intermediates whose reactions are controlled by the "SYN EFFECT".

(B) In Metallated Nitrosamines

In order to investigate the reasons for the unusual acidity of protons alpha to a nitrosamino function[16], we subjected the bridged biaryl nitrosamine, 5, to a variety of H-D exchange reactions in alcoholic media[17]. Compound 5 has four non-equivalent benzylic protons which are well resolved in its 100 MHz ^{1}H spectrum.

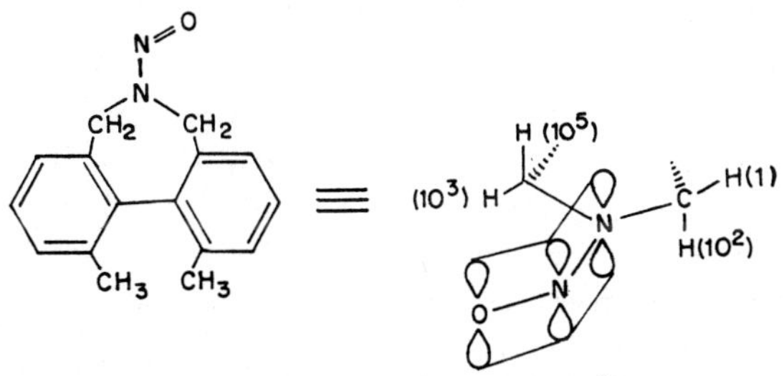

By monitoring the disappearance of each peak with time during the base-catalyzed exchange in t-butanol-O-d, their relative exchange rates could be measured. They are given in brackets beside the appropriate protons in the stereoformula for 5. Thus each axial proton exchanged 100 times faster than its equatorial partner and each syn proton exchanged 1000 times faster than the anti proton in the same (axial equatorial) environment. Because the half-life for interconversion of the syn and anti environments was only two hours (via N-N bond rotation) the accuracy of the quoted rates was in some runs only ±20%. While the axial-equatorial rate ratio of 100 was anticipated as a reflection of the importance of stereoelectronic control during the exchange[18], the observed 1000-fold rate difference for syn vs anti protons was totally unexpected.

To explore the cause of the preferential syn exchange, the relative rates were also measured in the presence of crown ether. The observation that 1.1 equivalent of dicyclohexyl-18-crown-6 caused only a slight (2 to 3 fold) decrease in rates of exchange of both the syn(equatorial) and anti(axial) protons was a clear indication that chelation was not a factor contributing to the faster syn exchange. The best explanation available to us at that time was an effect of orbital symmetry. Since the anion could be considered, as shown in the canonical form 6, to be isoelectronic

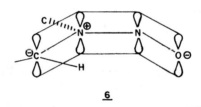

6

with butadiene dianion, its HOMO has the symmetry property which permits an attractive interaction across the termini of the four atom 6 pi electron fragment. This is a paraphrase of the argument originally expressed by Hoffmann and Olofson[19] to account for the unusual relative stabilities of dialkoxy and dihaloethylenes. To complement our studies on the H-D exchange we also studied the stereochemistry of the methylation of alpha-nitrosamino carbanions. The lithiation of 5 using LDA (lithium diisopropylamide) in THF (tetrahydrofuran) gave a deep red solution which decolorized on addition of methyl iodide to yield a single alkylation product having the methyl group in place of the syn-axial proton[20]. This product slowly equilibrated to a mixture of syn and anti axial methyl derivatives. For a more detailed study of the alkylation reaction (21) the conformationally biased* 4-phenyl-N-nitrosopiperidine

---

* The phenyl group at the 4 position of the piperidine ring would be expected to have an effect similar to that in phenylcyclohexane in which the axial conformation is populated to less than 1% (see reference 22).

7, was chosen. Lithiation of 7 using LDA in THF at 0° followed by cooling to -78°, then addition of methyl iodide and warming 15 minutes later to room temperature, hereafter simply stated as "standard conditions", gave methylated 8 in good yield. The methyl group was found to be predominantly syn (3:1) and exclusively axial. It was assumed the anti axial derivative arose from isomerization (N-N rotation) of the product, see [2], and no attempts were made to avoid this process during work-up.

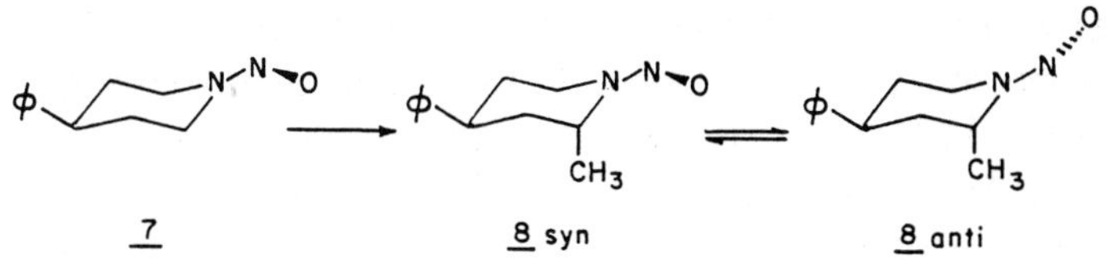

| 7 | 8 syn | 8 anti | [2] |

A second consecutive alkylation under standard conditions gave, as the only product, the 2,6-diaxial dimethylpiperidine 9. Thus the second alkylation was also completely stereoselective in yielding only axial substitution product in spite of the strain accompanying the introduction of a second axial group. A closer examination of the factors involved in this highly stereoselective alkylation was achieved by determining the relative stabilities of the observed product 9, and the alternative alkylation product, 10 and 11, shown in [3].

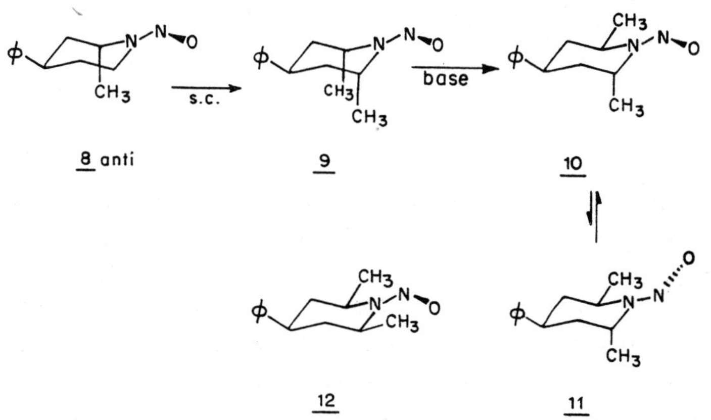

Equilibration of 9 with its isomers 10, 11 and 12 was attempted by heating it at 90° in DMSO (dimethylsulfoxide) containing potassium t-butoxide. Starting with either 9 or 10, a 25:75 ratio of 9 to 10 was attained after 24 hours with no detectable amounts of either 11 or 12 being formed. This provides a value for ΔG

for the equilibrium 9 $\rightleftarrows$ 10 of -0.8 kcal/mole. Since no 11 (the rotamer of 10) coul
be detected in a spectrum of pure isolated 10 it must be at least 2.8 kcal less stab
than 10 and thus at least 2.0 less stable than 9. Clearly the second alkylation
requires formation of the syn anion from 8 (anti) because formation of a syn anion
from 8 (syn) would suffer a severe A[1,3] strain of over 4.4 kcal/mol (21,23). Alkyla
tion then takes place exclusively trans to the phenyl permitting the formation of 9
in preference to the much less stable 11. That not even a trace of 11 is formed
during alkylation of 8 suggests that stereoelectronic factors[18] may be reinforcin
the steric preference for axial attack. Concurrent with our work on the stereochemistry of the alkylation of nitrosamines, Seebach and his coworkers carried out
extensive studies on reactions involving lithiated nitrosamines which established th
synthetic utility of the method[24].

(C) In Enolate Equivalents

As alluded to earlier, the same syn-axial methylation and 2,6-diaxial dimethylati
was again observed in the reactions of the hydrazones[25,26], oxime ethers[25] and
imines[27] of 4-t-butylcyclohexanone. In the case of the configurationally stable oxi
ether the methylation was shown to be $\geq$ 99.5% stereoselective. Small amounts of a
second isomer were formed in the reactions of hydrazones (10%) and imines (7%), in
each instance their presence being attributable to epimerization of the product. Th
systems 1, 2, 3 and 7 have a marked preference for formation of a syn lithio derivative whose configuration then confers upon it two additional properties, regioselect
vity in the site of lithiation and a propensity for axial attack by an electrophile.
Steric effects appear to be responsible for the regioselectivity. Consider, for
example, the lithiation of the carbonyl derivative of 2-butanone 13 which could
produce isomers 14, 15 and 16.

That 16 is formed almost exclusively[28] in those derivatives where X=N(CH$_3$)$_2$[26] or
alkyl[28] can be attributed to steric destabilizations in 14 and 15, both of which
suffer an eclipsing of the methyl group at C$_1$ with a substituent on C$_2$. As long as
N-X is not configurationally stable, 16 can be formed. In the case of X=OR, lithiation can only produce 14 and 15. Evidence in support of this interpretation was
obtained by a study of the acidities of a few simple aldimines and ketimines[29]. The
relative pK's of imines 17, 18, 19 and 20 are given with the formulae shown on the
next page.

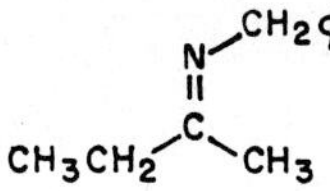

$$N\text{-}CH_2\phi$$

$$CH_3CH_2\text{-}C\text{-}CH_3$$

(pK = 0)

17

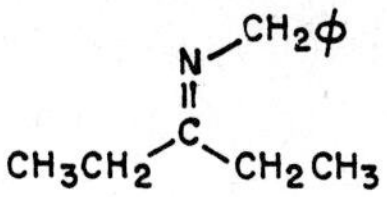

$$N\text{-}CH_2\phi$$

$$CH_3CH_2\text{-}C\text{-}CH_2CH_3$$

(pK = 2.0)

18

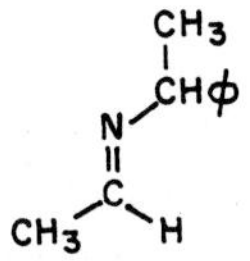

$$CH_3$$
$$CH\phi$$
$$N$$
$$CH_3\text{-}C\text{-}H$$

(pK = 0.3)

19

$$CH_3$$
$$CH\phi$$
$$N$$
$$CH_3CH_2CH_2\text{-}C\text{-}H$$

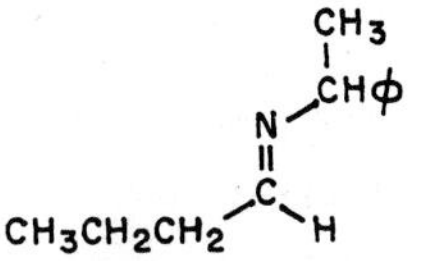

(pK = 0.4)

20

Thus the introduction of a methyl group into 17 results in a 2 unit increase in the pK of 18, whereas the introduction of a methyl group into 19 does not influence the pK of 20. This is consistent with the presence of a steric destabilization in 18-Li. Regardless of the configuration at the C_1-C_2 bond, it can be seen in the structures for (E) and (Z)18-Li that the methyl on the lithiated carbon encounters steric strain with either the ethyl in the (E) configuration or the benzyl group in the (Z) isomer.

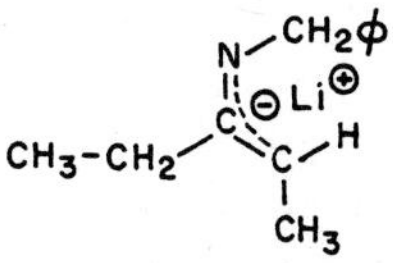

$$N\text{-}CH_2\phi$$
$$C\text{-}Li^{\oplus}$$
$$CH_3\text{-}CH_2\text{-}C\text{-}H$$
$$CH_3$$

or

$$N\text{-}CH_2\phi$$
$$C\text{-}Li^{\oplus}$$
$$CH_3CH_2\text{-}C\text{-}CH_3$$
$$H$$

$$N\text{-}CH_2\phi$$
$$C\text{-}Li^{\oplus}$$
$$H\text{-}C\text{-}H$$
$$CH_2CH_3$$

(E)-18-Li   (Z)-18-Li 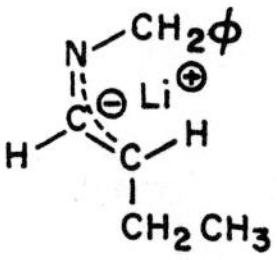   (E)-20-Li

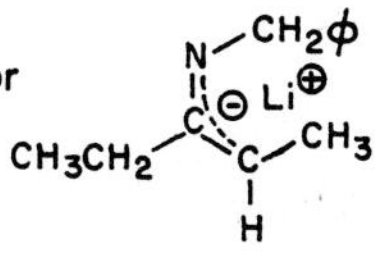

No such problem occurs in <u>17</u>-Li or <u>20</u>-Li so that the pK's are similar for <u>17</u>, <u>1</u> and <u>20</u>, the electronic effect of a methyl group being similar to hydrogen[30]. In summary then, the regioselectivity and stereoselectivity of "masked enolates" are clearly determined by the "syn effect".

III. ORIGIN OF THE SYN EFFECT

The most intriguing question, "What is the origin of the <u>syn</u> effect?" cannot ye be fully answered. The rather substantial number of laboratories involved in lithiations of nitrosamines, hydrazones, oximes and related derivatives were at one time equally divided in embracing chelation or orbital symmetry as the best explan- ation for the "<u>syn</u> effect". However, our observation[27] that lithioimines showed just as large a propensity for <u>syn</u> anion formation could not be accommodated by either rationale.

The imine behaviour also brought into question the validity of either explanati as applied to the other lithiated species. The most informative evidence on this theoretical problem has been provided by the <u>ab initio</u> MO calculations of Houk and his group[10]. They have calculated the fully optimized geometries and the relative energies of the two anions derived from N-methylacetaldimine. The results, in brackets beside the formulae, find the <u>syn</u> anion, <u>21</u>, to be more stable than the <u>anti</u> anion, <u>22</u>, by seven kcal/mole, a result in good qualitative agreement with experiment*, specifically that alkylation of N-isopropylpropionaldimine gave virtually exclusively the <u>syn</u> product (> 96%).

+7.0 kcal/mol

<u>21</u>   <u>22</u>

---

* The arguments presented in the literature[31] are qualitative in that the strain in lithiated aldimine was estimated to be at least two kcal/mol. We have now obtained direct evidence in support of this estimate[29]. The pK of the isobutyraldimine of α-phenethylamine was observed to be 2.5 units greater than the corresponding imine acetaldehyde, yet its lithio derivative is still 100% <u>syn</u>.

The geometries of 21 and 22 seem perfectly normal, an indication of the lack of any unusual intramolecular effects such as n-π^* interactions. The calculations were also carried out on the radical and cationic analogues of 21 and 22. For the radicals the anti isomer was slightly more stable than the syn and for the cations the stability of anti over syn became substantial. Additional probes of the effects of electron density[10] led to the conclusion that the most dominant influence of the syn-anti energy difference is a simple two-electron repulsion between the lone pair density at nitrogen and the pi electron density at C-1. More extensive calculations have been made in an attempt to include both the lithium cation and solvent. Inclusion of a bare lithium leads to a change in the relative energies, anti becoming more stable. As solvation of the lithium is added the syn again becomes the more stable for a σ-coordinated lithium, but not in the case of the pi coordinated lithium[32]. Until such calculations involving solvation can be fully optimized we must be satisfied with the qualitative indications of theory. So far, attempts to obtain a crystalline lithiated imine to provide experimental parameters to simplify such calculations have been unsuccessful. An impressive aspect of the calculations on 21 and 22 was their ability to account for the anomalous experiments in which endocyclic imines gave products arising from exclusive anti lithiation. The calculated geometry for 21 showed the N-C_2-C_1 bond angle to be 133°. Thus the endocyclic imines examined experimentally, 23, were constrained to much smaller values for this angle, 120° or less, thereby introducing considerable strain in the syn anion. Anti lithiation was attained by default. Qualitative support for this interpretation comes from pK measurements[29] on 24 and 23 (n=1). The anti protons of 23 were found to be 2 pK units less acidic than those of 24 (n=1) as would be expected for an anti versus a normal syn anion.

23 (n = 0, 1, 2)   24

IV. SPECTROSCOPIC STUDIES OF THE SYN LITHIATED INTERMEDIATES

The power of ^{1}H and ^{13}C nmr spectroscopy in ascertaining the stereochemistry of metal enolates has been clearly established by the work of House and coworkers[33-36]. That this stereochemistry exerts profound effects on the erythro-threo ratio of aldol condensation products is now well documented by the efforts of Dubois[37], Ireland[38], Heathcock[39], as well as by many more recent studies[1]. Such success notwithstanding, only a few studies involving direct spectroscopic examination of species belonging to

classes 1, 2 and 3 have been reported until recently[11,40,41]. In the first of these
Bergbreiter and Newcomb[11] were able to prove by ^{1}H nmr and trapping experiments that
the lithiation of the N,N-dimethylhydrazone of propionaldehyde gave only one intermediate under standard conditions, that having the E-syn stereochemistry as shown
in 25.

25   26

A drastic change in the stereochemistry was observed when the deprotonation was
carried out in THF containing 2.5 eq of hexamethylphosphoramide (HMPA). Under these
conditions the major lithio derivative was the Z-anti isomer 26 with 25 comprising
only 15% of the total. The configurations at C_1-C_2 were clearly indicated by ^3J
values of 12.5 and 7.5 Hz for 25 and 26. Trapping the 15.85 mixture of the two with
butyl iodide gave a 15:85 ratio of syn to anti hydrazones, thereby establishing the
configuration at nitrogen. The impact of such a change on the degree of asymmetric
syntheses was also examined in collaboration with Enders[42]. Enders' chiral hydrazone[42], whose formula appears in Table 3, was deprotonated to give an E-syn anion
whose alkylation gave, after hydrolysis, the chiral aldehyde of 82% optical purity.
The Z-anti isomer of the same hydrazone, generated as above in the presence of HMPA,
gave alkylated aldehyde of the opposite chirality in only 10% optical purity. Thus
reactions of the intermediates, which do not equilibrate under standard conditions,
are profoundly influenced by their stereochemistry. Indeed, in a related study, Meyers
and Williams,[43] showed that the optical yields obtained by lithiation and alkylation
of chiral ketone hydrazones were increased by heating the lithiated intermediate for
2 hours at 76° prior to alkylation. The higher optical yields were attributed to
promotion of a Z → E isomerization. The temperatures required are indicative of a
barrier to rotation about C_1 - C_2 which must have been > 23 kcal/mole. A much
smaller barrier has recently been measured by nmr[44] for two lithiated imines. For
the lithiated cyclohexyl and t-butyl imines of acetaldehyde, the barrier was found
to be only 17 kcal/mole.

Spectroscopic studies in our laboratories have focussed on the lithiated imines,
3, in which we have detected a very different rotational process as indicated by
variable temperature nmr[45]. The simplest of these, lithium 4-methyl-3-azapentenide,
27 exhibited normal ^{1}H and ^{13}C spectral characteristics at room temperature[46].
Unexpected, however, was the observation of two equally populated species at low
temperatures whose coalescence temperature of -20° for the formyl proton and -8°C

for the ^{13}C peaks due to C_2 indicate the presence of an interconversion process having $\Delta G^{\neq}$ = 12.7 kcal/mole.   That the two species are actually rotamers 27a and 27b,   arising from slow rotation about the $N-C_4$ bond was indicated by a number of facts. (a) Replacement of the formyl proton at C_2 by a methyl or an isopropyl group caused the population ratio to change to 2:1 in either case, an effect which is

27 a   27 b

not consistent with the alternative explanation of interconverting syn and anti forms. (b) Also inconsistent with this explanation was the observation of equal vicinal coupling constants for the protons at C_1 and C_2 (^3J = 7.5 and 14.5 Hz) in both species. Bergbreiter and Newcomb had earlier observed differences of at least 1 Hz in ^3J (trans) for the syn and anti isomers of lithiated acetaldehyde NN-dimethyl-hydrazone[11,42]. (c) Replacement of the N-isopropyl by N-(β)-phenethyl group resulted in only one observable species over the temperature range -70 to +30°C. (d) Crude ab initio calculations on the interconversion 27 (a) $\overrightarrow{\leftarrow}$ 27 (b) are in agreement, yielding a value for the barrier of 11.1 kcal/mole[45] although failing by 3 kcal/mole to reproduce the observed population ratio. Such qualitative agreement by calculations, which were not fully optimized during rotation, is good support for the proposed rotational equilibrium.

The presence of a barrier to rotation at the $N-C_4$ bond is consistent only with an anion having the syn configuration and also required the barrier to inversion at nitrogen to be at least 12.7 kcal/mole. Other evidence from our laboratory provides a maximum value for this barrier to syn-anti interconversion. In an experiment in which anti acetaldimine in THF was placed in a ^{13}C nmr tube, lithiated at -78°C with LDA, transferred immediately after addition of CH_3I to the ^{13}C probe at -65°, and spectra accumulated over the next ten minutes, only signals from syn lithiated product could be observed. This limits the half life of initially formed anti lithiated imine to <5 min at -65°, indicative of an inversion barrier of < 14.5 kcal/mole.

Additional evidence that the lithiated imines have the syn configuration is also provided by direct nmr observation.   The ^{13}C shieldings for the C_4 carbons of the lithiated aldimines are particularly informative. These ^{13}C signals exhibit an upfield shift of about 9 ppm on lithiation[28]. This shift reflects the change in configuration at the C_2-N from anti in the aldimine 28 to syn in the lithio derivative 28-Li thereby introducing a γ-gauche shielding interaction between C_1 and C_4. In neutral imines the γ-shielding effect was less, varying from 5.8 ± 0.7 ppm in

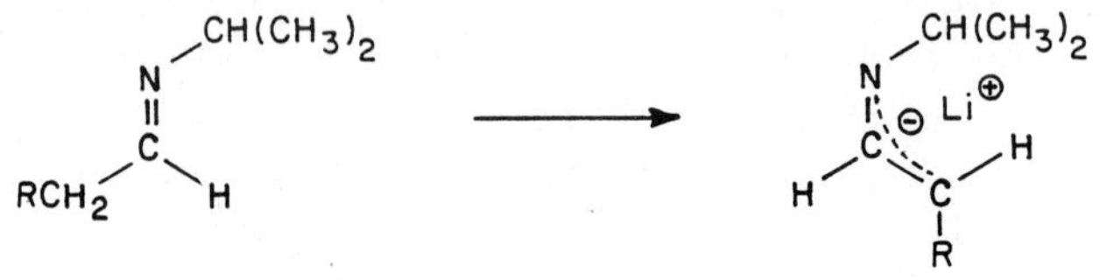

28   28-Li

aldimines to 9 ± 2.7 ppm in ketimines.[46] Even in the case of N-t-butyl ketimines there
is evidence that the lithio derivatives have the syn configuration[28].

Bergbreiter and Newcomb have also made extensive use of nmr techniques in inves-
tigating the regioselectivity in the deprotonation of hydrazones. Contrary to the
first report that the deprotonation took place at the anti carbon of 3-pentanone
N,N-dimethylhydrazone[47] they showed from studies on the hydrazone 29 stereoselective
labelled with ^{13}C, that deprotonation occurred equally at the syn and anti carbons.[49]
Treatment of 29 with lithium diethylamide in THF at 0°C yielded the two syn anions as
indicated by the formation of equal amounts of the differently labelled syn products
as shown in [4]. The same conclusion was reached independently in our laboratory[48]
Lithium diisopropyl amide was more selective showing a 70:30 preference for syn
deprotonation and greater selectivity was found at lower temperatures[49a]. These

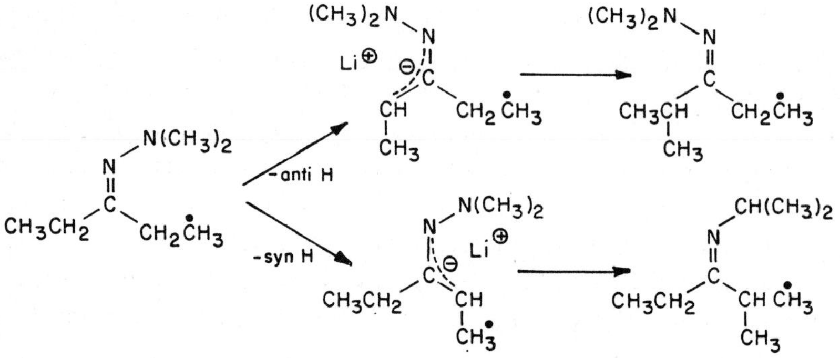

[4]

results are in accord with the expectation that an exothermic deprotonation should
proceed via transition states which only partly reflect the stabilities of the
lithiated intermediates. Most recently, application of the same technique to assess
the deprotonation of the imines of 3-pentanone also showed little regioselectivity[49]
for that process.

V. CONFIGURATIONAL STABILITIES OF THE INTERMEDIATES

The metallation of the three classes of carbonyl derivatives can be described as having the ability to occur by all the available pathways outlined in [5], the specific route being determined by the magnitude of the inversion barrier at nitrogen

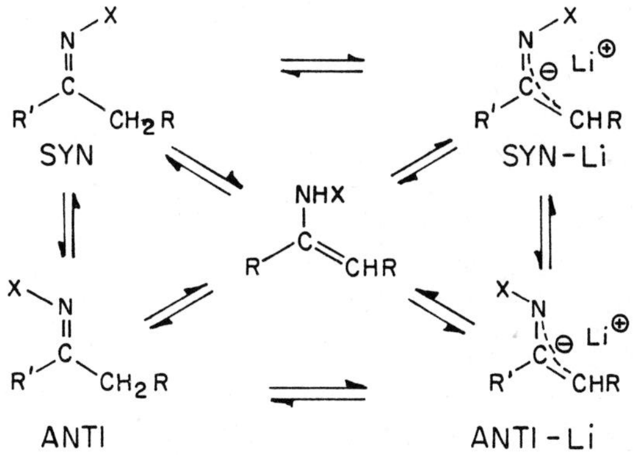

[5]

in conjunction with the metallation conditions. Several definitive statements about the deprotonation mechanisms can be made on the basis of existing experimental evidence coupled with our knowledge of the magnitude of the inversion barriers in the various carbonyl derivatives. These barriers representing the thermal inversion barrier, are summarized in Table 1. For all compounds there also exists, in principle, a lower barrier for an acid or a base-catalyzed inversion.

(a) Except for derivatives of propionaldehyde all observable lithiated inter-mediates have existed exclusively in the syn configuration. This indicates in general a low inversion barrier for the equilibrium SYN-Li ⇄ ANTI-Li with SYN-Li more stable. However, lithiated propionaldehyde N,N-dimethylhydrazone and lithiated propionaldehyde N-cyclohexyl imine are exceptions in that deprotonation in THF con-taining 2 equivalents of HMPA results in syn-anti ratios of 15:85 and 44:56 respec-tively[11].   The equilibrium shown in [5a] could only be achieved for the lithiated imine, as the solution of lithiated hydrazone decomposed. At equilibrium the syn-anti ratio was 7:1. That some anti isomer is present in this case need not be

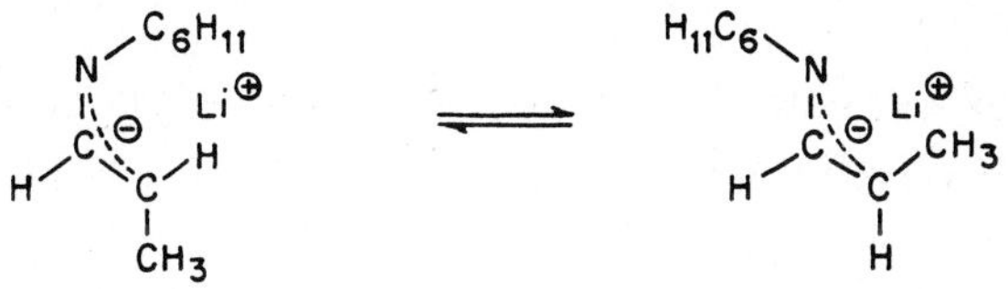

[5a]

interpreted as an indication of a diminished magnitude for the syn effect. Rather, this syn isomer, which has the E configuration at the C-C bond is not as strongly favored over the anti-isomer, since the anti form has the Z configuration at the C-C bond, to strongly enhance its stability [13,50]. The significance of the role of HMPA is also of importance and is currently under investigation in our laboratory.

(b) Deprotonation of a symmetrical ketone derivative will usually proceed via direct abstraction of a <u>syn</u> proton[27,51] although in certain cases appreciable amount of <u>anti</u> abstraction can take place[49], even becoming predominant[49d].

(c) Deprotonation of unsymmetrical carbonyl derivatives will tend to greater sy deprotonation in forming the regioselectively favored isomer as the more stable pre-cursor will have the smaller group <u>syn</u>.

TABLE 1

Inversion Barriers for Carbonyl Derivatives

	$\Delta G^{\ddagger}$ (kcal/mole)	Temp. (°C)	Approximate half-life at 25°	Reference
X = OCH_3	> 35	25	>> 100 hr.	a
= ONa	32	25	∿ 100 hr.	b
= $N(CH_3)_2$	23-26	> 140	1-24 hr.	c
= CH_3	24-26	25	2-24 hr.	d
α-Li imine	12-14	-70	1 sec	e

a. D.Y. Curtin, E.J. Grubbs and C.G. McCarty, J. Am. Chem. Soc., 88 (1966) 2775.

b. E.J. Grubbs, D.R. Parker and W.D. Jones, Tetrahedron Lett., (1973) 3297.

c. H.O. Kalinowsky, H. Kessler, D. Liebfritz and A. Pfeffer, Chem. Ber, 106 (1973), 1023; G.J. Karabatsos and S. Stassinopoulous, Tetrahedron, 32 (1976) 1147.

d. W.B. Jennings and D.R. Boyd, J. Am. Chem. Soc., 94 (1972) 7187.

e. See discussion, p. 13. For a general view of structural effects on such barriers, see H. Kessler, Tetrahedron, 30 (1974) 1861.

(d) Deprotonation of aldehyde hydrazones and aldimines, both of which exist exclusi ly in the <u>anti</u> configuration[52,25] must therefore involve <u>anti</u> deprotonation. Whether this involves direct formation of <u>anti</u>-Li and inversion or transformation v its tautomer to the <u>syn</u> carbonyl derivative cannot be determined.

(e) The <u>anti</u> configurational isomer of aldoximes has been reported resistant to deprotonation (LDA at 0°)[53] yet we were able to achieve complete deprotonation to fc <u>syn</u> lithio derivative by heating the reaction mixture at room temperature[29]. Report

of the conversion of the _anti_ form of ketoximes into isomerized _syn_ lithio derivatives have also been described[51,54,81] as established by the reaction sequence [6]
which presumably involves initial _anti_ deprotonation.

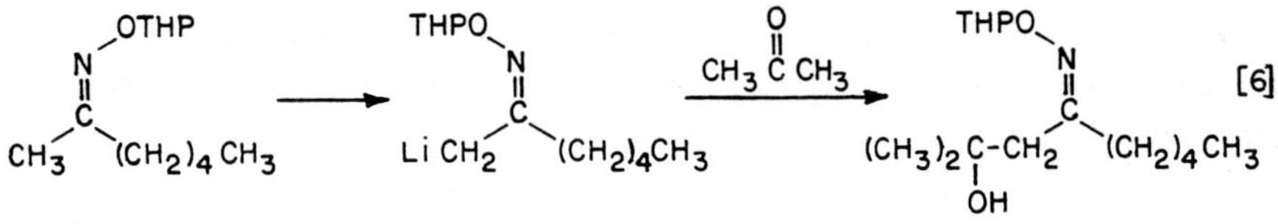

[6]

VI. HYDRAZONES AND THEIR DERIVATIVES

The first metallation alpha to the C=N bond of a hydrazone was described by
Hauser and his co-workers[55]. Treatment of the phenyl hydrazone of acetophenone with
two equivalents of potassium amide formed an intermediate dianion which could then
be converted to its dibenzyl derivative as in [7][55a]. Ensuing studies showed that

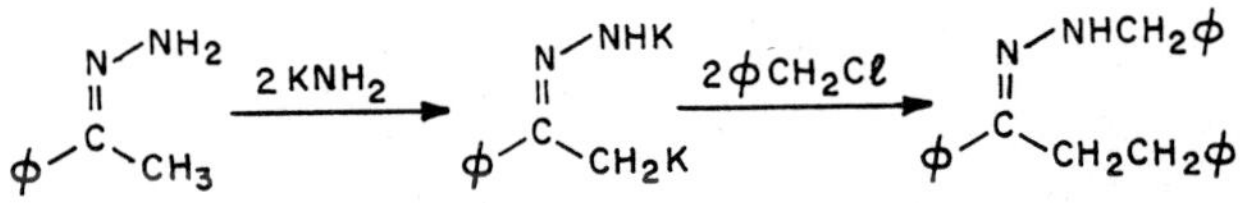

[7]

hydrazones could also be dimetallated with two equivalents of butyllithium and
reacted with ketones[54b] and esters[55c]. Trilithiated hydrazones were also generated
although direct proof of their intermediacy was lacking[55c].

A few years later Stork and Benaim showed that α-β unsaturated hydrazones could
be lithiated (with LDA) and alkylated at the alpha position in good yield[56].

A different base and solvent system, that involving lithium diethylamide in the
mixed solvent system, benzene-HMPA-THF, has been utilized extensively by Henri
Normant's group. They have studied metallations of imines and oximes as well as
hydrazones using this system to which they refer as producing "activated lithium
amides"[57]. Under their conditions, aldehyde hydrazones could be metallated and
alkylated in good yield[58]. Only when both R_1 and R_2 are alkyl groups does the
reaction fail, with elimination to the nitrile taking place (see [8]).

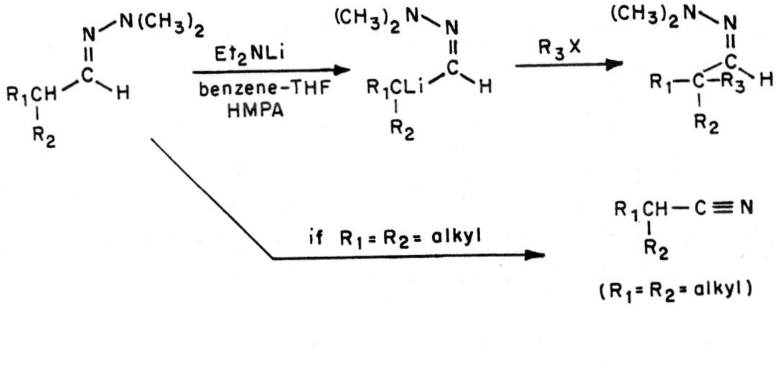

[8]

By far the most complete investigation of hydrazones as synthetic intermediates
has been conducted by Corey and Enders, who examined a wide range of reactions of
this enolate equivalent[59]. Preparation and lithiation of N,N-dimethylhydrazones of
aldehydes and ketones, their reactions with alkyl halides[59a], trimethylsilyl
chloride[59b], aldehydes and ketones[59c], disulfides[59d] and acid chlorides[59e], as well
as several methods of cleavage[59f,g] were all examined in considerable detail. In
addition, the regioselectivity of metallation and the stereochemistry of alkylation
were studied. Their work, and its further elaboration by Enders' group, has shown
that the combination of LDA with THF results in excellent yields of electrophilic
substitution products and this appears to be the method of choice for hydrazone
transformations. A few examples chosen from their work are given on the next page
along with their reported yields of pure isolated product. Particularly notable are
the demonstrated syntheses of a 1,4-diketone and an alpha-alkoxyketone.

Another area in which hydrazones have been shown to have a particular use is in
the synthesis of heterocycles. Specific examples of the synthesis of pyrazoles have
been described by Beam and co-workers[60].

Aziridines can be synthesized from hydrazonio derivatives using the route [9]
which was developed by Laurent and his co-workers[61]. The yields are generally 70%
or greater, a result which establishes the method as superior to an earlier similar

[9]

approach which involved the reaction of oximes with two equivalents of a Grignard
reagent[62].

Vinyl Anions from Tosylhydrazones

A very interesting and useful variant on the Bamford-Stevens reaction was
reported simultaneously by Shapiro[63a] and by Friedman and Schecter and coworkers[64].
When a tosylhydrazone is treated with two or more equivalents of alkyllithium in an
aprotic solvent, loss of lithium tosylate and nitrogen produces a vinyl anion by the
sequence shown in [10]. Parallel to the behaviour of other hydrazones, it has been
established that the syn alpha proton is abstracted preferentially (63b,c, 65) and,

Examples of Hydrazones in Synthesis

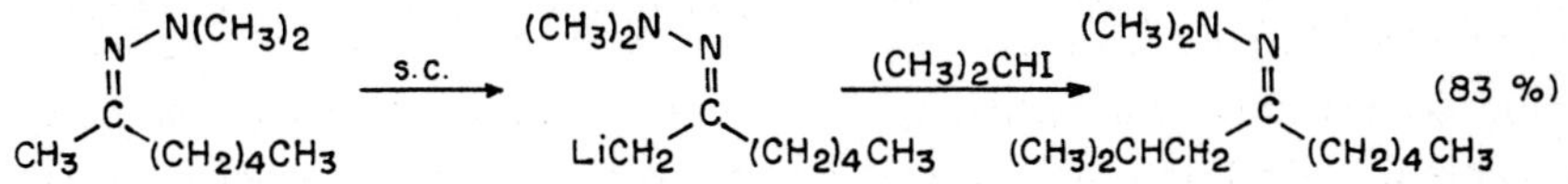

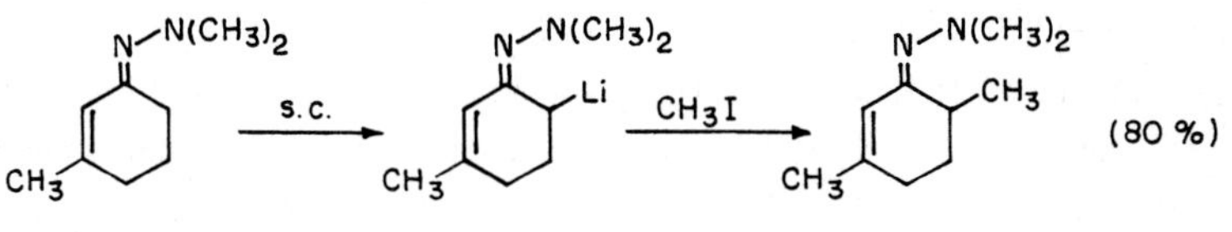

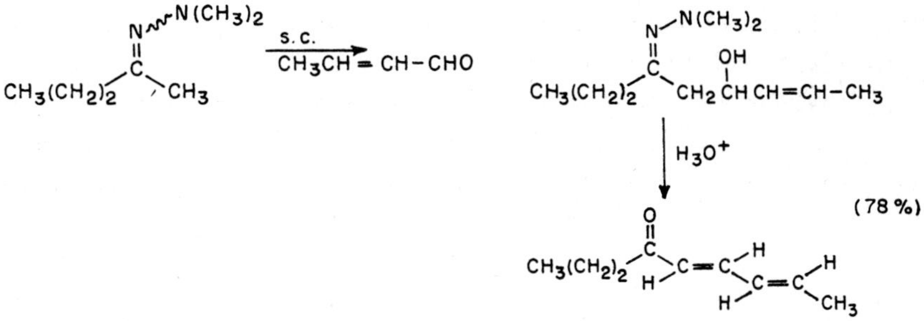

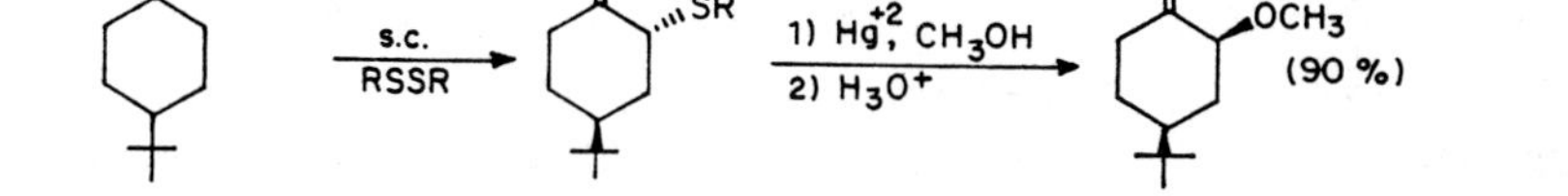

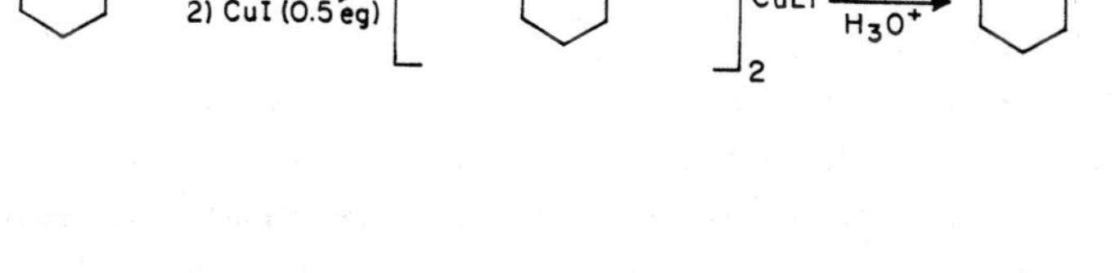

First reaction sequence:

$CH_3-C(=N-N(CH_3)_2)-(CH_2)_4CH_3$ $\xrightarrow{\text{s.c.}}$ $(CH_3)_2N-N=C(CH_2Li)-(CH_2)_4CH_3$ $\xrightarrow{(CH_3)_2CHI}$ $(CH_3)_2N-N=C((CH_3)_2CHCH_2)-(CH_2)_4CH_3$ (83 %)

Second reaction sequence (cyclohexenone hydrazone):

$\xrightarrow{\text{s.c.}}$ Li intermediate $\xrightarrow{CH_3I}$ product (80 %)

Third reaction sequence:

$CH_3(CH_2)_2-C(=N-N(CH_3)_2)-CH_3$ $\xrightarrow[CH_3CH=CH-CHO]{\text{s.c.}}$ $CH_3(CH_2)_2-C(=N-N(CH_3)_2)-CH_2-CH(OH)-CH=CH-CH_3$ $\xrightarrow{H_3O^+}$ (78 %)

$CH_3(CH_2)_2-C(=O)-CH=CH-CH=CH-CH_3$

Fourth reaction sequence:

$CH_3-C(=N-N(CH_3)_2)-CH_3$ $\xrightarrow[I_2(0.5\ eq)]{buLi}$ $CH_3-C(=N-N(CH_3)_2)-CH_2-CH_2-C(=N-N(CH_3)_2)-CH_3$ $\longrightarrow$ 2.5-hexanedione (90 %)

Fifth reaction sequence (t-butylcyclohexanone hydrazone):

$\xrightarrow[RSSR]{\text{s.c.}}$ SR intermediate $\xrightarrow[\text{2) } H_3O^+]{\text{1) } Hg^{+2},\ CH_3OH}$ OCH_3 product (90 %)

Sixth reaction sequence (methylcyclohexanone hydrazone):

$\xrightarrow[\text{2) CuI (0.5 eq)}]{\text{1) s.c.}}$ [dimer]$_2$ $\xrightarrow[H_3O^+]{CuLi,\ MVK}$ product with $-CH_2CH_2C(=O)CH_3$ (82 %)

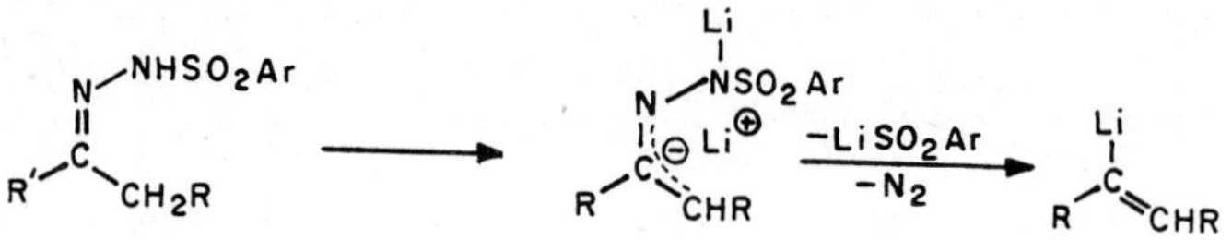

$$[10]$$

in unsymmetrical derivatives, the reaction is regioselective in forming the less substituted olefin exclusively[63a,e]. The vinyl anion has also been generated efficiently in TMEDA solvent [66a] and a triisopropylbenzenesulfonyl group has been shown to provide better yields by blocking the competitive ortho metallation reaction[66b]. Trapping of the vinyl anion has been achieved with many different electrophiles resulting in a variety of substituted olefins including vinyl alde-hydes[67], vinyl silanes[68,69], vinyl germanes and stannanes[69], as well as conjugated acids[70], ethers[71] and alpha-methylene lactones[72]. Application of this decomposition reaction to an α-thio derivative has been shown to provide a new method for a 1,2 carbonyl transposition[73]. Combination of an alkylation of the tosylhydrazone dianion with reconversion of the product to a new dianion, followed by decomposition and further substitution provides a useful synthetic sequence[74,75] as shown in [11]

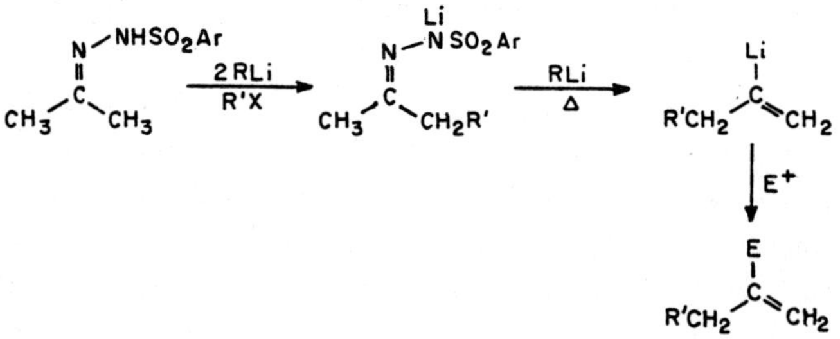

$$[11]$$

VII. OXIMES AND THEIR DERIVATIVES

That oximes and oxime tosylates can undergo highly stereoselective reactions has been known for some time. For example the oxime or the oxime tosylate of dibenzylketone produces a cis aziridine on treatment with lithium aluminum hydride[7] Later investigations[77,78] proved that the reaction proceeded via syn proton abstrac-tion and subsequent reduction of an azirine intermediate as shown in [12]. That the

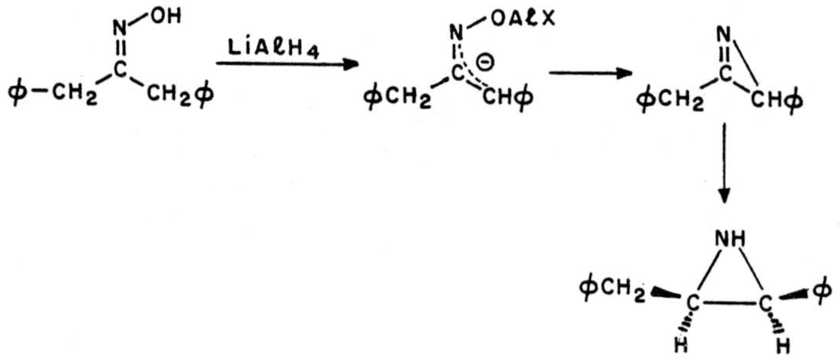

$$[12]$$

O-methyl ether of the same oxime exhibited a preference for abstraction of the syn proton in an H-D exchange was later disclosed by Spencer and Leong[79]. Shortly thereafter, a flurry of papers appeared[80,81,82,25] which established the synthetic potential of lithiated oximes and oxime ethers as enolate equivalents. In particular, it was shown that their lithiation was completely stereoselective occurring at the syn carbon only and subsequent electrophilic substitutions were achieved in high yield[80,81]. In the lithiation and methylation of cyclohexanone oxime ethers, the stereoselectivity was very high, the only detectable products being those resulting from syn-axial methylation[25,82]. A few examples which illustrate the syn and the axial stereoselectivity are shown below in [13] and [14].

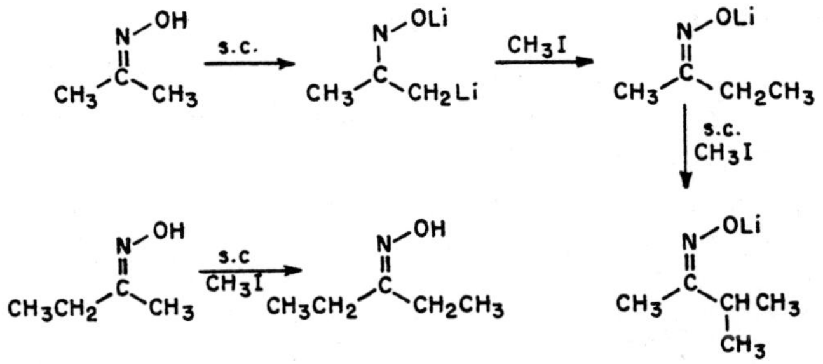

As [13] shows, the syn selectivity allows one to carry out alpha substitution of the unsymmetrical ketone at either carbon by the appropriate reaction sequence. The syn-axial course of alkylation was followed [14] even for the introduction of a highly strained second alkyl group. Another prominent feature of oxime reactivity as shown in [14] is that lithiation will not occur at a tertiary carbon[80,81,25]. Thus the normal conditions of metallation at a syn carbon[51] will fail if the carbon bears two alkyl groups.

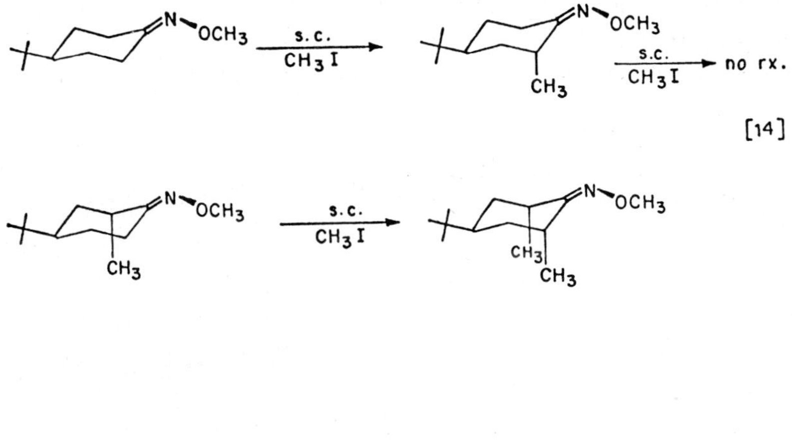

However, the use of more vigorous metallation conditions[81] or equilibration of the anion with unreacted starting material[54] can provide metallation at an anti-carbon as in [15].

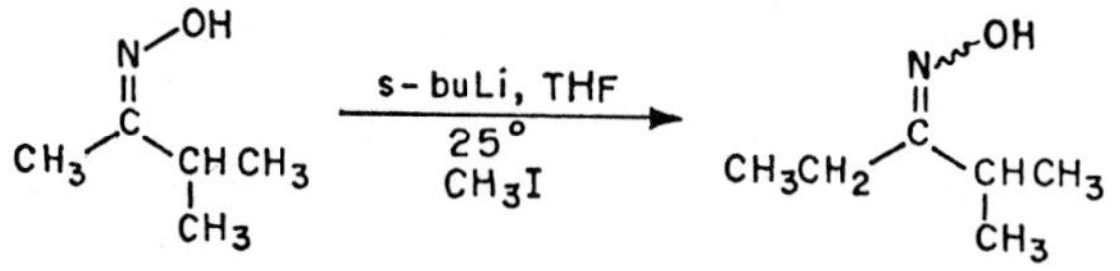

[15]

Thus the high barrier to inversion of the oxime function provides high configurational stability to this class. Forcing conditions either provide a base-catalyzed equilibration of starting materials or anti-lithiation to allow a subsequent more readily achieved inversion.

Applications in Isoxazole Synthesis

Oximes have been found to serve as starting materials for the synthesis of 3,5-disubstituted isoxazoles[83] and isoxazolines[84] by the reaction sequences in [16]. This method, though giving less than 50% yields of isoxazoles due to quenching of

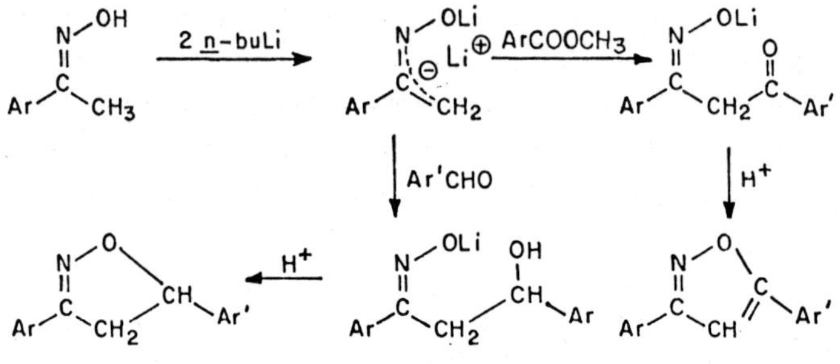

[16]

lithiated oxime by the acidic proton of the acylation product, until recently had provided the best route to this type of isoxazole[85]. A major improvement in isoxazole methodology was recently described by Barber and Olofson[54]. By using dimethylformamide as the electrophile to avoid quenching, and by selection of the oxime of desired stereochemistry, metallation at the desired syn carbon determined the type of isoxazole formed. As illustrated below in [17] and [18], the reactions of the E and Z oxime stereoisomers produce mono- and dialkylisoxazoles respectively.

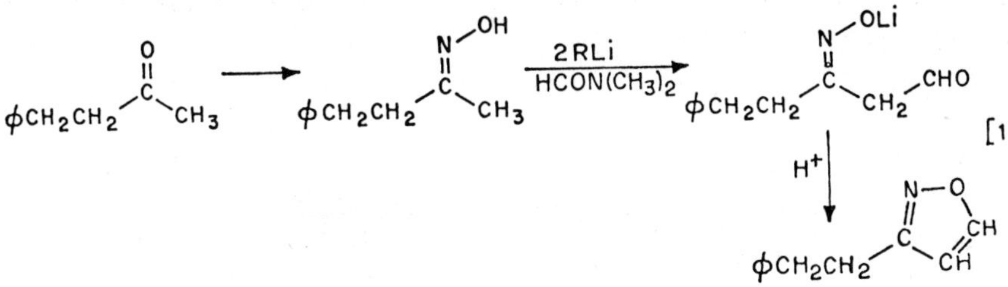

$$\phi CH_2CH_2 \overset{O}{\underset{||}{C}} CH_3 \longrightarrow \phi CH_2CH_2 \overset{N^{OH}}{\underset{||}{C}} CH_3 \xrightarrow[\text{HCON(CH}_3)_2]{2RLi} \phi CH_2CH_2 \overset{N^{OLi}}{\underset{||}{C}} CH_2 {}^{CHO}$$

$$\Big\downarrow H^+$$

$$\phi CH_2CH_2 \overset{N-O}{\underset{C}{\bigtriangleup}} \overset{CH}{\underset{CH}{}}$$

[17]

$$CH_3 \overset{N^{OH}}{\underset{||}{C}} CH_3 \xrightarrow[\phi CH_2Br]{s.c.} CH_3 \overset{N^{OLi}}{\underset{||}{C}} CH_2CH_2\phi \xrightarrow[H CON(CH_3)_2]{RLi} CH_3 \overset{N^{OLi}}{\underset{||}{C}} \underset{\underset{CH_2\phi}{|}}{CH} {}^{CHO}$$

$$\Big\downarrow H^+$$

$$CH_3 \overset{N-O}{\underset{C}{\bigtriangleup}} CH_2\phi$$

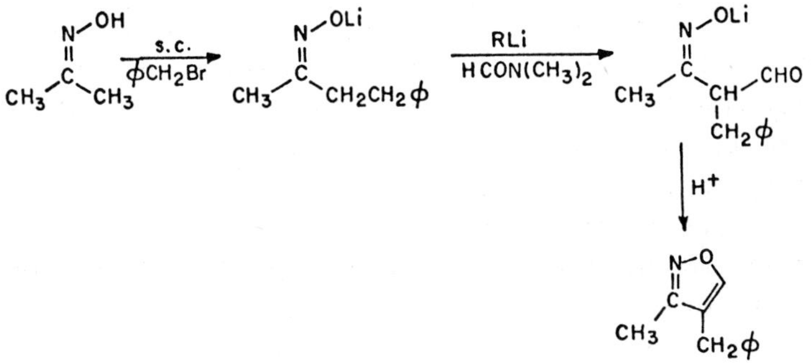

[18]

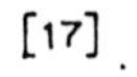

Oxygenation Reactions

Potentially a very useful reaction of oximes, and also of imines, was developed recently in Normant's laboratory[86]. Treatment of a dilithiated ketoxime (or a lithiated imine) with compressed air gave, after hydrolysis, the hydroxy ketone in yields which varied from 30 to 60%. As in the sequences [17] and [18], the site of hydroxylation in [19] should be controllable by selecting the appropriate oxime stereochemistry.

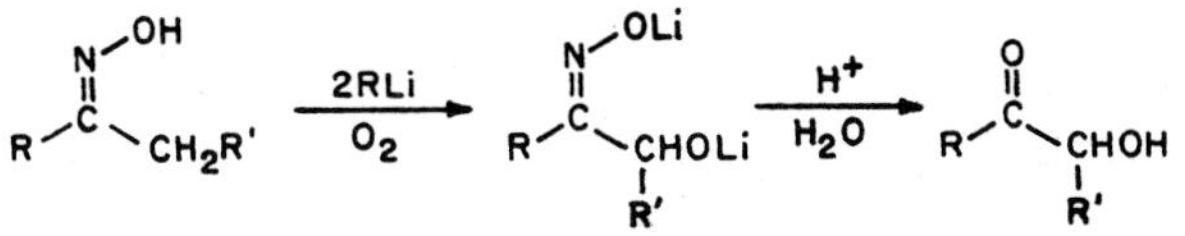

$$R \overset{N^{OH}}{\underset{||}{C}} CH_2R' \xrightarrow[O_2]{2RLi} R \overset{N^{OLi}}{\underset{||}{C}} \underset{\underset{R'}{|}}{CHOLi} \xrightarrow[H_2O]{H^+} R \overset{O}{\underset{||}{C}} \underset{\underset{R'}{|}}{CHOH}$$

[19]

A few sulfur analogues of oximes, specifically 5-phenylthioximes or sulfenimines have been reported to undergo lithiation and give electrophilic substitution products[87] in high yield (80-95%). However, the configurational stability of the nitrogen in these lithiated intermediates is low, thus making the reaction non-stereoselective as with hydrazones. Consequently, reaction of thioximes does not appear to possess any particular advantages.

Reactions of Isoxazolines

The ability of isoxazolines to serve as latent enones has been developed as a synthetic method by Jäger and coworkers[88]. When an isoxazoline is treated with an alkyllithium, metallation occurs at both alpha positions with the syn site slightly favored over the anti (2:1). The lithiation and alkylation of isoxazolines proceed at low temperature[88,89] to produce new isoxazolines which are then convertible to enones. The available reaction sequences appear in [20].

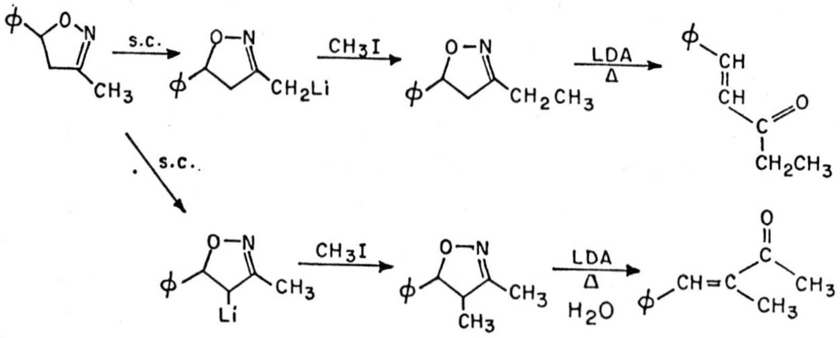

[20]

It is interesting to note that the syn deprotonation is achievable on isoxazolines but not in pyrrolines which give only anti lithiation[10]. A further unusual result was the successful lithiation and alkylation at a methine carbon[89] as in [21] in contrast to the failure of such reactions in acyclic oximes[81,25].

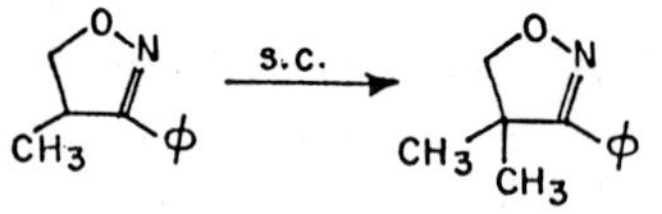

[21]

The difference is readily attributable to the fact that lithiation introduced little change in the steric interactions in a cyclic oxime ether, but causes appreciable $A^{1,3}$ strain in the acyclic systems.

Reactions of 1,2 Oxazines

The six-membered heterocycle, 3-methyl-4H-5,6-dihydro-1,2-oxazine is also convertible to an enone[90a] as indicated in [21a]. Since Shatzmiller has found conditions for regioselectivlely lithiating and alkylating the 1,2-oxazine at either the methyl at C-3 or the 4-methylene carbon in excellent yields[90b], this heterocyclic system provides a potentially useful route to α-methylene ketones.

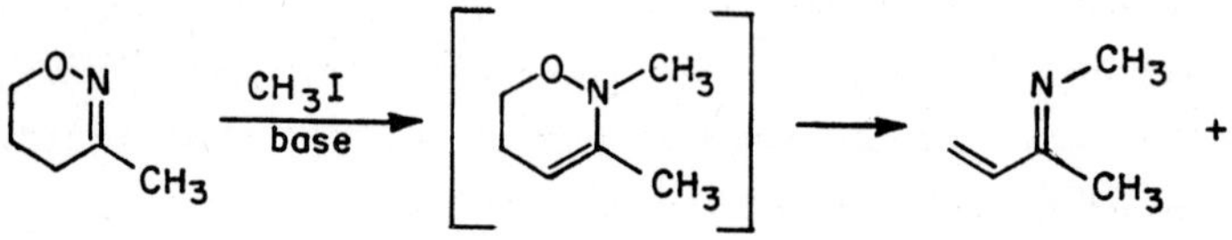

Advantages of Oximes and Oxime Ethers

The fact that hydrazones and imines are not configurationally stable under the conditions of lithiation results in the formation of the thermodynamically most stable <u>syn</u> lithio derivative, i.e., lithiation occurs at the least substituted carbon. In contrast, oximes and oxime ethers may be metallated under conditions which preserve the stereochemistry at nitrogen during the <u>syn</u> lithiation process. This allows a free choice of lithiation at the desired carbon regardless of whether it is primary or secondary. On other occasions, when desired, equilibration can be accomplished at higher temperatures as was described earlier on page 19.

VIII. IMINES

Chronologically, imines were the first of the carbonyl derivatives to be examined as synthetically useful intermediates. Essentially simultaneously, their utility as "masked enolates" was being examined by the research groups of Wittig[91] and Stork[92], both of whom reported their deprotonation and subsequent reactions with ketones[91] and alkyl halides[92] respectively. In addition to Wittig's report on the production of aldols[93], only a few scattered examples of the use of these imines appeared up to 1968. Büchi utilized this method, referred to by Wittig as the "directed aldol condensation", in the synthesis of β-sinensals[94] and nuciferal[95], and Tarbell's group showed that the bromomagnesium derivative of ketimines was significantly more reactive towards ethylene oxide than metal enolates[96]. Starting in 1969, the first of many papers on the reactions of lithiated imines was published by Henri Normant and coworkers[97]. One notable feature of their studies was the use of a new base-solvent combination, specifically lithium diethylamide in a HMPA-benzene-THF mixed solvent system for which the authors eventually coined the term "hyperbasic medium"[98]. At about the same time Evans[99] demonstrated the utility of ketimines in the synthesis of Δ^2-tetrahydropyridines as illustrated in [22]. Here he initiated the use of LDA in THF, which, as for hydrazones, has become the most popular of the

[22]

metallation conditions. Of great synthetic importance is the regioselectivity of metallation, which is the same as for hydrazones, occurring at the least substituted carbon. Normant's group was the first to draw attention to this feature in describing a synthesis of dihydrojasmone[100]. Their other contributions on the use of lithiated imines include reactions with epoxides,[101,102] iodine[101] and 2,3-dihalo-propenes[100,102]. They also described lithiation and alkylation of isobutyraldimine,[103,104] a reaction which had earlier been thought to be unattainable[91], yet independently achieved in the same year in House's[105] laboratory. This lithiation at a methine carbon, as shown in [23], is unique to aldimines and ketimines as the

$$
\begin{array}{c}
\underset{\underset{\underset{(CH_3)_2CH}{|}}{\overset{N^{C_6H_{11}}}{\underset{\parallel}{C}}}}{}{\overset{}{\diagdown}}{H}
\end{array}
\xrightarrow[\substack{THF-HMPA \\ -benzene \\ C\ell(CH_2)_nBr}]{(Et)_2NLi}
\quad
\underset{\underset{(CH_2)_n C\ell}{|}}{\overset{N^{C_6H_{11}}}{\underset{\parallel}{C}}}
\quad [23]
$$

corresponding hydrazones give elimination[58] while oximes[80] and oxime ethers[27] are not metallated. Other examples from Normant's group include the reactions of imine of pyruvaldehyde acetals[106,107] and alkylation reactions utilizing bromoacetals to provide 1,4-dicarbonyl compounds[108].

Ambient Character of Imines

Although, in principle, all enolate equivalents discussed in this article have the potential to react at either carbon or nitrogen, only the imines have displayed any ambident character. Only a few electrophiles have been observed to react at nitrogen, specifically, - acid chlorides[91c], anhydrides[109], silyl chlorides and sil triflates[110] and alcohols[111], as well as certain reactive alkyl halides in a mixed solvent system[112].

The site of electrophilic attack in lithiated α-β unsaturated aldimines also has been determined independently by several research groups[110b,113,114]. Attack o an alkyl halide takes place at the alpha carbon and interestingly, trimethylsilyl-chloride attacks the gamma carbon only[110b].

Control of the site at which the substituent enters has been achieved for the reaction of lithiated imines with silyl halides. Studies on the effects of the substituent at nitrogen[115a,b] have provided efficient routes to both C- and N-silyl derivatives as seen in [24]. Lithiated t-butylimines undergo exclusive attack at carbon. The C-silyl derivative can then be converted completely to the N-silyl enamine[115c]

$$
\underset{CH_3CH_2}{\overset{N^{C(CH_3)_3}}{\underset{\parallel}{C}}}{\diagdown}H
\xrightarrow[(CH_3)_3 M-C\ell]{s.c.}
\underset{CH_3-CH-M(CH_3)_3}{\overset{(CH_3)_3 C \diagdown N}{\underset{\parallel}{C}}}{\diagdown}H
$$

M = Si or Sn

[24]

$$
\underset{\underset{Si(CH_3)_3}{|}}{\underset{CH_3-CH}{\overset{N^{C(CH_3)_3}}{\underset{\parallel}{C}}}}{\diagdown}H
\xrightarrow{(CH_3)_3 SiC\ell}
\underset{CH_3CH}{\diagup}{=}CH-NSi(CH_3)_3
$$

The C-silylation of aldimines has recently been shown to provide a useful viny cation equivalent[116]. For example, as [24a] shows, the cyclohexylimine of acetalde hyde can be converted into the disubstituted olefin using silylation, with t-butyld methylsilyl chloride, alkylation and, after hydrolysis, nucleophilic attack at the carbonyl and stereoselective desilylation to either the _cis_ or _trans_ ethylene derivative.

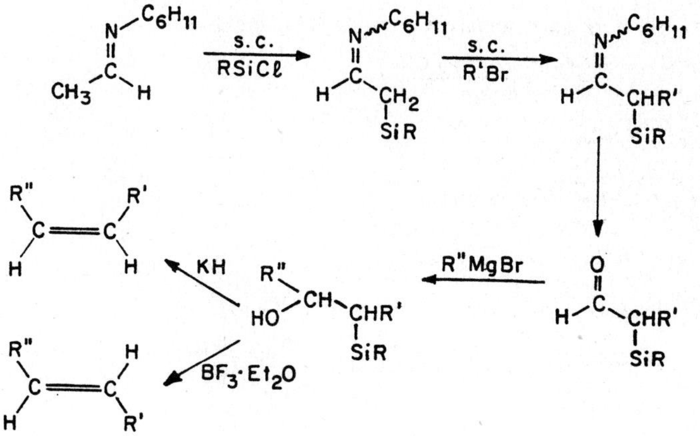

[24a]

Lithiation of Imines

(A) _Direct_. As mentioned before, the use of LDA in THF appears to be the most popular metallation method since it is both cheap and efficient. For details on the metallation experiment and also on the methods of synthesis and hydrolysis of imines the reader is referred to recent full papers[47,117,118]. The use of organolithiums for the proton abstraction step is usually complicated by competitive addition to the C=N bond[119]. The possibility that proton abstraction by a strong base might lead to formation of a 2-azaallyl anion has been thoroughly studied by Ahlbrecht and Farnum[120]. They found exclusive formation of the 3-azaallyl (or 1-azaallyl) anion except in the one instance shown below in [25]. For this imine formation of the 3-azaallyl isomer is strongly disfavored by two strong steric

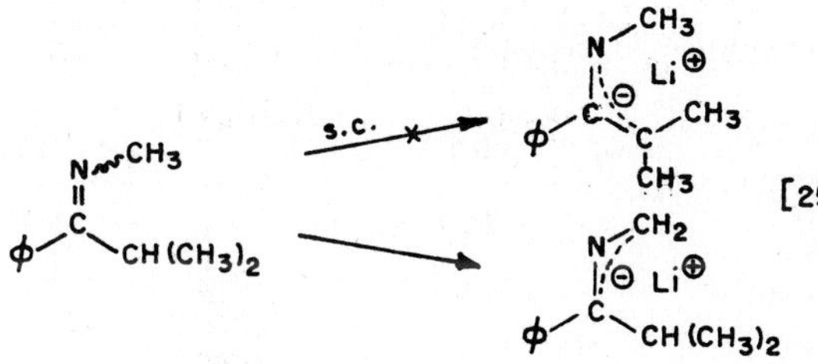

[25]

repulsive interactions, one between the phenyl and methyl groups at C_1 and C_2, and another between the two methyl groups on C_1 and nitrogen.

(B) <u>Indirect</u>. In order to achieve lithiation at the site opposite to that dictated by the regioselectivity of imine deprotonation, two research groups independently developed similar yet complementary solutions to the problem. Wender and coworkers showed that both allyl imines[121] and α-β unsaturated imines[122] could be transformed into metallated imines in the three ways illustrated in [26] and [27]. The yields of subsequent alkylation products were generally above 80%.

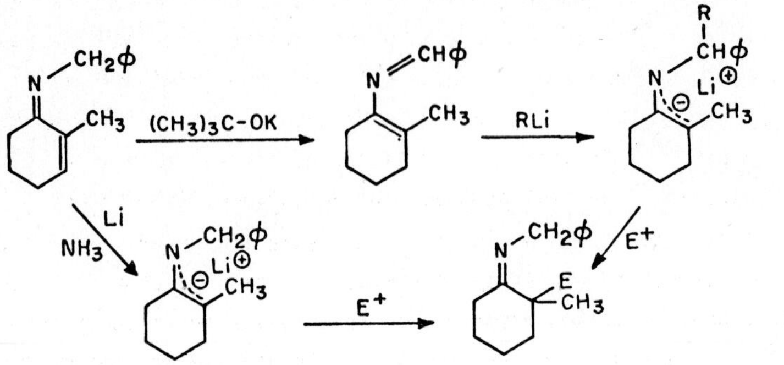

[26]

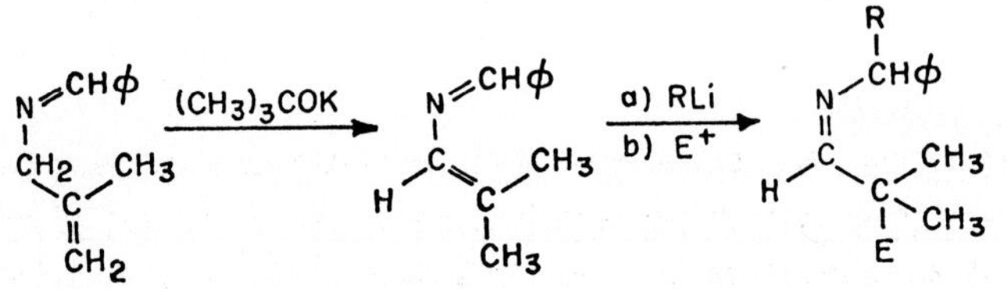

[27]

The variant approach of Martin and coworkers[123] used the condensation of a ketone with a phosphonate-substituted imine to form the α-β unsaturated imine.
Addition of an alkyllithium and an electrophile results in the synthetic transformation [28] which the authors refer to as a geminal acylation-alkylation reaction. T

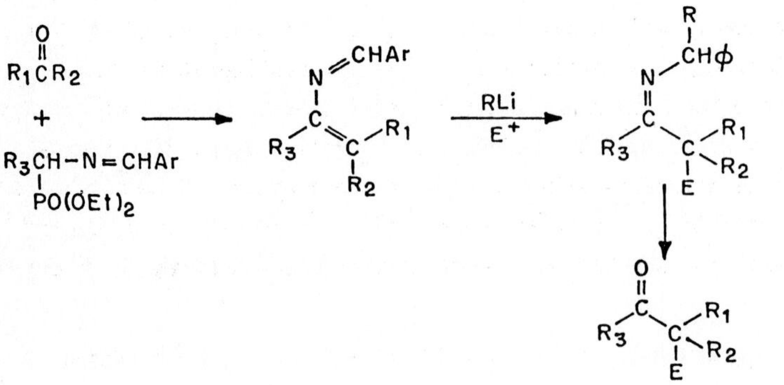

[28]

yields are high and provide, in conjunction with the synthetic sequences of Wender, a significant improvement in the methods of synthesis of tetrasubstituted carbon atoms. This general topic has been reviewed by Martin[124].

A number of reports have dealt with other applications of imines in synthesis. They have been observed to undergo Michael addition to chalcones in good yields, but to cyclohexanones in poor yields[125]. The lithiated imine of cyclohexanone is sufficiently reactive towards oxetane to provide hydroxypropyl derivatives in satisfactory yield[126]. In general, lithiated imines were found to react readily with oxygen[87] in yields comparable to those obtained with lithiated oximes, both reactions contributing a reasonable alternative to Vedejs' method of introducing an OH alpha to a carbonyl group[127]. Applications of lithiated imines in terpene synthesis[128,129], pyridine synthesis[130] and the synthesis of α-β unsaturated ketones[131,132,133] have been described. Additional examples of the synthetic potential of lithiated imines include the preparation of hexahydroindoles[152], tetrahydrofurans[153], 1,3 dianils[154], the use of copper salts to promote selective allylation[155] and a synthesis of multi-striatin[156]. A review of the reactions of lithiated imines and of enamines has recently appeared[157].

Advantages of Imines

In comparison with oximes and their derivatives, imines offer a distinct advantage in their ease of preparation, more rapid metallation and facile hydrolysis. With respect to hydrazones the above characteristics are more or less comparable as is their regioselectivity, their stereoselectivity and their ability to give high yields. The choice of the most appropriate carbonyl derivative for a given task might well be determined by the requisite conditions for hydrolysis. Imines can be readily hydrolysed over the small pH range 4 to 7 while hydrazones can be cleaved over the range 4 to 9, or by methylation and acid hydrolysis[59e] or, also by ozonolysis[141].

Advantages of imines over hydrazones include their ability to be lithiated at a tertiary carbon, either directly[103,104] or indirectly[121,122,123]. In such situations hydrazones either fail to metallate or, in the case of aldehyde hydrazones, undergo an elimination to afford nitriles[103]. In all likelihood, the lithiated imines have a stronger thermodynamic preference for the syn configuration than do the hydrazone anions, since conditions under which anti lithiated hydrazones were formed produced less anti imine[11b]. Such a penchant for syn lithiation will mean a stronger tendency for the imine anions to maintain this stereochemical influence even under adverse reaction conditions. Of great importance for both classes of carbonyl derivatives is their well-established, yet complementary, successes in the field of asymmetric synthesis.

IX. ASYMMETRIC SYNTHESIS WITH CARBONYL DERIVATIVES

(A) Cyclohexanones and other Cycloalkanones

At the time that research efforts were being concentrated on the use of chiral imines and hydrazones in asymmetrical synthesis, a high level of sophistication had already been developed in this general area by Meyers and his group using chiral oxazolines[134]. It is thus not too surprising, though still most laudable, that in a span of five years the asymmetric induction achieved in the majority of syntheses of chiral aldehydes and ketones surpassed the 90% level. The pioneer experiment in this area was a report by Horeau and Méa-Jacheet in 1968[135]. By treating the isobornyl imine of cyclohexanone with ethyl magnesium bromide and alkylation of the resultant anion with methyl iodide, they were able to obtain a 58% yield of the hydrolysis product, 2-methylcyclohexanone whose optical purity (hereafter referred to as enantiomeric excess, ee) was 72%. Since this value had to reflect a minimum of 86:[for the ratio of attack on the two faces of the alpha carbon, this observation conti buted the first evidence that metallated imines had a syn configuration.

Subsequent studies have been able to increase the ee of this methylation to up to 87% and to much higher levels for other alkylations. Much of the research which has focussed on this particular reaction can serve as a model for the approach to b taken, in general, towards elucidation of the factors leading to high levels of asymmetric synthesis. To illustrate some of the effects of reaction conditions as well as of the structure of the chiral substituent, a number of results for the methylation and other alkylations of cyclohexanone are summarized in Table 2. It is clear that the maximum ee has essentially been attained independently by three different groups[141,142,117]. Another feature of Table 2 shows that changes in alky ting agent, solvent or counterion can all alter the resultant ee and, as is borne o by additional studies, not in any predictable way. There is one concept, originally put forward by Meyers[139], that appears to be generally valid. He proposed that use a chelating substituent attached to the imine nitrogen to enhance the rigidity of t metallated intermediate should be an important component of any successful asymmetr synthesis. So far those hydrazone and imine derivatives which give the best ee's h embodied this structural feature. Even if this principle is valid, it must be recog nized that the intimate features of the transition state will remain a matter of co jecture until the interaction of the anion with its counterion can be elucidated (s Section III).

TABLE 2

Alkylations of Chiral Derivatives of Cyclohexanone

Structure of R-N in 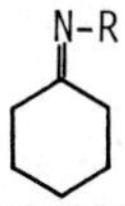 $\overset{N-R}{\bigcirc}$	Base (in THF)	Alkylating Agent	εε(% Yield) [Abs. Conf.]	Ref.
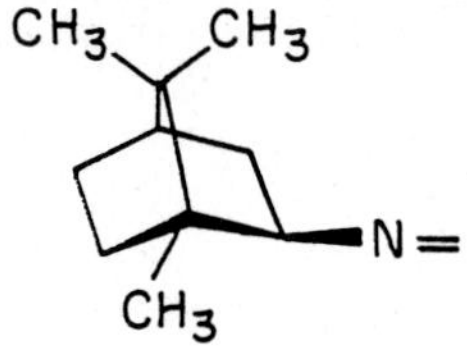 (camphor-derived) CH_3, CH_3 ... CH_3 —N=	i-C$_3$H$_7$MgBr	CH$_3$I	72[a] (58) [S]	135
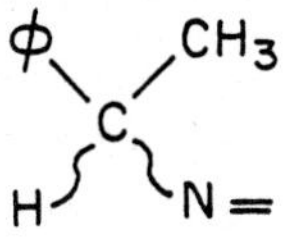 φ, CH$_3$ / H, N=	i-C$_3$H$_7$MgBr	CH$_3$I	<10 (high)	135
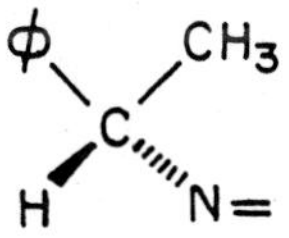 φ, CH$_3$ / H, N=	LDA	CH$_3$I	26 (42) [R]	136
"	LDA + MgBr$_2$	CH$_3$I	52[a] [R]	137
"	LDA	(CH$_3$)$_2$SO$_4$	13[a] [R]	137
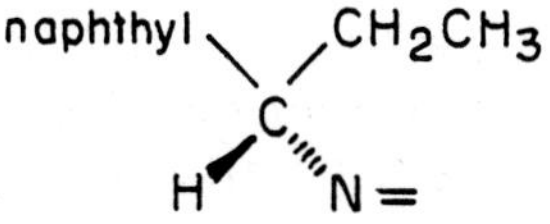 naphthyl, CH$_2$CH$_3$ / H, N=	LDA + TMEDA	CH$_3$I	67[b]	138
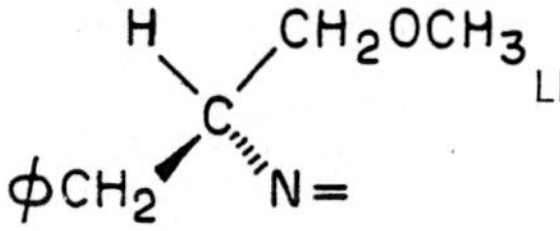 H, CH$_2$OCH$_3$ / φCH$_2$, N=	LDA	CH$_3$I	87 (65) [S]	117
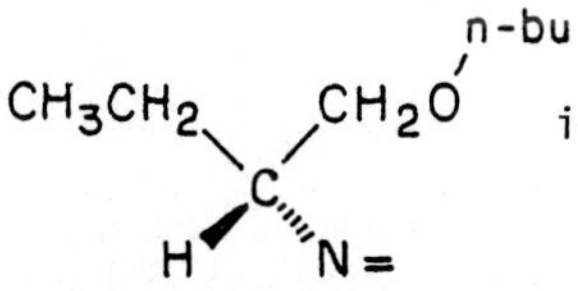 CH$_3$CH$_2$, CH$_2$O–n-bu / H, N=	i-C$_3$H$_7$MgBr	CH$_3$I	81 (50) [R]	140

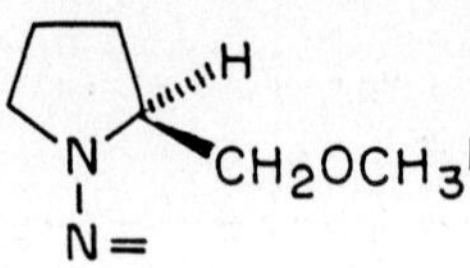 Pyrrolidine-CH₂OCH₃ (N=)	LDA	$(CH_3)_2SO_4$	99	[R] (80)	141
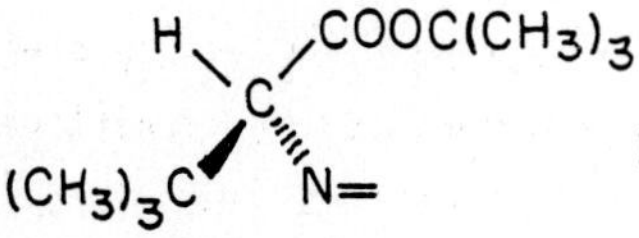 H–C(COOC(CH₃)₃)((CH₃)₃C)(N=)	LDA	CH_3I	97	[S]	142
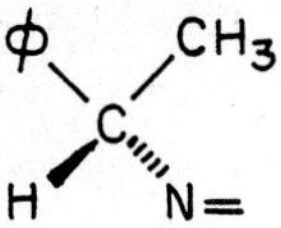 φ–C(CH₃)(H)(N=)	LDA	CH_3CH_2I	19ᵃ	[S]	137
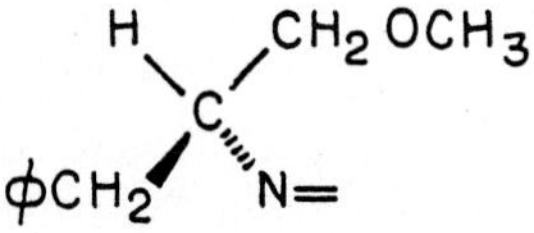 φCH₂–C(H)(CH₂OCH₃)(N=)	LDA	CH_3CH_3I	94	[S] (82)	117
"	LDA	$CH_3CH_2CH_2I$	99	[S] (76)	117
"	LDA	$CH_2=CH-CH_2Br$	99	[R] (80)	117
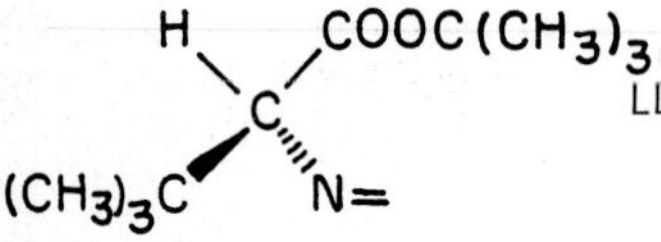 H–C(COOC(CH₃)₃)((CH₃)₃C)(N=)	LDA	$CH_3CH_2CH_2I$	97	[S]	142
"		$CH_2=CH-CH_2Br$	84	[R]	142
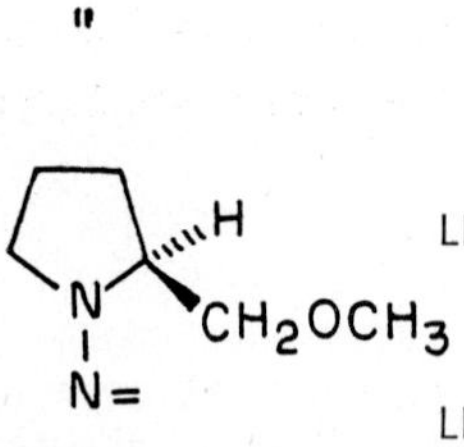 Pyrrolidine-CH₂OCH₃ (N=)	LDA	$CH_3CH_2CH_2I$	86	[R] (73)	141
"	LDA	$CH_2=CHCH_2Br$	73	[S] (67)	141

ᵃCalculated from diastereomer ratio of methylated imines. ᵇSame as ᵃ, except the experiment was done on resolved imine, but failed to yield optically active ketone, presumably due to racemization at the imine carbon.

In two recent papers, Meyers' group describes some very interesting results on chiral cycloalkanone imine alkylations[117,118]. They have successfully carried out dimethylation as well as sequential methylation and propylation to product chiral 2,6-dialkylcyclohexanones in 85% optical purity[118]. Optical yields diminished in the alkylation of the seven and 8-membered ketimines, as well as for 1-tetralone and 1-indanone imines. In studies similar to those of the hydrazones discussed earlier[43] an intriguing influence of the configuration of the C_1-C_2 bond in the lithiated imine on the chirality of 2-methylcyclododecanone and 2-methylcyclopentadecanone was clearly shown by a careful study of the stereochemistry of the lithiated imine inter-mediate[118]. Deprotonation of the C-12 or C-15 imines produced 100% of the E isomer which on reflux in THF rearranged 100% to the Z isomer. Methylation of the E inter-mediates produced methyl derivatives having the S configuration, while the Z inter-mediates gave products of opposite chirality, e.g., as shown in [29].

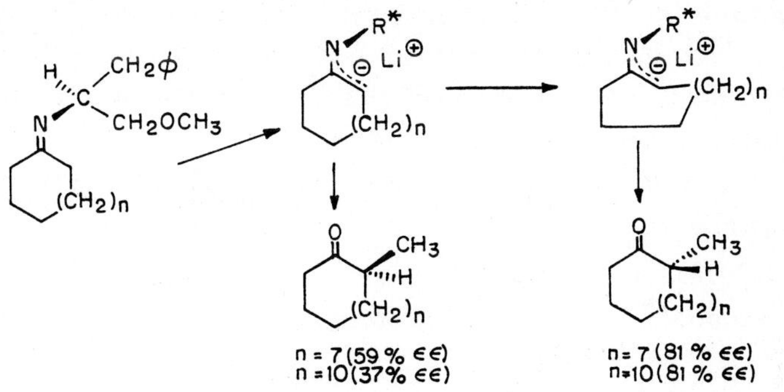

[29]

n = 7 (59 % EE)
n = 10 (37% EE)

n = 7 (81 % EE)
n = 10 (81 % EE)

In contrast, the alkoxyimine of cyclodecanone gave only a fair degree of asymmetric synthesis (ee ∿ 30% for methylation) and the same optical antipodes (S) were obtained with or without the period of reflux.

(B) Synthesis of Acyclic Aldehydes and Ketones

Direct alkylation of acyclic aldehyde derivatives has been studied by three groups and some of the results are given in Table 3 as illustrative examples of varia tions in ee with reaction parameters. Again, the extent of asymmetric synthesis fluctuates considerably with minor changes in reaction conditions. Additional studie on the methylation of octanal with imines derived from 5 different terpene alkanol- amines[145] failed to provide any real improvements. In stark contrast to the unim- pressive aldehyde results, an acyclic ketone has been synthesized in optically pure form. This remarkable accomplishment of Enders and Eichenauer[146] provided the pure [S] enantiomer of 4-methylheptan-3-one, an alarm pheromone of certain ants. Interest ingly, the effect of solvent in this alkylation was marked in that the ee decreased from 100% in ether to 85% in THF and to 20% in THF-HMPA mixture. Meyers' group was equally impressive in attaining high asymmetric induction (ee = 98%) in the synthesis of 3-methyl-heptane-4-one[43]. Details of their investigation of the alkylation of acyclic ketimines have been described[118].

(C) Conjugate Additions

Considerable success has been achieved by Kenji Koga and coworkers[147] who have studied a variety of conjugate additions of organometallics to chiral alpha-beta unsaturated imines. Their first report described the successful addition of alkyl and aryl magnesium bromides to a chiral imine of crotonaldehyde, referred to by them as the imine of tertiary leucine. Their excellent results are summarized in [30]. Malonates were also added[148].

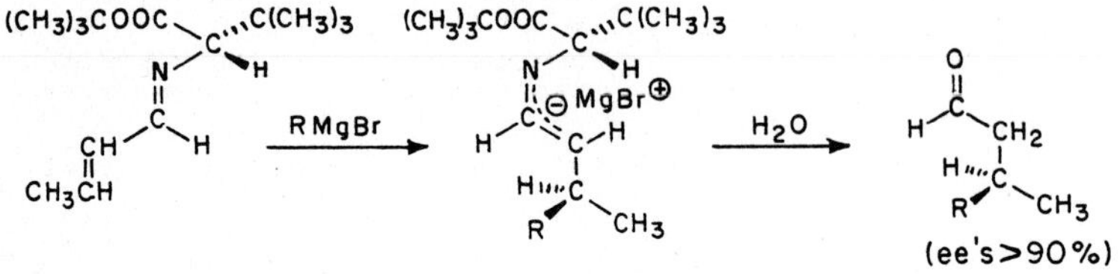

$$(ee's > 90\%) \qquad [30]$$

Equally as selective were the conjugate additions of Grignards to the same chira imines of cyclopentenone and cyclohexenone, both of which gave optical purities of more than 90%[149]. They subsequently showed that tandem addition and alkylation as shown below in [31] could by choice of appropriate conditions yield either the cis o the trans cycloalkene carboxaldehyde in variable yields with the same high enantio- meric excess[150].

TABLE 3

Alkylations of Chiral Aldehyde Derivatives

Structure

$$RCH_2$$
$$C{=}N$$

R=	R'=	Base (Additive)	Alkylating Agent	ee(% Yield) [Chirality]	Ref.
CH_3CH_2-	SAMP[c] [S]	LDA	CH_3I	62 (65) [R]	141
CH_3	MMAPE[d] [S]	LDA	$n-C_6H_{13}I$	42 (46) [S]	144
CH_3	$-CH-CH_2\emptyset$ $CH_2(OCH_2CH_2)_2OCH_3$ [S]	LDA	$n-C_6H_{13}I$	39 (70) [R]	144
CH_3	SAMP [S]	LDA	$n-C_6H_{13}I$	87 (61) [S]	143
CH_3	α-phenethyl [S]	LDA (HMPA + $MgBr_2$)	$n-C_6H_{13}I$	69 (87) [S]	137
CH_3	[S]	LDA	$n-C_6H_{13}I$	43[b]	137
CH_3	[S]	LDA ($MgBr_2$)	$n-C_6H_{13}I$	51[b]	137
$n-C_6H_{13}$	$-CHCH_2\emptyset$ $CH_2(OCH_2CH_2)_2OCH_3$	LiTMP	CH_3I	58 (70) [S]	144

[a]The ee and yields quoted represent the pure isolated aldehyde except for [b]calculated from diastereomer ratio as measured by [13]C. [c]SAMP refers to the carbonyl derivative of [S]-1-amino-2-methoxymethylpyrrolidine. [d]MMAPE refers to the carbonyl derivative of α-methoxymethyl-β-phenethylamine.

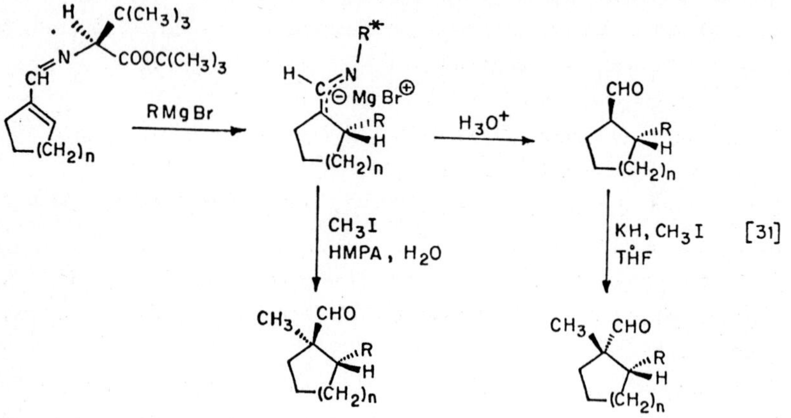

A different type of conjugate addition which involved adding a "metallated" chiral imine to acrylate and acrylonitrile was recently reported by de Jeso and Pommier[151]. This involved a stannyl imine which in the case cited below [32] gave a good yield of chiral product.

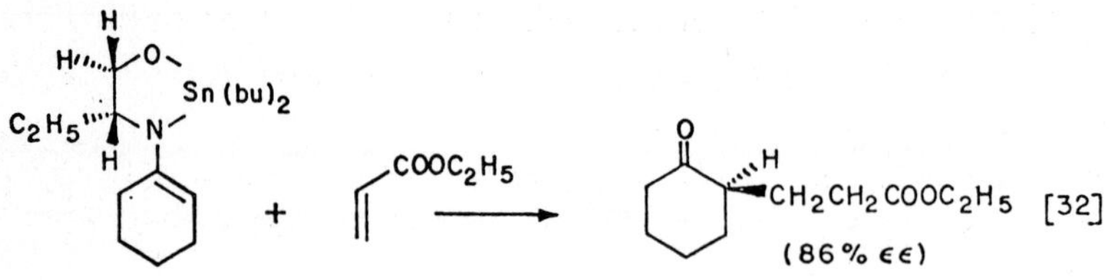

X. CONCLUSIONS

This extensive accumulation of experience with enolate equivalents has clearly established the diverse and complementary capabilities of lithiated hydrazones, oximes and imines in synthesis. Equipped with the knowledge of their individual advantages, the competent organic chemist can, like a piper, call the tune to which the appropriately chosen molecules must dance!

XI. REFERENCES

1. C.H. Heathcock, Chapter 4, this monograph.

2. J.C. Stowell, "Carbanions in Organic Synthesis", J. Wiley & Sons, N.Y. 1979.

3. R.L. Augustine, "Carbon Carbon Bond Formation", Vol. 1, Ch. 2, Marcel Dekker, N.Y. 1979.

4. E. Negishi, "Organometallics in Organic Synthesis", Wiley, N.Y. 1980.

5. P.A. Bartlett, Tetrahedron, 36 (1980) 2.

6. E.M. Kaiser, J. Organomet. Chem., 203 (1980) 1; 183 (1979) 1.

7. J. D'Angelo, Tetrahedron, 32 (1976) 2979.

8. D. Seebach and K.H. Geiss, "Organolithium Compounds in Organic Synthesis" in Journal of Organometallic Chemistry Library, Vol. 1, Ed. D. Seyferth, Elsevier, N.Y. 1976.

9. There are, in fact, several exceptions, two of them involving endocyclic imines which undergo lithiation at the anti position only (10) and endocyclic isoxazolines which give some anti lithiation (88). These are discussed in Section III. Exceptions in acyclic hydrazones and imines of propionaldehyde have been reported by Bérgbreiter et al. (11). Their special characteristics are discussed in Section V.

10. K.N. Houk, R.W. Strozier, N.G. Rondan, R.R. Fraser and N. Chuaqui-Offermanns, J. Am. Chem. Soc., 102 (1980) 1429.

11. (a) M. Newcomb and D.E. Bergbreiter, J. Chem. Soc. Chem. Commun., (1977) 486; (b) J.Y. Lee, T.J. Lynch, D.T. Mao, D.E. Bergbreiter and M. Newcomb, J. Am. Chem. Soc., 103 (1981) 6215.

12. (a) P. Caubere and M.F. Hochu, Bull. Soc. Chim Fr., (1968) 459; (b) T.J. Prosser, J. Am. Chem. Soc., 83 (1961) 1701; (c) C.C. Price and W.H. Snyder, J. Am. Chem. Soc., 83 (1961) 1773.

13. (a) A. Schouteeten and M. Julia, Tetrahedron Lett., (1975) 607; (b) G. de Saqui-Sannes, M. Riviere and A. Lattes, Tetrahedron Lett., (1974) 2073.

14. K. Geiss, B. Seuring, R. Pieter and D. Seebach, Angew. Chem. Int. Ed., 13 (1974) 479.

15. S. Bank, J. Am. Chem. Soc., 87 (1965) 3245.

16. L.K. Keefer and C.H. Fodor, J. Am. Chem. Soc., 92 (1970) 5747.

17. (a) R.R. Fraser and L.K. Ng, J. Am. Chem. Soc., 98 (1976) 5895; (b) R.R. Fraser and Y.Y. Wigfield, Tetrahedron Lett., (1971) 2515.

18. R.R. Fraser and P.J. Champagne, J. Am. Chem. Soc., 100 (1978) 657.

19. R. Hoffmann and R.A. Olofson, J. Am. Chem. Soc., 88 (1966) 943.

20. R.R. Fraser, G. Boussard, I.D. Postescu, J.J. Whiting and Y.Y. Wigfield, Can. J. Chem., 51 (1973) 1109.

21. R.R. Fraser and T.B. Grindley and S. Passannanti, Can. J. Chem., 53 (1975) 2473.

22. E.L. Eliel, N.L. Allinger, S. Angyal and J.W. Morrison, Conformational Analysis Interscience Inc., New York 1965, p. 44, 440.

23. Y.L. Chow, C.J. Colon and J.N.S. Tam, Can. J. Chem., 46 (1968) 2821.

24. D. Seebach and D. Enders, Angew. Chem. Int. Ed., 14 (1975) 15.

25. R.R. Fraser and K.L. Dhawan, J. Chem. Soc. Chem. Commun., (1976) 674.

26. (a) E.J. Corey and D. Enders, Tetrahedron Lett., (1976) 3; (b) E.J. Corey and D. Enders, Chem. Ber., 111 (1978) 1337.

27. R.R. Fraser, J. Banville and K.L. Dhawan, J. Am. Chem. Soc., 100 (1978) 7999.

28. R.R. Fraser and N. Chuaqui-Offermanns, Can. J. Chem., 59 (1981) 3007.

29. R.R. Fraser and N. Chuaqui-Offermanns, unpublished observations.

30. G. Lamaty in "Isotopes in Organic Chemistry", Vol. 2, E. Buncel and C.C. Lee, Ed., Elsevier, Amsterdam, 1976, p. 49 presents a summary of evidence that the relative rates of base-catalyzed H-D exchange at the methyl and methyl ene groups in 2-butanone is between 0.8 and 1.0.

31. R.R. Fraser and J. Banville, J. Chem. Soc., Chem. Commun., (1979) 47.

32. K.N. Houk, and R.W. Strozier, unpublished calculations.

33. H.O. House, W.F. Fischer, Jr., M. Gall, T.E. McLaughlin and N.P. Peet, J. Org. Chem., 36 (1971) 3429.

34. H.O. House, R.A. Auerbach, M. Gall and N.P. Peet, J. Org. Chem., 38 (1973) 514.

35. H.O. House, A.V. Prabhu and W.V. Phillips, J. Org. Chem., 41 (1976) 1209.

36. H.O. House, D.S. Crumrine, A.Y. Teranishi and H.D. Olmstead, J. Am. Chem. Soc., 95 (1973) 3310.

37. J.E. Dubois and P. Fellmann, Tetrahedron Lett., (1975) 1225.

38. R.E. Ireland, R.H. Mueller and A.K. Willard, J. Am. Chem. Soc., 98 (1976) 2868.

39. W.A. Kleschick, C.T. Buse and C.H. Heathcock, J. Am. Chem. Soc., 99 (1977) 247.

40. K.G. Davenport, M. Newcomb and D.E. Bergbreiter, J. Org. Chem., 46 (1981) 3142.

41. H. Ahlbrecht, E.O. Düber, D. Enders, H. Eichenauer and P. Weuster, Tetrahedron Lett., (1978) 3691.

42. K.G. Davenport, H. Eichenauer, D. Enders, M. Newcomb and D.E. Bergbreiter, J. Am. Chem. Soc., 101 (1979) 5654.

43. A.I. Meyers and D.R. Williams, J. Org. Chem., 43 (1978) 3245.

44. J.Y. Lee, T.J. Lynch, D.E. Bergbreiter and M. Newcomb, private communication.

45. R.R. Fraser, N. Chuaqui-Offermanns, K.N. Houk and N.G. Rondan, J. Organomet. Chem., 206 (1981) 131.

46. R.R. Fraser, J. Banville, F. Akiyama and N. Chuaqui-Offermanns, Can. J. Chem., 59 (1981) 705.

47. M.E. Jung and T.J. Shaw, Tetrahedron Lett., (1977) 3305.

48. M.E. Jung, T.J. Shaw, R.R. Fraser, J. Banville and K. Taymaz, Tetrahedron Lett. (1979) 4149.

49. (a) D.E. Bergbreiter and M. Newcomb, Tetrahedron Lett., (1979) 4145; (b) J.K. Smith, D.E. Bergbreiter and M. Newcomb, J. Org. Chem., 46 (1981) 3157

(c) A. Hosomi, A. Shirahata, Y. Araki and H. Sakurai, J.Org.Chem. 46(1981)4631;

(d) A. Hosomi, Y. Araki and H. Sakurai, J. Am. Chem. Soc., 104 (1982) 2081.

50. M. Schlosser and J. Hartmann, J. Am. Chem. Soc., 98 (1976) 4676.

51. (a) H.E. Ensley and R. Lohr, Tetrahedron Lett., (1978) 1415; (b) R.E. Lyle, H.M. Fribush, G.G. Lyle and J.E. Saavedra, J. Org. Chem., 43 (1978) 1275.

52. (a) G.J. Karabatsos and R.A. Taller, Tetrahedron, 24 (1968) 3923;

(b) G.J. Karabatsos and J.J. Lande, Tetrahedron, 24 (1968) 3907; (c) J. Hine and C.Y. Yeh, J. Am. Chem. Soc., 89 (1967) 2669.

53. M. Bellasoued, F. Dardoize, Y. Frangin and M. Gaudemar, J. Organomet. Chem., 165 (1979) 1.

54. G.N. Barber and R.A. Olofson, J. Org. Chem., 43 (1978) 3015.

55. (a) F.E. Henoch, K.G. Hampton and C.R. Hauser, J. Am. Chem. Soc., 89 (1967) 463;

(b) R.M. Sandifer, S.E. Davies and C.F. Beam, Synth. Commun., 6 (1976) 339;

(c) C.F. Beam, R.M. Sandifer, R.S. Foote and C.R. Hauser, Synth. Commun., 6 (1976) 5.

56. G. Stork and J. Benaim, J. Am. Chem. Soc., 93 (1971) 5938.

57. H. Normant, T. Cuvigny and D. Reisdorf, C.R. Acad. Sci. Ser. C, 268 (1969) 521.

58. T. Cuvigny, J.R. Le Borgne, M. Larchevêque and H. Normant, Synthesis (1976) 237.

59. (a) E.J. Corey and D. Enders, Tetrahedron Lett., (1976) 11; (b) E.J. Corey, D. Enders and M.G. Bock, Tetrahedron Lett., (1976) 7; (c) E.J. Corey and S. Knapp, Tetrahedron Lett., (1976) 4687; (d) D. Enders and P. Weuster, Tetrahedron Lett., (1978) 2853; (e) E.J. Corey and D. Enders, Chem. Ber., 111 (1978) 1362.

60. (a) C.F. Beam, C.W. Thomas, R.M. Sandifer, R.S. Foote and C. Hauser, Chem. Ind. (London), (1976) 487; (b) C.F. Beam, R.S. Foote and C.R. Hauser, J. Chem. Soc., C, (1971) 1658.

61. (a) S. Arseniyadis, A. Laurent and P. Mison, Bull. Soc. Chem. Fr.,(1980)II 246;

(b) G. Alvernhe, S. Arseniyadis, R. Chaabouni and A. Laurent, Tetrahedron Lett., (1975) 355.

62. (a) J. Hoch, K.N. Campbell, B.K. Campbell, L.G. Hess and I.J. Schaffner, J. Org. Chem., 9 (1944) 184; (b) K.N. Campbell, Org. Synthesis, Coll. Vol., 3 (1955) 148.

63. (a) R.H. Shapiro and M.J. Heath, J. Am. Chem. Soc., 89 (1967) 5734;

(b) R.H. Shapiro and K.J. Kolonko, J. Org. Chem., 43 (1978) 1404;

(c) M.F. Lipton and R.H. Shapiro, J. Org. Chem., 43 (1978) 1409;

(d) R.H. Shapiro, M.F. Lipton, K.J. Kolonko, R.L. Buswell and L.A. Capuano, Tetrahedron Lett., (1975) 1811; (e) R.H. Shapiro, Org. Reactions, 23 (1975) 405.

64. G. Kaufman, F. Cook and H. Schechter, J. Bayless and L. Friedman, J. Am. Chem. Soc., 89 (1967) 5736.

65. (a) W.G. Dauben, G.T. Rivers, W.T. Zimmerman, N.C. Yang, B. Kim and J. Yang, Tetrahedron Lett., (1976) 2951; (b) W.G. Dauben, G.T. Rivers and W.T. Zimmermar J. Am. Chem. Soc., 99 (1977) 3414.

66. (a) J.E. Stemke and F.T. Bond, Tetrahedron Lett., (1975) 1815;
    (b) A.R. Chamberlin, J.E. Stemke and F.T. Bond, J. Org. Chem., 43 (1978) 147.

67. P.C. Traas, H. Boelens and H.J. Takken, Tetrahedron Lett., (1976) 2287.

68. T.H. Chan, A. Baldassare and D. Massuda, Synthesis (1976) 801.

69. R.T. Taylor, C.R. Degenhardt, W.P. Melega and L.A. Paquette, Tetrahedron Lett., (1977) 159.

70. J.E. Stemke, A.R. Chamberlin and F.T. Bond, Tetrahedron Lett., (1976) 2947.

71. C.A. Bunnell and P.L. Fuchs, J. Am. Chem. Soc., 99 (1977) 5184.

72. (a) R.M. Adlington and A.G.M. Barrett, J. Chem. Soc. Chem. Comm., (1978) 1071;
    (b) R.M. Adlington and A.G.M. Barrett, J. Chem. Soc. Chem. Comm., (1979) 1122;
    (c) R.M. Adlington and A.G.M. Barrett, Tetrahedron, 37 (1981) 3935.

73. S. Kano, T. Yokomatsu, T. Ono, S. Hibino and S. Shibuya, J. Chem. Soc. Chem. Commun. (1978) 414.

74. A.R. Chamberlin and F.T. Bond, Synthesis (1979) 44.

75. F.T. Bond and R.A. Dipietro, J. Org. Chem., 46 (1981) 1315.

76. K. Kitahonoki, K. Kotera, Y. Matsukawa, S. Miyazaki, T. Okada, H. Takahashi and Y. Takano, Tetrahedron Lett., (1965) 1059.

77. (a) H. Tanida, T. Okada and K. Kotera, Bull Chim Soc., Japan, 46 (1973) 934;
    (b) K. Kotera, T. Okada and S. Miyazaki, Tetrahedron, 24 (1968) 5677.

78. (a) J.C. Philips and C. Perianayagam, Tetrahedron Lett., (1975) 3263;
    (b) G. Ricart and D. Couturier, C.R. Acad, Sci. Ser. C., 284 (1977) 941.

79. T.A. Spencer and C.W. Leong, Tetrahedron Lett., (1975) 3889.

80. W.G. Kofron and M.-K. Yeh, J. Org. Chem., 41 (1976) 439.

81. M.E. Jung, P.A. Blair and J.A. Lowe, Tetrahedron Lett., (1976) 1439.

82. R.E. Lyle, G.G. Lyle, J.E. Saavedra, H.M. Fribush, J.L. Marshall, W. Lijinsky and G.M. Singer, Tetrahedron Lett., (1976) 4431.

83. (a) C.F. Beam, C.D. Dyer, R.A. Schwarz and C.R. Hauser, J. Org. Chem., 35 (197 1806; C.F. Beam, R.S. Foote and C.E. Hauser, J. Heterocyclic Chem., 9 (1972) 18
    (c) R.M. Sandifer, L.M. Schaffner, W.M. Hollinger, D.C. Reames and C.F. Beam, J. Heterocyclic Chem., 13 (1976) 607.

84. C.A. Park, C.F. Beam, E.M. Kaiser, R.J. Kaufman, F.E. Henoch and C.R. Hauser, J. Heterocyclic Chem., 13 (1976) 449.

85. M. Perkins, C.F. Beam, M.C.D. Dyer and C.R. Hauser, Org. Synthesis, 55 (1976) 3

86. T. Cuvigny, G. Valette, M. Larchevêque and H. Normant, J. Organomet. Chem., 15! (1978) 147.

87. F.A. Davis and P.A. Mancinelli, J. Org. Chem., 43 (1978) 1797.

88. H. Grund and V. Jäger, Liebigs Ann. der Chemie, (1980) 80; (b) V. Jäger and H. Grund, Angew. Chem. Int. Ed., 15 (1976) 50.

89. V. Jäger and W. Schwab, Tetrahedron Lett., (1978) 3129.

90. (a) B. Hardegger and S. Shatzmiller, Helv. Chim. Acta, 59 (1976) 2765;
     (b) B. Hardegger and S. Shatzmiller, J. Am. Chem. Soc., 103 (1981) 5916.

91. (a) G. Wittig, H.D. Frommeld and P. Suchanek, Angew. Chem., 75 (1963) 987;
     (b) G. Wittig and H.D. Frommeld, Chem. Ber., 97 (1964) 3548; (c) G. Wittig
     and H. Reiff, Angew. Chem. Int. Ed. 7 (1968) 7.

92. G. Stork and S. Dowd, J. Am. Chem. Soc., 85 (1963) 2178.

93. G. Wittig and P. Suchanek, Tetrahedron, 22 (1966) 347.

94. G. Buchi and H. Wüest, Helv. Chim. Acta, 50 (1967) 2440.

95. G. Buchi and H. Wüest, J. Org. Chem., 34 (1969) 1122.

96. W.E. Harvey and D.S. Tarbell, J. Org. Chem., 32 (1967) 1679.

97. T. Cuvigny and H. Normant, C.R. Acad. Sci. Ser. C., 268 (1969) 1380.

98. (a) T. Cuvigny and H. Normant, Bull. Soc. Chim. Fr., (1970) 3976; (b) T. Cuvigny,
     M. Larchevêque and H. Normant, C.R. Acad. Sci. Ser. C, 277 (1973) 511.

99. D.A. Evans, J. Am. Chem. Soc., 92 (1970) 7594.

100. T. Cuvigny, M. Larchevêque and H. Normant, Tetrahedron Lett., (1974) 1237.

101. M. Larchevêque, G. Valette, T. Cuvigny and H. Normant, Synthesis, (1975) 256.

102. T. Cuvigny, M. Larchevêque and H. Normant, Liebigs Ann. Chem., (1975) 719.

103. (a) T. Cuvigny, J.F. Le Borgne, M. Larchevêque and H. Normant, J. Organomet.
     Chem., 70 (1974) C5 .

104. J.F. Le Borgne, J. Organomet. Chem., 122 (1976) 123.

105. (a) H.O. House, W.C. Liang and P.D. Weeks, J. Org. Chem., 39 (1974) 3102;
     (b) P. Groenewegen, H. Kallenberg and A. Van der Gen., Tetrahedron Lett.,
     (1978) 491.

106. T. Cuvigny and H. Normant, Synthesis (1977) 198.

107. T. Cuvigny, M. Larchevêque, H. Normant and N. Naulet, Synthesis, (1978) 390.

108. J.F. Le Borgne, T. Cuvigny, M. Larchevêque and H. Normant, Tetrahedron Lett.,
     (1965) 1379.

109. J. Thomas, J. Organomet. Chem., 101 (1975) 249.

110. W. Oppolzer and W. Fröstl, Helv. Chim. Acta. 58 (1975) 587; (b) H. Ahlbrecht
     and D. Liesching, Synthesis, (1976) 746; (c) H. Ahlbrecht and E.O. Düber,
     Synthesis (1980) 630.

111. (a) R. Knorr and P. Löw, J. Am. Chem. Soc., 102 (1980) 3241; (b) R. Knorr,
     A. Weiss, P. Löw and E. Räpple, Chem. Ber., 113 (1980) 2462.

112. G.J. Heiszwolf and H. Kloosterziel, Rec. Trav. Chim., 89 (1970) 1217.

113. K. Takabe, H. Fujiwara, T. Katagiri and J. Tanaka, Tetrahedron Lett., (1975)
     1237.

114. G.R. Kieczykowski, R.H. Schlessinger and R.B. Sulsky, Tetrahedron Lett., (1976)
     597.

115. (a) J.M. Brocas, B. de Jeso and J.-C. Pommier, J. Organomet. Chem., 120
     (1976) 217; (b) B. de Jeso and J.-C. Pommier, J. Organomet. Chem., 122

(1976) Cl; (c) M. Fourtinon, B. de Jeso and J.-C. Pommier, J. Organomet. Chem 193 (1980) 165.

116. P.F. Hudrlik and A.K. Kulkarni, J. Am. Chem. Soc., 103 (1981) 6251.

117. A.I. Meyers, D.R. Williams, G.W. Erickson, S. White and M. Druelinger, J. Am. Chem. Soc., 103 (1981) 3081.

118. A.I. Meyers, D.R. Williams, S. White and G.W. Erickson, J. Am. Chem. Soc., 10 (1981) 3088.

119. L. Hu, B. Mauze and L. Miginiac, C.R. Acad. Sci. Ser. C. 284 (1977) 195; cf. oximes, H.G. Richey, Jr., R.C. McLane and C.J. Phillips, Tetrahedron Lett., (1976) 233.

120. H. Ahlbrecht and W. Farnung, Chem. Ber., 110 (1977) 596.

121. P. A. Wender and J.M. Schaus, J. Org. Chem., 43 (1978) 782.

122. P.A. Wender and M.A. Eissenstat, J. Am. Chem. Soc., 100 (1978) 292.

123. (a) S.F. Martin, G.W. Phillips, T.A. Puckette and J.A. Colpret, J. Am. Chem. Soc., 102 (1980) 5866; (b) S.F. Martin and G.W. Phillips, J. Org. Chem., 43 (1978) 3792.

124. S.F. Martin, Tetrahedron, 36 (1980) 419.

125. L. Horrichon-Guigon and S. Hammerer, Tetrahedron, 36 (1980) 631.

126. P.F. Hudrlik and C.-N. Wan, J. Org. Chem., 40 (1975) 2963.

127. E. Vedejs, D.A. Engler and J.E. Telschow, J. Org. Chem., 43 (1978) 188.

128. D.A. McCrae and L. Dolby, J. Org. Chem., 42 (1977) 1607.

129. K. Takabe, H. Fujiwara, T. Katagiri and J. Tanaka, Tetrahedron Lett., (1975) 1239.

130. K. Takabe, H. Fujiwara, T. Katagiri and J. Tanaka, Tetrahedron Lett., (1975) 4375.

131. G.R. Kieczykowski, C.S. Pogonowski, J.E. Richman and R.H. Schlessinger, J. Org Chem., 42 (1977) 175.

132. W. Nagata and Y. Hayase, J. Chem. Soc. C, (1969) 460.

133. R.M. Jacobson, R.A. Raths and J.H. McDonald III, J. Org. Chem., 42 (1977)2545

134. A.I. Meyers, Rev. Pure Applied Chem., 51 (1979) 1255.

135. D. Méa-Jacheet and A. Horeau, Bull. Soc. Chim. Fr., (1968) 4571.

136. M. Kitamato, K. Hiroi, S. Terashima and S. Yamada, Chem. Pharm. Bull. Japan, 22 (1974) 459.

137. R.R. Fraser, F. Akiyama and J. Banville, Tetrahedron Lett., (1979) 3929.

138. R.R. Fraser and F. Akiyama, unpublished observations.

139. A.I. Meyers, D.R. Williams and M. Druelinger, J. Am. Chem. Soc., 98 (1976) 3032.

140. J.K. Whitesell and M.A. Whitesell, J. Org. Chem., 42 (1977) 377.

141. D. Enders and H. Eichenauer, Chem. Ber., 112 (1979) 2933.

142. S. Hashimoto, N. Komeshima, S. Yamada and K. Koga, Chem. Pharm. Bull. Japan, 27 (1979) 2437.

143. D. Enders and H. Eichenauer, Tetrahedron Lett., (1977) 191.

144. A.I. Meyers, G.S. Poindexter and Z. Brich, J. Org. Chem., 43 (1978) 892.

145. A.I. Meyers, Z. Brich and G.W. Erickson, J. Chem. Soc. Chem. Commun. (1979) 566.

146. D. Enders and H. Eichenauer, Angew. Chem. Int. Ed., 18 (1979) 397.

147. (a) S. Hashimoto, S. Yamada and K. Koga, J. Am. Chem. Soc., 98 (1976) 7450;
      (b) S. Hashimoto, S. Yamada and K. Koga, Chem. Pharm. Bull. Japan, 27 (1979) 771.

148. S. Hashimoto, N. Komeshima, S. Yamada and K. Koga, Tetrahedron Lett., (1977) 2907.

149. S. Hashimoto, H. Kogen, K. Tomioka and K. Koga, Tetrahedron Lett., (1979) 3009.

150. H. Kogen, K. Tomiaka, S. Hashimoto and K. Koga, Tetrahedron Lett., (1980) 4005;
      H. Kogen, K. Tomiaka, S. Hashimoto and K. Koga, Tetrahedron, 37 (1981) 3951.

151. B. de Jeso and J.-C. Pommier, Tetrahedron Lett., (1980) 4511.

152. P. Bercot and A. Horeau, C.R. Acad. Sci. Ser. C., 272 (1971) 1509.

153. K. Takabe, N. Nagaoka, T. Endo and T. Katagiri, Chem. and Ind., (1981) 540.

154. R. Knorr, A. Weiss and H. Polzer, Tetrahedron Lett., (1977) 459.

155. M. Yamaguchi, M. Murakami and T. Mukaiyama, Chem. Lett., (1979) 957.

156. G.T. Pearce, W.E. Gore and R.M. Silverstein, J. Org. Chem., 41 (1976) 2797.

157. P.W. Hickmott, Tetrahedron, 38 (1982) 3363.

CHAPTER 3

THE FORMATION AND TRANSFORMATIONS OF ALLENIC-α-ACETYLENIC CARBANIONS

R. EPSZTEIN

Laboratoire de Chimie Organique Biologique, Université Paris-Sud, bâtiment 420,
91405 Orsay Cédex, France

CONTENTS

I. INTRODUCTION

At the outset of this article we wish to emphasize that, contrary to what might be thought from its title, the article does not deal with two different kinds of chemical species. As will be shown, allenic and α-acetylenic carbanions, $\underline{1}$ and $\underline{2}$, are in fact the limiting resonance structures of the mesomeric carbanion $\underline{3}$. The structures of these species will be considered in detail subsequently. In the

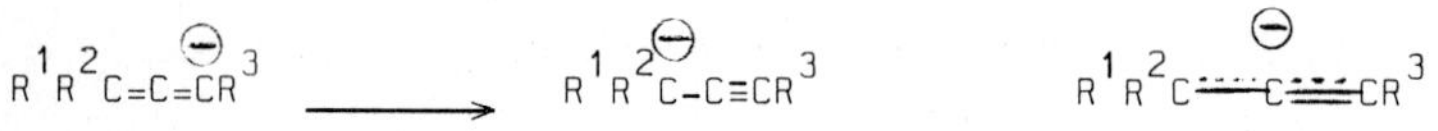

$$R^1R^2C=C=\overset{\ominus}{C}R^3 \longrightarrow R^1R^2\overset{\ominus}{C}-C\equiv CR^3 \qquad R^1R^2\overset{\ominus}{C}\cdots C\cdots CR^3$$

$$\underline{1} \qquad\qquad \underline{2} \qquad\qquad \underline{3}$$

following section, for convenience and brevity, they will generally be referred to simply as allenic anions.

Although their essential structure was elucidated less than thirty years ago, allenic carbanions have been used as a synthetic tool for more than fifty years. Their importance grew as time elapsed and they are of great utility in synthesis in various areas such as steroidal and other kinds of hormones, vitamins, acetylenic aminoacids, potential anticancer agents and so on. As organometallics they can be used as such, with the advantage of being more reactive than their saturated analogu

However they are also capable of being involved in more elaborate reactions in which their unsaturated character participates.

Allenic carbanions have already been reviewed in several articles.[1] One might wonder then about the value of adding a new survey to the previous ones. To our knowledge, however, no comparative study of the properties of these compounds, either prepared in different ways or carrying different metallic counterions, has been made hitherto. We are uncertain whether we have altogether succeeded in this difficult task.

II. METHODS OF PREPARATION OF ALLENIC CARBANIONS

The chemistry of allenic carbanions is intimately tied to the manner in which they are prepared. Thus, a common practice in effecting synthetic transformations of the carbanion is to carry out the reaction in situ, that is where the organometallic is not isolated. This is in part because of their sensitivity to air and moisture and, in part, due to their general instability at ambient temperature, being transformed to other products. Therefore the approach taken in this article, which separates their manner of preparation from their reactions, is one of convenience rather than of fundamental significance.

There are various ways of preparing allenic carbanions, the more commonly used being : (a) the reaction of a propargylic or an allenic halide with a metal, such as magnesium, zinc, aluminium, cadmium or sometimes sodium; (b) the deprotonation of an acetylene or allene by means of a strong alkali metal base. Other methods will also be mentioned although most of these are of lesser synthetic interest.

(A) Reaction of a propargylic or allenic halide with a metal

The first example of the preparation of an allenic carbanion by reacting a halide with a metal was reported by Salzberg and Marvel in 1928.[2] These authors prepared the Grignard of tris-t-butylethynylmethyl bromide $\underline{4}$ in ether solution.

$$(t\text{-}BuC{\equiv}C)_3CBr \qquad R\overset{4}{C}{\equiv}CC\overset{2}{R}R^3Br \qquad R^1R^2C{=}C{=}CH_2 \qquad R^1C{\equiv}CCH_2R^2$$

$$\underline{4} \qquad\qquad \underline{5} \qquad\qquad\qquad \underline{6} \qquad\qquad\qquad \underline{7}$$

$$a)\ R^2 = H \qquad b)\ R^2 = CO_2H$$

Somewhat later Marvel and coworkers[3] obtained from various tertiary acetylenic bromides $\underline{5}$ the corresponding Grignards or sodium derivatives. These gave rise on carbonation to carboxylic acids which were thought to be acetylenic. However, Ford, Thompson and Marvel[4] reinvestigated this work and noted that the carbonation products were not acetylenic but allenic. Newman and Wotiz,[5] closely followed by Lappin,[6] prepared the Grignards from ω-substituted primary propargyl bromides 21 and showed that their carbonation as well as protonation products were mixtures of allenes 6 and acetylenes 7 (R^2=H, COOH). Prévost, Gaudemar and Honigberg[7] prepared the Grignard of propargyl bromide itself and noted also that on carbonation it yields a mixture of acetylenic and allenic acids. However, addition to aldehydes and ketones led only to the acetylenic alcohol 8. Gaudemar[8] extended this work to 3-bromobut-1-yne 9 and obtained a very similar result. Almost at the same time

$$HC\equiv CCH_2CR^1R^2OH \qquad HC\equiv CCHMeBr \qquad HC\equiv CCH_2Br \qquad BuC\equiv CCH_2Br$$

$$\underline{8} \qquad\qquad\qquad \underline{9} \qquad\qquad\qquad \underline{10} \qquad\qquad\qquad \underline{11}$$

Wotiz and coworkers[9] prepared magnesium derivatives from various primary, secondary and tertiary propargylic bromides 10-14 which on protolysis or carbonation yielded either a mixture, in the case of 10-12, or only allenes in the case of 13 and 14.

$$HC\equiv CCHBuBr \qquad BuC\equiv CCHMeBr \qquad BuC\equiv CCMe_2Br \qquad Me_3SiC\equiv CCH_2X$$

$$\underline{12} \qquad\qquad\qquad \underline{13} \qquad\qquad\qquad \underline{14} \qquad\qquad\qquad \underline{15} \quad (X = Cl, Br)$$

Gaudemar[10] also described the preparation of aluminium and zinc derivatives of propargyl bromide. Later Komarov, Shostakovski and Astaf'eva[11] prepared the Grignard reagent from 1-trimethylsilyl 3-chloropropyne 15, and very recently Karaev, Guseinov and Akhundov[12] obtained the corresponding bromide. Eiter and coworkers[13] prepared its zinc derivative while Daniels and Paquette[14] the aluminium derivative. Propargyl bromide has been converted into its cadmium derivative.[15,16]

The vinylogue of propargyl bromide, 1-bromopent-2-en-4-yne 17a has also

$$HC\equiv CCR^3=CR^2CHR^1X \qquad\qquad\qquad\qquad BrMgC\equiv CCH_2MgBr$$

17   19

a) $R^1 = R^2 = R^3 = H$

b) $R^1 = Me$, $R^2 = R^3 = H$ $R^1CBr=C=CHR^2$

c) $R^1 = R^3 = H$, $R^3 = Me$ 20

d) $R^1 = H$, $R^2 = R^3 = Me$ 18

e) $R^1 = R^2 = R^3 = Me$ $R^1C\equiv CCHR^2Br$

f) $R^1 = Me$, $R^2 - R^3 = (CH_2)_4$ 21

g) $R^1 - R^2 = (CH_2)_4$, $R^3 = Me$

been converted into its Grignard and zinc derivatives[23] and later some of its homologues 17b-g.[24] Among these, the macrocyclic bromide 18 has to be mentioned.[18]

From the Grignard prepared from propargyl bromide, upon treatment with ethylmagnesium bromide, the di-Grignard 19 was obtained.[8,17]

Finally we shall mention the preparation of organometallics from allenic bromides. In all cases the allene 20 yields the same derivative as the acetylene 21.[19-22]

(B)   Metalation of allenes and acetylenes

This section will be divided in two parts. The first will be devoted to allenic and acetylenic hydrocarbons. The second will deal with heterosubstituted compounds. Silylated hydrocarbons will be included in the first part although the ease and direction of the proton abstraction may be influenced by the silyl group. Metalation of terminal acetylenes will be reported only when more than one metal atom is bound.

(1)   Metalation of allenic and acetylenic hydrocarbons.   Removal of propargylic and allenic protons was performed for the first time by Bourguel[25] who caused a triple bond to migrate, from the "inside" of internal acetylenes to a terminal position by heating with sodamide. This prototropic rearrangement doubtless proceeds via the transient formation of various sodium derivatives in equilibrium, with the evolution towards the more stable terminal carbanion, that is the acetylide. Such a transformation was also achieved by Vaughn.[26]

Many years later, Eberly and Adams[27] followed by West, Carney and Mineo,[28] succeeded in the treatment of but-1-yne and buta-1,2-diene with an excess of butyllithium in hexane which yielded polylithiated derivatives. The polymetalation of various acetylenes was then studied.[29-38,43-49] Among these investigations the dimetalation of 1-alkynes is of particular interest as will be seen later.

Polymetalation of allene was performed by Jaffe[35] while Brandsma and Mugge[39] observed that phenyl allene gives a dilithioderivative. It should also be mentioned here that whereas most internal acetylenes can be polymetalated, but-2-yne gives only a monolithioderivative.[37] This brings us to the interesting possibility of preparing monometalated rather than polymetalated derivatives.

Corey and Kirst[31] obtained a monolithio derivative from 1-trimethylsilyl-propyne 22 in ether in the presence of tetramethylethylenediamine (TMEDA).

$$Me_3SiC\equiv CMe \qquad R^1_3SiC\equiv CCH_2SiR^2_3$$
$$\underline{22} \qquad\qquad\qquad \underline{23}$$

Jaffe[35] showed that allene could be easily monometalated with BuLi in THF-hexane at -50°. Linstrumelle and coworkers[40,41,50] lithiated di- and trisubstituted allenes. Creary[51] metalated an allene using, instead of a carbon base, a dialkyl lithium amide. Vinyl allenes were metalated by Goré and coworkers[52] who studied the effect of the reaction time and solvent on the orientation of deprotonation.

Finally, in 1981 Yamakado and coworkers[53] metalated $\alpha'\beta'$-disilylated propyne 23, while Zakharin[54] prepared the lithio-derivative of an allenylcarborane.

(2) <u>Metalation of heterosubstituted acetylenes and allenes.</u> In 1961 Pourcelot Cadiot and Willemart[78] reported the facile rearrangement of propargylic thioethers 24 to propynylic thioethers 25 in the presence of sodium alkoxide or hydroxide.

$$HC\equiv CCH_2SR \qquad MeC\equiv CSR \qquad HC\equiv CCHMeSR \qquad CH_2=C=CMeSR$$

| 24 | 25 | 26 | 27 |

Under the same conditions the secondary thioether 26 yields the allenic derivative 27. Here, as in the Bourguel reaction,[25] the rearrangement proceeds via a transient sodium derivative. Brandsma, Wijers and Arens[55] actually reported the metalation of acetylenic thioethers 28 using sodamide in liquid ammonia. Mantione and Alves[57]

$$RCH_2C\equiv CSEt \qquad R^1C\equiv CCH_2OR^2 \qquad \emptyset C\equiv CCH_2OMe \qquad C_5H_{11}C\equiv CCH_2OTHP$$

| 28 | 29 | 30 | 31 |

showed that propargylic ethers[29] can also be lithiated with BuLi in ether in the cold, whereas the ether 30, under these conditions, yields a dilithio derivative.[5] Corey and Terashima[60] followed by Mercier, Epsztein and Holand[61] lithiated the tetrahydropyranyl ether of oct-2-yn-1-ol 31 with BuLi in THF at low temperature. The latter workers also metalated the silylated ether 32 as well as the propargyli

$$Me_3SiC\equiv CCH_2OTHP \qquad Me_3SiC\equiv CCH_2N\begin{array}{c}CH_2CH_2\\ \diagup \qquad \diagdown\\ \diagdown \qquad \diagup \\ CH_2CH_2\end{array}O \qquad Me_3SiC\equiv CCHRN=CH\emptyset$$

| 32 | 33 | 34 |

amine 33, but with the latter proton abstraction was more difficult. This observation was confirmed by Epsztein and Mercier[62] who compared the metalating ability of several propargylic amines carrying different substituents on the nitrogen atom

Interestingly, Metcalf and Casara[64] reported the remarkably easy metalation of the Schiff base 34 (R = H) and even that of the carbamate 35.

$$Me_3SiC\equiv CCH_2NHCO_2Bu\text{-}t \qquad MeSC\equiv CCH_2OMe \qquad MeSC\equiv CCH(OEt)_2$$

| 35 | 36 | 37 |

Other examples of the metalation of propargylic ethers were given by Carlson and coworkers[67] with the preparation of lithio derivatives from the monothiodieth 36 and the thioether acetal 37. Dipropargylic ethers 38 were lithiated by Huché and Cresson[63] while Reich and Shah[66] obtained the dilithio derivative of the selenoether 39.

$R^1C\equiv CCR^2_2OCH_2C\equiv CR^3$ $\quad$ $\emptyset SeCH_2C\equiv CH$ $\quad\quad$ $R^1CH=C=CHOR^2$

$\quad\quad$ 38 $\quad\quad\quad\quad\quad\quad\quad$ 39 $\quad\quad\quad\quad\quad\quad$ 40

Allenic ethers have also been lithiated. Hoff, Brandsma and Arens[56] metalated the allenic ethers 40 and thioethers 43, while Mantione and Alves[59] lithiated the mono and disilylated compounds 41 and 42. Clinet and Linstrumelle[68] metalated substituted allenic ethers 44 with BuLi as well as with lithium amides and,

$t-BuOCH=C=CMeSiMe_3$ $\quad\quad$ $t-BuOCH=C=C(SiMe_3)_2$ $\quad$ $R^1CH=C=CHSR^2$

$\quad\quad$ 41 $\quad\quad\quad\quad\quad\quad\quad\quad$ 42 $\quad\quad\quad\quad\quad\quad\quad$ 43

finally, Braverman and co-workers[69] obtained a lithium derivative from the diallenic sulfone 45.

$\quad\quad\quad\quad$ $CH_2=C=CROMe$ $\quad\quad\quad\quad$ $(Me_2C=C=CHR)_2SO_2$

$\quad\quad\quad\quad\quad\quad$ 44 $\quad\quad\quad\quad\quad\quad\quad\quad$ 45

(C) $\quad$ Other methods of preparation of allenic carbanions

$\quad\quad$ It has been shown recently[70] that elimination of alcohol from the acetylenic diethers 46 by means of an excess of alkali amide in liquid ammonia, which gives rise to the enyne 48,[71] proceeds via the allenic anion 47. This species is unstable under the conditions used but it can be trapped by ethyl bromide leading to 49.

$\quad\quad\quad\quad\quad\quad\quad\quad\quad$ $\ominus$

$ROCH_2C\equiv CCH_2OR$ $\quad\quad$ $ROC=C=CHCH_2OR$ $\quad\quad$ $HC\equiv CCH=CHOR$ $\quad\quad$ $ROCEt=C=CH_2OR$

$\quad\quad$ 46 $\quad\quad\quad\quad\quad\quad\quad$ 47 $\quad\quad\quad\quad\quad\quad\quad$ 48 $\quad\quad\quad\quad\quad\quad\quad$ 49

$\quad\quad$ Under different conditions, i.e. by using BuLi in ether, Mantione and Alves[72] obtained from diethers 50, after elimination of alcohol, a lithio derivative which they considered as being an equilibrium mixture of compounds 51 and 52.

$R^1OCHR^2C\equiv CCH_2OR^1$ $\quad\quad$ $R^2CH=CLi-C\equiv COR^1$ $\quad\quad$ $R^2CH=C=C=CLiOR^1$

$\quad\quad$ 50 $\quad\quad\quad\quad\quad\quad\quad\quad$ 51 $\quad\quad\quad\quad\quad\quad\quad$ 52

$\quad\quad$ In the same way, the mono- and dithiodiethers 53 yielded the alkylthio-lithio derivatives 54 and 55.

$R^3XCHR^2C\equiv CCH_2SR^1$ $\quad\quad$ $R^2CH=CLiC\equiv CSR^1$ $\quad\quad$ $R^2CH=C=C=CLiSR^1$

$\quad\quad$ 53 (X = O, S) $\quad\quad\quad\quad\quad\quad$ 54 $\quad\quad\quad\quad\quad\quad\quad$ 55

Atsumi and Kuwajima[73] treated furans 56 under the same conditions and obtained the dilithio derivatives 57.

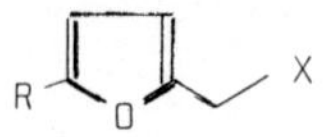

$$LiOCR=CHCH=C=CXLi$$

56   57   (X = $\emptyset Me_2Si$, $\emptyset CH=CH$)

Instead of elimination, 1,4-addition of lithium alkyls[74] and lithium dialkyl amides[75] to vinylacetylenes has been performed by Russian scientists leading to allenic derivatives 58 and 59. In the same way Grignards have been added to vinylacetylenes giving rise to allenic Grignards 60.[76]

Allenic lithio derivatives have been prepared as well from bromoallenes via metal halogen exchange on treatment with a lithium base.[40,77]

Finally, conversion of allenic lithium derivatives into zinc,[61] copper,[51,79] magnesium,[53] and tritium derivatives[80] has been achieved by treating them with the respective salts or alkoxides.

$$R^1CLi=C=CH_2R^2 \qquad R^1CLi=C=CHCH_2NR^2 \qquad R^1CMgBr=C=CHCH_2R^2$$

58   59   60

Other derivatives, for instance organotin and organolead compounds have been obtained similarly, but owing to their stability, they can hardly be likened to carbanions. Therefore they will be omitted in the present section.

III. STRUCTURE

(A) Foreword

Among the various methods of preparation of allenic carbanions, proton abstraction from an acetylene or an allene by a strong base can be differentiated from the other methods in that in this case there is often more than one site of attack. Exceptions arise when there is only one acidic proton or when a grouping which may favour elimination at a given carbon atom is present. Therefore, in general, more than one monometalated species, together with polymetalated ones are possible, as illustrated in the following scheme :

$$R^1CH_2C\equiv CCH_2R^2 \longrightarrow \begin{cases} R^1CH_2C\equiv CCHMR^2 \\ R^1CHMC\equiv CCH_2R^2 \\ R^1CHMC\equiv CCHMR^2 \quad \text{or} \quad R^1CM_2C\equiv CCH_2R^2 \quad \text{etc.} \end{cases}$$

$$R^1CH=C=CHR^2 \quad \left[\begin{array}{l} \longrightarrow \ R^1CH=C=CMR^2 \\ \longrightarrow \ R^1CM=C=CHR^2 \\ \longrightarrow \ R^1CM=C=CMR^2 \end{array} \right.$$

Of course, these examples, which are not exclusive, do not necessarily give the correct formulae of the lithio derivatives since each represents only one of the possible tautomeric structures.

The other methods are less ambiguous as, for example, in the case of halogen metal exchange.

$$R^1C\equiv CCHXR^2 \longrightarrow R^1C\equiv C\overset{\ominus}{C}HR^2$$

$$R^1CH=C=CXR^2 \longrightarrow R^1CH=C=\overset{\ominus}{C}R^2$$

or in the addition of a lithium base or a Grignard to a vinylacetylene :

$$RC\equiv CCH=CH_2 \xrightarrow{\ B^- \ } RCB=C=CHCH_2^{\ominus}$$

However, even when the center of attack is known there is still uncertainty as to the real structure of the organometallic as will be seen here.

Actually, prior to 1935, when Ford, Thompson and Marvel noted that allenes were obtained on allowing electrophiles to react with the organometallics generated from tertiary propargylic bromides[4] (see section IIA), no one had questioned the structures of these intermediates. Even many years later this "anomalous" result was still explained on the basis of a rearrangement occurring during the reaction.

It soon became obvious, however, as investigations in the field of propargylic organometallic derivatives proceeded, that the relationships between structure and chemical behaviour are not simple and, therefore, that the former could not be directly deduced from the latter. In 1951 Wotiz and co-workers[9] suggested that propargyl bromide's Grignard is a mixture of an allenic and acetylenic derivative in equilibrium but progress could be made on that matter only after the development of modern physico-chemical methods of investigation.

(B) <u>Carbanions prepared from propargylic halides</u>

In one of the first studies of this kind Gaudemar,[10] in 1956, studied the IR spectra of the organometallics derived from propargyl bromide. At that time little was known of the shifts in frequency brought about by the presence of a metal atom on the absorption of a vicinal unsaturated grouping. Therefore, though the author assigned the allenic structure <u>61</u> to the zinc and aluminium derivatives, he first concluded that the Grignard was acetylenic (<u>62</u>). This assumption was

$MCH=C=CH_2$	$HC\equiv CCH_2M$	$BrMgC\equiv CMe$	$HC\equiv CCHRBr$	$RC\equiv CCH_2Br$
<u>61</u>	<u>62</u>	<u>63</u>	<u>64</u>	<u>65</u>

supported by the observation that this compound rapidly disproportionates, yielding among other products the di-Grignard 19 and prop-1-ynyl magnesium bromide 63. This would imply the presence of an acidic hydrogen atom in the mono-Grignard which should hence be acetylenic. In contrast to the magnesium derivative, the zinc and aluminium species are fairly stable. Moreover, on addition of ethylmagnesium bromide to this Grignard, one equivalent of ethane is evolved whereas the aluminium derivative does not give rise to gas evolution and the zinc derivative, only to a small extent.

Subsequently it was found[81] that allylic organometallics absorb at lower frequencies than the corresponding hydrocarbon, the shift increasing with the electropositivity of the metal, and the correct assignments could than be inferred for propargyl bromide metallic derivatives. The zinc and aluminium compounds absorb as found before, at respectively 1900 and 1905 cm^{-1} whereas the frequency of the Grignard was found to be 1880 cm^{-1}, which means that all of these are allenic.[81]

Similarly, secondary propargylic bromides 64 in which the triple bond is terminal also lead to allenic organometallics. In contrast, compounds 65 with an internal triple bond yield a mixture of allenic and acetylenic metallic derivatives. Compound 66 could not be converted into its Grignard although it gave a zinc derivative 67 which was shown to be acetylenic.[81,82] Prévost and later Gaudemar[81,] explained these results by taking into account the charge distribution in the carbanion 68 as well as steric effects. With propargyl bromide there is no steric hindrance and only electronic factors play a role in favouring the binding of the metal to carbon 1. Similarly with compounds 69 steric factors act equally

on both carbon atoms. With secondary bromides 64 steric and electronic effects act in the same direction and attack at carbon 1 will again be favoured. In all three cases the metallic derivative is allenic. On the other hand, with compounds 65 electronic and steric factors act in opposite directions. The metallic derivative will then be a mixture of both allenic and acetylenic forms, except when steric hindrance is important enough to prevent completely the formation of the allenic derivative as in the case of 66. This expectation was confirmed later as can be seen in Tables 1 and 2.

$$R^1C{\equiv}CCHR^2Br$$

69

TABLE 1

IR data of allenic organometallics

$$R^1_1C{\equiv}CCR^2R^3Br \xrightarrow{\quad M \quad} R^1CM{=}C{=}CR^2R^3$$

| R^1 | R^2 | R^3 | (cm^{-1}) | | |
			Mg	Zn	Al
H	H	H	1685[83b]	1909[16]	1874[16]
H	Me	H	1892[86]	1913[16]	
H	n-Pr	H	1891[83b]	1915[86]	
H	i-Pr	H		1904[89]	
H	Me	Me	1905[88]		
Ø	Me	H	1891[83b]	1908[89]	

TABLE 2

IR data of acetylenic organometallics

$$RC{\equiv}CCH_2Br \xrightarrow{\quad M \quad} RCM{=}C{=}CH_2 \;+\; RC{\equiv}CCH_2M$$

| R | M | Solvent | (cm^{-1}) | | Ref. |
			C=C=C	-C≡C-	
Me	Mg	Ether	1891	2187	86
Et	-	-	1888	2183	83b
n-Pr	-	-	1891	2189	83b
n-Bu	-	-	1890	2179	86
n-C$_5$H$_{11}$	-	-	1889	2183	83b
Ø	-	-	1887	2159	83b
Me	Zn	THF	1912	2206	86
Et	-	-	1911	2199	86
n-Pr	-	-	1911	2205	89
n-C$_5$H$_{11}$	-	-	1911	2201	86
Ø	-	-	1900	2180	83a

Goré and co-workers[24] studied the IR spectra of the Grignards derived from vinylacetylenic halides 70 and found an allenic band at 1875-1880 cm^{-1} in agreement with the structure 71. Similarly the Grignard prepared from the macrocyclic bromide 18 was shown to possess the allenic structure 72.[18]

$$R^1CH=CR^2CR^3=C=CHMgBr$$

<u>71</u>

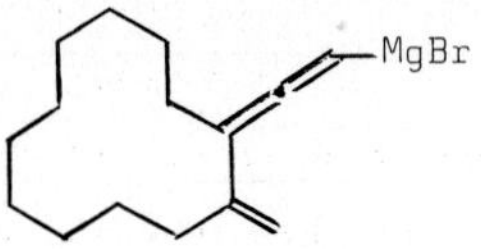

$$BrCH\ R^1CR^2=CR^3C\equiv CH$$

<u>70</u>

<u>72</u>

NMR studies on the Grignards obtained from various propargylic bromides[84,85] show in all cases a rapid equilibrium between the allenic and acetylenic forms. The presence of the latter could be confirmed by Moreau and Gaudemar[86] who prepared from deuterated propargyl bromide <u>73</u> the diol <u>74</u> which was free of deuterium. This result can be explained by the following scheme :

$$DC\equiv CCH_2Br \xrightarrow{Mg} BrMgCD=C=CH_2 \rightleftharpoons DC\equiv CCH_2MgBr$$

$$\underline{73} \qquad\qquad\qquad\qquad\qquad\qquad\qquad \downarrow EtMgBr$$

$$HOCMe_2C\equiv CCH_2CMe_2OH \xleftarrow{Me_2CO} BrMgC\equiv CCH_2MgBr$$

$$\underline{74}$$

The formation of the di-Grignard is thus achieved through hydrogen (in the present case deuterium) abstraction from the acetylenic form of the mono-Grignard of propargyl bromide.

The existence of this form also explains the disproportionation of this Grignard (see section IIIB) as shown by the scheme :

$$BrMgCH=C=CH_2 \rightleftharpoons HC\equiv CCH_2MgBr$$

$$\downarrow BrMgCH=C=CH_2$$

$$BrMgC\equiv CCH_2MgBr + \begin{cases} MeC\equiv CH \\ CH_2=C=CH_2 \end{cases}$$

$$\underline{19}$$

$$BrMgCH=C=CH_2 + MeC\equiv CH \longrightarrow MeC\equiv CMgBr + \begin{cases} MeC\equiv CH \\ CH_2=C=CH_2 \end{cases}$$

The di-Grignard <u>19</u> has been shown to be acetylenic since it absorbs at 1936 cm^{-1}.

(C) Carbanions obtained by means of lithiation

While structure determination of the organometallics derived from halides is relatively simple, in the case of metalation of allenes and acetylenes, in addition to the question concerning the nature of the multiple bond(s) in the metalated species, one has to consider as well the question of positional isomerism. IR spectroscopy was of great utility regarding the former, whereas for the latter NMR was the more useful tool.

As we have seen above, the first known examples of lithio derivatives obtained by deprotonation of acetylenes are the dimetalated compounds derived from terminal acetylenes[27,28] to which was ascribed structure 75 on the basis of their chemical behaviour. However, West and Jones[29] studied the IR spectra in hexane of the

$$LiC{\equiv}CCHLiR \qquad CH_2{=}C{=}CLi_2 \qquad , Me_3SiC{\equiv}CCH_2SiMe_3 \qquad Me_3SiC{\equiv}CCH(SiMe_3)_2$$

$$\underline{75} \qquad\qquad \underline{75a} \qquad\qquad \underline{76} \qquad\qquad \underline{77}$$

lithiated compounds of propyne and of its di- and trisilylated derivatives 76 and 77 and concluded that all are allenic. The results are shown in Table 3.

TABLE 3

IR data of lithiated derivatives of propyne and its silylated derivatives

Compound	(cm^{-1})	Compound	(cm^{-1})
$C_3H_2Li_2$	1870	$C_3HLi(SiMe_3)_2$	1870
C_3HLi_3	1770	$C_3Li(SiMe_3)_3$	1850
C_3Li_4	1675	$C_3Li_2(SiMe_3)_2$	1790

Klein and coworkers studied the mono and dilithiated derivatives of acetylenes[32-34] and noted, either on the basis of their chemical properties or through NMR data, that in the dilithio compounds both metal atoms are bound to the same propargylic carbon atom. In order to explain this rather unexpected behaviour, they proposed a "sesquiacetylenic" structure 78 for the carbanion, which was predicted by "ab initio" molecular orbital calculations to be the more stable.

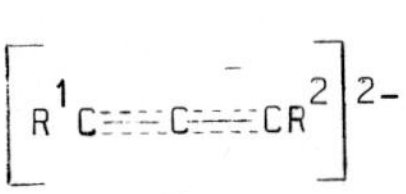

$$\left[R^1C{\cdots}C{\cdots}CR^2 \right]^{2-} \qquad \left[\begin{array}{c} R^1 \\ R^2 \end{array}\!\!>\!\!C{\cdots}C{\cdots}CLi \right]^{-} Li^{+}$$

$$\underline{78} \qquad\qquad\qquad \underline{79}$$

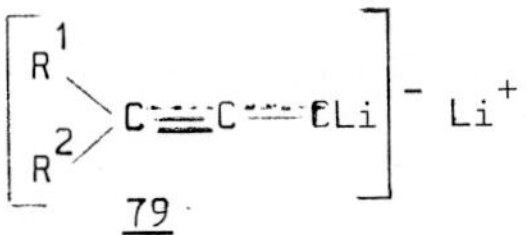

Subsequently Klein and Becker[38] noted that the IR parameters of phenylated dilithioalkynes are strongly dependent on the solvent, indicating that their structure varies as the solvent is changed. The IR data are given in Table 4.

TABLE 4

IR data of lithiated derivatives of 1-phenylpropyne

Compound	(cm^{-1})	Solvent
$\varnothing C\equiv CCHLi_2$	1895	hexane or ether
-	2080, 1895	THF
-	2045	HMPT
$\varnothing C\equiv CCLi_3$	1800	hexane or ether
-	2045	HMPT
$\varnothing C\equiv CCHLiMe$	1870	hexane or ether
$\varnothing C\equiv CCLi_2Me$	1795	- -
-	1795, 2000, 2040	- + TMEDA
$\varnothing C\equiv CCHLi\varnothing$	1870	- or ether
$\varnothing C\equiv CCLi_2\varnothing$	1790	-
-	1790, 2050	ether
-	2075	TMEDA

They hence concluded that in hexane or ether solution these compounds are allenic whereas in strongly dissociating solvents they possess the "sesquiacetyleni" structure. In less dissociating solvents both forms will be present.

Priester, West and Ling Chwang[43] taking into account the considerable batho-chromic shift ($180\ cm^{-1}$) which accompanies the transformation of monolithiated terminal acetylenes into dilithio derivatives, proposed the "propargylide" structure 79.

If the structure of polymetallic derivatives of allenes and acetylenes has been abundantly studied, a lesser interest was shown at the beginning to their monolithio derivatives. Klein and Brenner[32,33] recorded the NMR spectra of the monolithio derivatives of alk-4-ene-1-ynes 80 and noted that the abstracted proton

$$R^1C\equiv CCH_2CR^2=CHR^3$$
80

originated from the propargylic carbon. The upfield shift of the vinyl proton(s) shows that the resulting negative charge is delocalized onto the carbon atom to which the proton(s) is(are) attached. However the magnitude of the charge is less than in other cases such as pentadienyllithium or phenylallylsodium. The remaining propargylic proton is shifted to lower field due to a change in hybridization of this carbon atom.

Jaffe found that the monolithio derivative of allene absorbs at 1885 cm^{-1},[35] which is consistent with the allenic structure 81. With higher homologues of allene, i.e. 82, there are two different possible metalation sites. However, owing

$CH_2=C=CHLi$ $RCH=C=CH_2$ $Me_2C=C=CH_2$ $Me_2C=C=CHLi$

81   82   83   84

to steric effects, the terminal carbon atom is the preferred one and the lithiation is regioselective. Creary[51] prepared the monolithio derivative of 1,1-dimethyl-allene 8 and from NMR data in benzene solution (δ_{Me} = 1.33 ppm (doublet), δ_{CHLi} = 4.70 ppm (heptet), J = 4.8 Hz), and in ether where the data were very similar, assigned to it the allenic structure 84.

Finally, Ishigura[80] noted that titanium derivatives of 1-trimethylsilyl propyne 85 and -but-1-yne 86 absorb at 1898 and 2092 cm^{-1}, which implies that they are respectively allenic (87) and acetylenic (88).

$Me_3SiC\equiv CMe$ $Me_3SiC\equiv CCH_2Me$ $Me_3SiCM=C=CH_2$ $Me_3SiC\equiv CCHMMe$

85   86   87   88

With heterosubstituted lithium derivatives, physico-chemical data are difficult to obtain owing to the instability of these compounds at room temperature. However, the zinc derivatives are more amenable to such studies. A pertinent study has been made with the zinc derivatives of 3-methoxy-1-phenylpropyne 89 and 3-t-butoxy-1-trimethylsilylpropyne 90.[87] The IR spectrum of the former shows a band at 1895 cm^{-1} which is in agreement with the allenic structure 91, whereas

$\emptyset C\equiv CCH_2OMe$ $Me_3SiC\equiv CCH_2OBu\text{-}t$ $\emptyset CM=C=CHOMe$

89   90   91

the latter absorbs at 1900 and 2105-2140 cm^{-1} and hence is a mixture of both allenic and acetylenic derivatives 92 and 93.

$Me_3SiCM=C=CHOBu\text{-}t$ $Me_3SiC\equiv CCHMOBu\text{-}t$

92   93

(D) Conclusion

In conclusion, with few exceptions most of the monometalated derivatives of
acetylenes and allenes possess an allenic structure, regardless of their method
of preparation. However the nature of the counterion introduces a difference
as shown by NMR measurements. Lithio derivatives are purely allenic, whereas other
organometallics are mixtures of both forms. The case of a polymetalated compound
is different since the structure varies sometimes considerably with the metal;
thus the dimagnesium derivative of propyne 19 is acetylenic while its dilithio
derivative 75a is allenic. There is also evidence that the solvent can play a role
on the structure of these species.[38]

IV. REACTIONS

The first transformations effected on allenic carbanions are carbonation
and hydrolysis.[2-4] These reactions were performed with Grignards derived from
tertiary propargylic bromides which possess the allenic structure 94 but may
contain small amounts of their acetylenic isomer 95. All these precursors gave
rise to allenes 96 and 97.

$$BrMgCR^1=C=CR^2R^3 \qquad R^1C\equiv CCR^2R^3MgBr \qquad R^1CH=C=CR^2R^3 \qquad HO_2CCR^1=C=CR^2R^3$$

94   95   96   97

With the Grignards derived from the primary propargylic bromides 10 (allenic)
and 65 (a mixture), the analogous reactions[5-7,9] yielded mixtures of allenes 98
or 99 and acetylenes 100 or 101 (R^1 = H, alkyl). However, with aldehydes, the
allenic Grignard derived from 10 led only to acetylenic alcohols.[7,8] A similar

$$R^1CH=C=CH_2 \qquad HO_2CCR^1=C=CH_2 \qquad R^1C\equiv CMe \qquad R^1C\equiv CCH_2CO_2H$$

98   99   100   101

result was obtained from the secondary propargylic bromide 9,[8] the Grignard of
which is also known to be allenic.

These few examples chosen among the earlier studies show that the structures of the products may depend not only on the nature of the reactant but
also on the type of reaction. Like other organometallics, allenic carbanions
react with a great variety of electrophiles. These reactions can be divided into
two main types : addition to unsaturated bonds and substitution, both types being
illustrated in the above examples. The reactants used in the addition processes
can also be divided into two varieties : carbon-heteroatom double bonds which
include carbonyls, imines, carbon dioxide, carboxylic acids, esters, chlorides,
amides, thioketones, dithioesters, as well as carbon-carbon double or triple bonds.

The substitution reactants include a great number of compounds both organic
and inorganic. The possible paths followed in all these reactions can be illustrated in Scheme 1 where E is a general electrophile.

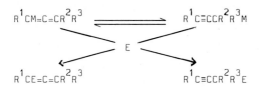

SCHEME 1

The mechanism depends on various factors, steric as well as electronic,
pertaining to the carbanion as well as the electrophile.

There is also a third kind of transformation, limited to a few examples,
which consists of a rearrangement of the carbanion. In all of these the starting
reagent carries a heteroatom which then plays the role of the electrophile.

We shall develop now these three types of reactions.

(A) Addition reactions

    (1) Addition to carbon-heteroatom double bonds

        (a) Addition to carbonyls

Among the reactions of allenic carbanions, addition to carbon-oxygen double
bonds has been the most widely studied. Due to the abundance of data in this field,
it seems important to us to start with a short summary.

We have already mentioned that purely allenic carbanions, that is those for
which IR spectroscopy does not indicate the presence of any acetylenic compound,
react with aldehydes to give acetylenic alcohols. This would mean that the reaction
is accompanied by rearrangement. However, bearing in mind that the presence of at
least traces of acetylenic organometallic has been proven in several instances,
one could also consider this intermediate to be the actual reactant. We will return
to this point subsequently. Actually, as the number of known examples increased,
on varying the structure of the starting acetylene or allene as well as that of
the carbonyl compound, and the nature of the metal and the solvent, it soon
appeared that the observed regiospecificity was far from general. As a matter of
fact, whereas with allenic aluminium compounds no allenic alcohol could be detected
in the reaction products, with other metals, namely zinc, magnesium and cadmium,
some of this product was found to be present. Later, when lithium derivatives
became readily available it was found that the allenic alcohols were formed
sometimes to significant extent, if not exclusively.

Moreover, when branched alcohols having the formula 102 are prepared by the
above method, they are found to be mixtures of the two possible diastereoisomers,

$R^1C\equiv CCHR^2CR^3R^4OH$ t-BuCOPr-i

$$\begin{array}{c} \text{t-Bu} \\ \text{i-Pr} \end{array}\!\!\!>\!\!C\text{-CH=C=CH}_2 \quad \underset{\text{OH}}{} \qquad (\text{t-Bu})_2\text{CO}$$

<u>102</u>   <u>103</u>   <u>104</u>   <u>105</u>

the proportions of which may vary widely depending on the same factors as does regioselectivity.

We shall now consider more in detail all these problems.

(i) <u>Regioselectivity</u>

As has been stated, allenic organoaluminium derivatives react with carbonyl compounds to yield only acetylenic alcohols[10,76,88,91-93,97,98,106] regardless of the nature of the solvent. Allenic Grignards behave similarly, except on reaction with hindered ketones.[16] Thus, allenylmagnesium bromide gives with 2,2,4-trimethyl-3-pentanone <u>103</u> in ether as well as in THF, in the presence or absence of HMPT, up to 10% of the allenic alcohol <u>104</u>. The amount of this compound is not appreciably affected by the polarity of the medium. With 2,2,4,4,tetramethyl-3-pentanone <u>105</u>, 20% of allenic alcohol <u>106</u> is obtained. It is also interesting to note that, whereas the Grignard <u>107</u>, derived from a tertiary propargylic bromide, affords

$(\text{t-Bu})_2C(OH)CH=C=CH_2$ $Me_2C=C=CHMgBr$ $MeCH(OH)CMe_2C\equiv CH$ $MeEtC=C=CHMgBr$

       <u>106</u>   <u>107</u>   <u>108</u>   <u>109</u>

the pure acetylenic alcohol <u>108</u>, upon reaction with acetaldehyde,[88] its higher homologue <u>109</u> yields a mixture of <u>110</u> and <u>111</u>[21a] when treated with acetone.

$HC\equiv CCMeEtCMe_2OH$ $HOCMe_2CH=C=CMeEt$ $R^1CH=CR^2CR^3=C=CHMgBr$

     <u>110</u>   <u>111</u>   <u>112</u>

$R^1CH=CR^2R^3C(C\equiv CH)CR^4R^5OH$ $\emptyset C\equiv CCHMeBr$ $\emptyset C\equiv CCHMeCMe_2OH$ $Me_2C=C=CHCu$

       <u>113</u>   <u>114</u>   <u>115</u>   <u>116</u>

Finally, vinylallenic Grignards <u>112</u> prepared from the vinylogue <u>17</u>[23] of propargyl bromide ($R^1 = R^2 = R^3 = H$) and its homologues[24] give rise with aldehydes and ketones to the vinylacetylenic alcohols <u>113</u>.

Concerning zinc derivatives, from <u>114</u> pure acetylenic alcohol <u>115</u>[83a] is obtained whereas <u>10</u> gives, even with aldehydes (except chloral), small amounts of allenes, while ketones give rise to larger proportions of these products as shown in Table 5.[16] The influence of solvent on the extent of formation of the

$HC\equiv CCMe_2CMe_2OH$ $R^1C\equiv CEt$ $R^1C\equiv CCHMeCHR^2OH$ $R^1C\equiv CMe$

     <u>117</u>   <u>118</u>   <u>119</u>   <u>120</u>

allenic alcohol is also important, mainly with hindered ketones as can be seen in Table 6. However in ether/THF the zinc derivative of trimethylsilylpropargyl bromide **15** on reaction with acetophenone yielded the acetylenic alcohol **102** $(R^1 = Me_3Si, R^2 = H, R^3 = Me, R^4 = \emptyset)$.[13]

 With cadmium compounds, which give altogether poor yields, the amount of allenic alcohol is important even with relatively unhindered ketones (table 7).[16]

TABLE 5

Addition of $CH_2=C=CHZnBr$ to carbonyl compounds in THF

Carbonyl compound	Allenic alcohol %	Total yield %
CCl_3CHO	0	44
MeCHO	5	75
i-PrCHO	5	62
MeCOMe	12	74
MeCOBu-t	20	69
$(Et)_2CO$	25	60
$(i-Pr)_2CO$	54	49
t-BuCOPr-i	20	14
$(t-Bu)_2CO$		0
$\emptyset$CHO	7	65
MeCO$\emptyset$	14	54
$\emptyset_2CO$	52	48

TABLE 6

Addition of $CH_2=C=CHZnBr$ to 3-pentanone and 2,4-dimethyl-3-pentanone

Solvent	$HO(Et)_2CCH=C=CH_2$ %	$HO(i-Pr)_2CCH=C=CH_2$ %
Dioxane	12	23
THP	13	40
THF	18	36
DME	14	38
THF + DMSO	14	36
THF + DMF	20	45
THF + HMPT	25	56
DMSO	22	45
HMPT	28	58

TABLE 7

Addition of $CH_2=C=CHCdBr$ to ketones

Solvent	ketone	allenic alcohol %	total yield %
HMPT	$(Et)_2CO$	68	14
THF + HMPT	$(Et)_2CO$	75	15
THF + HMPT	AcMe	26	17

It is also interesting to note that the allenic copper derivative 116 yields with acetone the acetylenic alcohol 117.[51] Similarly titanium compounds[80] derived from the acetylenic hydrocarbons 118 via their lithium derivatives also give with aldehydes the acetylenic alcohols 119 almost exclusively. Only when $R^1 = \emptyset$, does cyclohexylcarboxaldehyde yield the expected alcohol contaminated with 6% of its allenic isomer. Table 8a shows these results which are compared with those obtained from the corresponding lithium derivatives.

TABLE 8a[80]

Addition of $MCR^1=C=CHMe$ to aldehydes R^2CHO in THF

R^1	R^2	M	acetylenic alcohol %	total yield %
Me_3Si	C_6H_{11}	Ti	> 99	69
Me_3Si	$\emptyset$	"	"	79
Et	C_6H_{11}	"	"	42
Et	C_6H_{11}	Li	84	
$\emptyset$	C_6H_{11}	Ti	94	89
$\emptyset$	C_6H_{11}	Li	93	
$\emptyset$	$\emptyset$	Ti	> 99	92
$\emptyset$	$\emptyset$	Li	98	

Now with organometallics derived from primary propargylic bromides 65 $(R \neq H)$, regardless of the nature of the metal, a mixture of acetylenic and allenic alcohols, 122 and 121, is always obtained.[81,83a,99,100]

$HOCR^2R^3CR^1=C=CH_2$ $R^1C\equiv CCH_2CR^2R^3OH$ $R^1C\equiv CMe$ $HOCHR^2CR^1=C=CH_2$

121   122   123   124

It is interesting to mention, however, that titanium derivatives[80] of compounds 123 yield with aldehydes the allenic alcohols 124 almost exclusively (See Table 8b).

TABLE 8b
Addition of the metalated derivative of $R^1C\equiv CMe$ on aldehydes R^2CHO

R^1	R^2	M	acetylenic alcohol %	total yield %
Me_3Si	C_6H_{11}	Ti	1	93
Me_3Si	C_6H_{11}	Li	58	
Me_3Si	C_6H_{11}	Mg	56	
Me_3Si	C_6H_{11}	Zn	75	
Me_3Si	Ø	Ti	1	87
Me_3Si	Ø	Li	65	
Me	C_6H_{11}	Ti	1	9U
Me	C_6H_{11}	Li	53	
Me	Ø	Ti	1	80
Me	Ø	Li	49	

It is also important to note the peculiar behaviour of the organometallics derived from the silylated propargylic halides 15. Their aluminium derivatives yield with carbonyls allenic alcohols 125,[14] whereas their zinc and magnesium derivatives give rise to the acetylenic alcohols 126.

$$R^1R^2C(OH)C(SiMe_3)=C=CH_2 \qquad\qquad Me_3SiC\equiv CCH_2CR^1R^2OH$$

125   126

Finally, with heterosubstituted zinc[61,62,94,95] and titanium[80] derivatives which certainly are allenic, the reaction is quite regioselective. Thus from propargylic ethers 29 acetylenic hydroxyethers 128a were obtained whereas the propargylic amines 127 yielded acetylenic aminoalcohols 128b except when R ≠ Ø, in which case most of the product was shown to be allenic.[87]

$$R^1C\equiv CCH_2NR^2_2 \qquad\qquad R^1C\equiv CCHYCR^3R^4OH \qquad \begin{array}{l} a) \ Y = OR^2 \\ b) \ Y = NR^2_2 \end{array}$$

127   128

With a few exceptions the above results are quite consistent in that a definite relationship seems to exist between steric effects and regioselectivity. The lithio derivatives exhibit a less simple behaviour as will be seen. The lithiated derivatives of the silylated compounds 22[35] and 129[97] lead with carbonyl derivatives to the respective acetylenic alcohols 126 and 130. Using the lithioallenes 58 as the reactants, whereas with aldehydes the acetylenic alcohol 119 was favoured for R^2 = Me[80] (Table 8a), with ketones the allenic alcohols 131 were obtained.[71]

$Me_3SiC{\equiv}CCHMe_2$ $Me_3SiC{\equiv}CCMe_2CR^1R^2OH$ $HOCR^3R^4CR^1{=}C{=}CHR^2$

     129   130   131

Lithiodimethylallene 84 on reaction with acetone was reported by Patrick and coworkers to give rise to the allenic alcohol 132.[77]

$HOCMe_2CH{=}C{=}CMe_2$ $HC{\equiv}CCMe_2CH\emptyset OH$

     132   133

However Linstrumelle and Michelot[40] obtained with benzaldehyde the acetylenic alcohol 133. Creary[51] added the same reagent to various carbonyls and obtained the results shown in Table 9, which are in close agreement with those of the latter workers.[40] There is a slight difference, compared to the former,[77] which might be due to a difference in the reaction medium. Furthermore, it can be seen here that, in contrast with the other organometallics, parameters other than steric seem to play an important role.

     Also to be mentioned here is the reaction of the lithio derivative of the disilylated compound 23 (R^1 = Me) with carbonyls which yields olefins.[53] This reaction, known as the Peterson olefination, proceeds via the lithium alkoxides 134 (Scheme 2). It has been performed with various aldehydes as well as with cyclohexanone, as shown in table 10. It appears to be always regiospecific except when one of the R^2's is bulky and the carbonyl compound is cyclohexylcarboxaldehyde It is interesting to note that in this case, with the corresponding Grignard, the reaction is still highly regioselective.

23 $\xrightarrow{\text{BuLi}}$ $Me_3SiCLi{=}C{=}CHSiR^2_3$ $\xrightarrow{R^3R^4CO}$ $Me_3SiC{\equiv}CCHSiR^2_3$ $\longrightarrow$ $Me_3SiC{\equiv}CCH{=}CR^3R^4$

                                                $R^3R^4{\overset{|}{C}}OLi$

                                          134

             SCHEME 2

TABLE 9

Reaction of dimethylallenyllithium 84 with carbonyls

Carbonyl compound	Acetylenic alcohol %	Total yield %
HCHO	100	
ØCHO	100	45
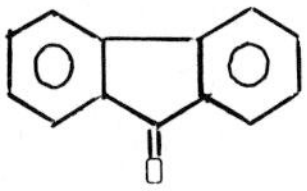 (fluorenone)	100	
MeCHO	92	36
ØCOMe	92	71
ØCOEt	80	76
ØCOPr-i	0	90
ØCOBu-t	0	100
$Ø_2CO$	49	
Me_2CO	19	53
MeCOEt	11	80
MeCOBu-t	0	91
Et_2CO	0	89
$(t-Bu)_2CO$	0	100
$(CH_2)_5CO$	12	73
$p-MeOC_6H_4COMe$	75	62
$Me_2C=CHCOMe$	49	63

A certain number of lithio derivatives of heterosubstituted acetylenes and allenes have also been reacted with carbonyl compounds. Mantione and coworkers[57,101] obtained from the propargylic secondary ethers 135 furan derivatives arising from the cyclization of the allenic alcohols 136. From the silylated allenic ethers 137

$$ØC≡CCHR^1OR^2 \qquad HOCR^3R^4CØ=C=C\ R^1OR^2 \qquad t-BuOCH=C=CRSiMe_3$$

135   136   137

(R = H) the allenic alcohols 138[59] were prepared, whereas methoxyallene yielded the alcohols 139.[56b]

On the other hand the silylated thioether 140 led to the acetylenic alcohols 141.[102]

TABLE 10

Reaction of $Me_3SiC\equiv CCH_2SiR^2_3$ with carbonyl compounds

Carbonyl compound	R^2_3Si	metal	Z-olefin %	total yield %
$(CH_2)_5CO$	Me_3Si	Li		83
$(CH_2)_5CO$	Me_3Si	Mg		88
$(CH_2)_5CO$	Et_3Si	Mg		93
$n\text{-}C_5H_{11}CHO$	Me_3Si	Li	75	77
$n\text{-}C_5H_{11}CHO$	Me_3Si	Mg	88	75
$n\text{-}C_5H_{11}CHO$	Et_3Si	Li	86	78
$n\text{-}C_5H_{11}CHO$	Et_3Si	Mg	97	89
$n\text{-}C_5H_{11}CHO$	$t\text{-}BuMe_2Si$	Mg	> 98	65
$(CH_2)_5CHCHO$	Me_3Si	Li	89	69
$(CH_2)_5CHCHO$	Me_3Si	Mg	95	84
$(CH_2)_5CHCHO$	Et_3Si	Li	91	88
$(CH_2)_5CHCHO$	Et_3Si	Mg	96	96
$(CH_2)_5CHCHO$	$t\text{-}BuMe_2Si$	Li	92	55*
$(CH_2)_5CHCHO$	$t\text{-}BuMe_2Si$	Mg	97	75
$\varnothing CHO$	Me_3Si	Li	50	53
$\varnothing CHO$	Me_3Si	Mg	75	76
$\varnothing CHO$	Et_3Si	Li	50	63
$\varnothing CHO$	Et_3Si	Mg	67	78
$\varnothing CHO$	$t\text{-}BuMe_2Si$	Mg	50	55
$\varnothing CH=CHCHO$	Me_3Si	Mg	67	88
$\varnothing CH=CHCHO$	$t\text{-}BuMe_2Si$	Mg	88	90

* A relatively large amount of the regioisomeric cumulene was produced in this particular case.

$t\text{-}BuOCH=C=C(SiMe_3)CMe_2OH$ $CH_2=C=C(OMe)CR_2OH$ $Me_3SiC\equiv CCH_2SBu\text{-}t$

138 139 140

$Me_3SiC\equiv CCH(SBu\text{-}t)CR^1R^2OH$ $C_5H_{11}C\equiv CCHOTHP$ $\overset{|}{\varnothing CHOH}$

141 142

With the primary propargylic ether **31**, however, addition to benzaldehyde led to a mixture of the expected acetylenic hydroxyether **142** and the dihydrofuran **143**, which could be proven to be the cyclization product of the allenic hydroxy-ether **144**.[61]

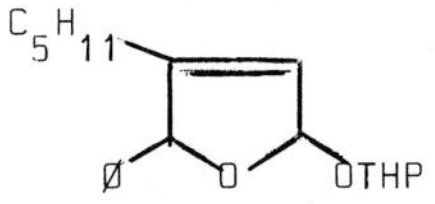

143 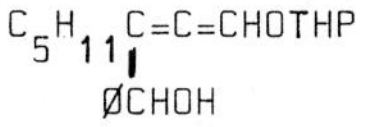   **144**

It was shown later[103] that all the reactions performed with lithiated primary propargylic ethers with aldehydes and ketones give a mixture of the acetylenic derivative and the isomeric dihydrofuran or, in some cases, the allenic alcohol.

A study was then initiated in order to determine the influence of various parameters on orientation in this reaction.[103] The results are shown in Tables 11-13. They indicate that steric factors seem to be less important than electronic factors on orientation.

TABLE 11

Reaction of lithiated $\emptyset C \equiv CCH_2 OMe$ with benzaldehyde in different solvents

Solvent	$\emptyset C \equiv CCH(OMe)CH\emptyset OH$ %	total yield %
THF + TMEDA	84	60
THF	78	66
Et_2O	37	60
hexane + TMEDA	21	45
hexane	25	15

TABLE 12

Reaction of lithiated $\emptyset C\equiv CCH_2OMe$ with various carbonyl compounds in THF

Carbonyl	Acetylenic alcohol %	total yield %
$\emptyset_2 CO$	84	70
$\emptyset CHO$	78	66
t-BuCHO	70	70
MeCO$\emptyset$	60	65
$(CH_2)_5 CO$	35	73
MeCOBu-t	35	70

TABLE 13a

Reaction of lithiated $RC\equiv CCH_2OMe$ with benzaldehyde

R	Acetylenic alcohol %	total yield %
$\emptyset$	78	66
Me	57	70
$n-C_5H_{11}$	63	70
t-Bu	80	75
Me_3Si	57	65

TABLE 13b

Reaction of lithiated $\emptyset C\equiv CCH_2OR$ with benzaldehyde

R	Acetylenic alcohol %	total yield %
Me	78	66
Et	74	70
t-Bu	77	55
THP	80	65

(ii) Stereochemistry

On reacting the magnesium and aluminium compounds derived from 3-bromo but-1-yne $\underline{9}$ with aldehydes, Chodkiewicz and coworkers[104,105] obtained a mixture of diastereoisomers $\underline{145}$ with generally a predominance of the threo $\underline{145-t}$. The

HC≡CCHMeCHROH

$\underline{145}$

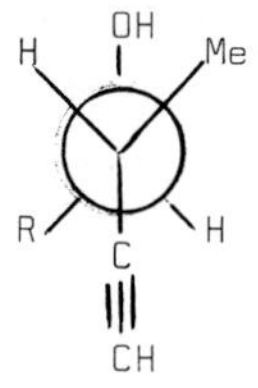

145-e 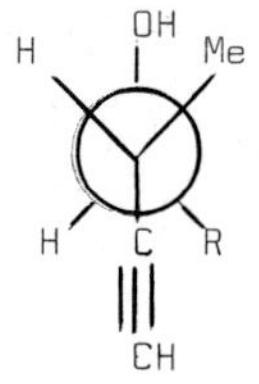 145-t

results are shown in Table 14 where it can be seen that the stereoselectivity increases with the size of R. Addition to acetophenone is also stereoselective, although in this case the configurations have not been ascribed (see Table 15). It should be noted that neither with benzaldehyde nor with its substituted derivatives (p-methyl, 2,4,6-trimethyl and p-nitro benzaldehyde), stereoselectivity could not be observed.

TABLE 14

Addition of MCH=C=CHMe to aldehydes RCHO

R	Metal	threo alcohol %	total yield %
Me	Mg	76	62
Me	Al	56	75
Et	Mg	82	54
i-Pr	Mg	90	50
i-Pr	Al	90	80
t-Bu	Mg	95	75
C_6H_{11}	Mg	94	80
CCl_3	Mg	64	100

TABLE 15

Addition of MCH=C=CHMe to acetophenone

Metal	Diastereoisomer I %	total yield %
Mg	80	90
Zn	89	90
Al	95	80

Comparable results were obtained recently by Yamakado and coworkers[53] who converted the lithiated derivatives of bis-silylated compounds 23 into olefins on addition to carbonyls. The reaction proceeds as already stated via the inter- mediate lithium alkoxides 134 which then undergo syn elimination. It can then be seen in Table 10 that the addition reaction is stereoselective in favour of the threo isomer, which gives rise to the Z olefin. The selectivity depends on the R^3Si moiety and the carbonyl substituents. When lithium is replaced by magnesium the stereoselectivity is enhanced. It is interesting to note that with benzaldehyde there is no stereoselectivity when the lithio derivative is used, whereas that with the Grignard is moderate.

Titanium derivatives also give rise to a significant stereoselectivity in favour of the threo isomer. Table 16 shows that this is more important than with lithium and even magnesium derivatives.[80]

TABLE 16

Addition of $MCR^1=C=CHMe$ to aldehydes R^2CHO

R^1	R^2	Metal	threo alcohol %	total yield %
Me_3Si	C_6H_{11}	Li	62	48
Me_3Si	C_6H_{11}	Mg	65	42
Me_3Si	C_6H_{11}	Ti	89	69
Me_3Si	Ø	Ti	84	79
Ø	C_6H_{11}	Li	80	67
Ø	C_6H_{11}	Ti	99	89
Et	C_6H_{11}	Ti	91	42

With heterosubstituted allenic carbanions stereoselectivity was also observed, with lithium[103] as well as with zinc[103] and titanium[80] counterions. In the case of lithium, solvent has an important influence on the stereo- selectivity (Table 17).

It can be seen that in hexane addition to benzaldehyde gives an equal amount of both diastereoisomers. When the polarity of the medium increases, however, the proportion of the erythro isomer increases. (It should be noted that due to the presence of an heterosubstituted grouping on one of the asymmetric carbon atoms, the definition of the threo and erythro diastereoisomers is reversed).

TABLE 17

Addition of lithiated $\emptyset C \equiv CCH_2OMe$ to benzaldehyde

Solvent	erythro alcohol %	total yield %
Hexane	52	15
Hexane + TMEDA	58	45
Et_2O	64	60
THF	66	66
THF + TMEDA	79	60

TABLE 18

Addition of lithiated $\emptyset C \equiv CCH_2OMe$ to carbonyl compounds in THF

Carbonyl	erythro alcohol %	total yield %
$\emptyset CHO$	66	66
t-BuCHO	86	70
$MeCO\emptyset$	75	65
MeCOBu-t	31	70

TABLE 19

Addition of lithiated $RC \equiv CCH_2OMe$ to benzaldehyde in THF

R	erythro alcohol %	total yield %
$\emptyset$	66	66
Me	69	70
$n-C_5H_{11}$	66	70
t-Bu	61	75
Me_3Si	53	65

TABLE 20

Addition of lithiated $\emptyset C\equiv CCH_2OR$ to benzaldehyde in THF

R	erythro alcohol %	total yield %
Me	66	66
Et	66	70
t-Bu	76	55
THP	75	65

The influence of the carbonyl substituents is also important, as can be seen in Table 18. However there does not seem to exist a simple rule which would explain how stereoselectivity varies with their respective size and it is interestir to note the inversion of selectivity when going from an aromatic to an aliphatic ketone. In contrast with these data, the influence of the acetylene substituent and of the ether group is not very important (Tables 19 and 20).

With zinc derivatives the stereoselectivity is generally higher, as indicated in Tables 21-24,[103] and with 17-ketosteroids the reaction is almost stereospecific. Epsztein and coworkers in reporting these results erroneously assigned to the isolated compounds the threo configuration,[94a] on the basis of Horeau's method for t determination of absolute configuration.[109] It could be shown, however, that this method gives incorrect results with acetylenes and the products were proven to be the erythro compounds.[94b]

TABLE 21

Addition of the zinc derivative of $\emptyset C\equiv CCH_2OMe$ to carbonyls R^1COR^2

R^1	R^2	erythro alcohol %	R^1	R^2	erythro alcohol %
Me	H	78	Et	Me	53
$n\text{-}C_7H_{15}$	H	82	i-Pr	Me	57
t-Bu	H	92	$\emptyset(CH_2)_2$	Me	60
$\emptyset CH_2$	H	76	t-Bu	Me	65
Cl_3C	H	90	$p\text{-}MeOC_6H_4$	Me	18
$p\text{-}MeOC_6H_4$	H	78	$\emptyset$	Me	19
$\emptyset$	H	80	$\emptyset$	Et	20
$p\text{-}ClC_6H_4$	H	75	t-Bu	$\emptyset$	70
$p\text{-}NO_2C_6H_4$	H	73	$\emptyset$	$C\equiv CSiMe_3$	40
$\emptyset CH=CH$	H	87	$\emptyset CH=CH$	Me	38
$\emptyset C\equiv C$	H	90	Me	$C\equiv C\emptyset$	40

In Table 21 is indicated the influence of the carbonyl substituents on the stereochemistry. With the exception of aryl ketones the erythro isomer is always predominant and the stereoselectivity is generally related to the relative size of both substituents of the carbonyl group.

TABLE 22

Addition of the zinc derivative of $RC \equiv CCH_2OMe$ to benzaldehyde

R	erythro alcohol %	total yield %
Ø	80	78
Me	90	70
$n-C_5H_{11}$	88	70
t-Bu	80	80
Me_3Si	75	80

TABLE 23

Addition of the zinc derivative of $ØC \equiv CCH_2OR$ to benzaldehyde

R	erythro alcohol %	total yield %
Me	80	78
Et	76	80
t-Bu	74	88
Me_3Si	71	60
THP	65	75

TABLE 24

Addition of the zinc derivative of $Me_3SiC \equiv CCH_2OR$ to benzaldehyde

R	erythro alcohol %	total yield %
Me	75	80
Ø	77	93
t-Bu	73	90
THP	64	65

TABLE 25

Addition of the zinc derivatives of $R^1C\equiv CCH_2OR^2$ to carbonyls

R^1	R^2	erythro alcohol % with		
		MeCOBu-t	t-BuCHO	MeCOMe
$\emptyset$	Me	65	92	19
$\emptyset$	THP	76	95	22
Me_3Si	Me	70		23
Me_3Si	THP	77		34

TABLE 26

Addition of $MCR^1=C=CHOR^2$ to aldehydes R^3CHO in THF

R^1	R^2	R^3	M	erythro alcohol %	total yield %
Me_3Si	THP	C_6H_{11}	Zn	71	51
Me_3Si	THP	C_6H_{11}	Ti	90	67
Me_3Si	THP	$n-C_5H_{11}$	Ti	88	65
Me	THP	C_6H_{11}	Li	81	44
Me	THP	C_6H_{11}	Zn	88	48
Me	THP	C_6H_{11}	Ti	94	81
Me	THP	$n-C_5H_{11}$	Ti	95	76
Me	CMe_2OMe	$n-C_5H_{11}$	Ti	93	57
Me	THP	C_6H_{11}	Ti	95	59
Me	THP	$n-C_5H_{11}$	Ti	95	68

In the latter case, however, there is a small but definite decrease in selectivity as the substituent decreases in size, when benzaldehyde is used. In contrast, in the case of hindered ketones or aldehydes, the stereoselectivity increases with the size of the ether grouping as shown in Table 25.

Finally with titanium derivatives of heterosubstituted compounds[80] the stereoselectivity is also higher than with zinc derivatives (See Table 26).

(iii) Mechanism

In order to explain the formation of β-acetylenic alcohols from allenic organometallics it can be assumed that they arise either directly with concomitant rearrangement, or from the acetylenic form which must be present in the

reaction medium, at least in small quantities. Very early, however, the allenic
structure was chosen as the true one.[10,18,81,83a,99] The mechanism was then
thought to be cyclic, that is an SE1' process as shown on Scheme 3. Later Chodkiewicz
and co-workers[104,105] proposed instead an SE2' mechanism (Scheme 4) in order to
explain the stereoselectivity of the addition of Grignards derived from **3**-bromobut-
1-yne **9** to aldehydes. The eclipsed transition state which is stabilized through

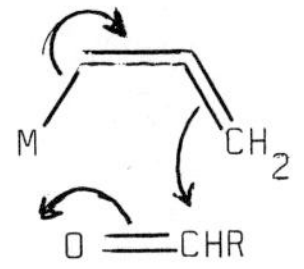

| SCHEME 3 | SCHEME 4 | SCHEME 5 |

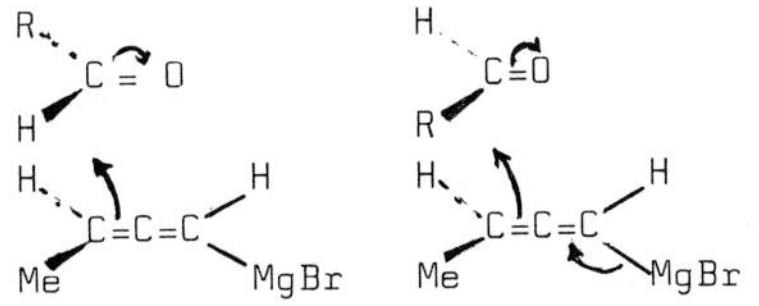

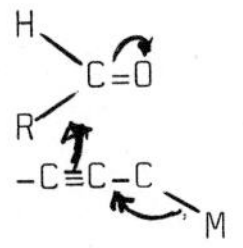

overlap of the orbitals of the allene and of the carbonyls can exist in two configurations. The more stable one in which the larger substituent of the carbonyl
lies close to the smaller substituent of the allene gives rise to the threo isomer.
On the other hand, as shown by Felkin and co-workers,[107] for the allylic series
the SE2 process will be slower than the SE2' and hence Chodkiewicz rejected the
acetylenic organometallic as a possible precursor.

An SE2' mechanism was also taken in account later in order to explain stereoselectivity in similar cases.[24,61,80,87]

Now, regarding the formation of allenic alcohols, one might assume for the
reasons given above that here also an SE2' process takes place from the acetylenic
organometallic, according to Scheme 5. Actually, such a process has not been
proven so far, to our knowledge. It is pertinent in this regard to consider a
study by Lequan and Guillerm[108] who carried out the addition to chloral of
acetylenic and allenic organotin derivatives **146** and **146a**. As these species are
rather stable, conditions could be found under which isomerization during reaction

$$HC \equiv CCHR^1 SnR^2_3 \qquad R^2_3 SnCH=C=CHR^1 \qquad Cl_3 CCH(OH)CH=C=CHR^1 \qquad HC \equiv CCHR^1 CH(OH)CCl_3$$

| **146** | **146a** | **147** | **147a** |

would not take place. However, both reactions proceeded with complete rearrangement;
146 led to the allenic alcohol **147**, whereas **146a** yielded the acetylenic alcohol **147a**.

Since tin compounds show appreciable stability they can hardly be considered
as typical carbanions. It seems reasonable, nevertheless, to assume that the
above example provides evidence in favour of an SE2' mechanism also in the case
of other allenic/acetylenic organometallics. The reaction of an allenic carbanion
with a carbonyl compound can then be summarized by Scheme 6. The proportions of

the acetylenic and allenic alcohols should then be determined by the position of the equilibrium between the two organometallics, and by the relative rates of their addition to the carbonyl function.

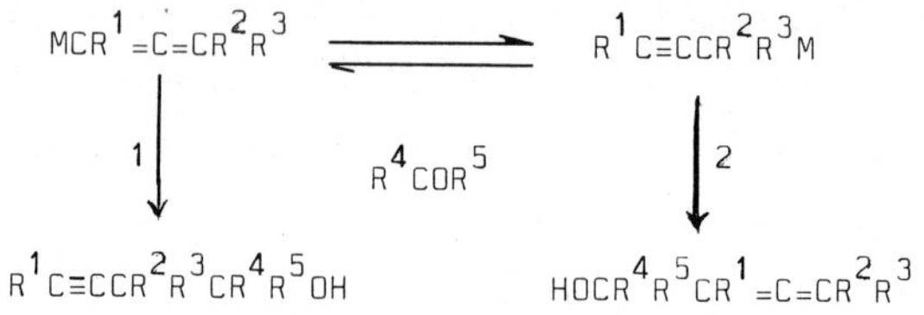

$$MCR^1{=}C{=}CR^2R^3 \quad \rightleftharpoons \quad R^1C{\equiv}CCR^2R^3M$$

$$1 \downarrow \qquad R^4COR^5 \qquad \downarrow 2$$

$$R^1C{\equiv}CCR^2R^3CR^4R^5OH \qquad\qquad HOCR^4R^5CR^1{=}C{=}CR^2R^3$$

SCHEME 6

As has already been stated, the position of the equilibrium is mainly controlled by the nature of the substituents R^1, R^2 and R^3. Thus the organometallic will be predominantly allenic, except when $R^2 = R^3 = H$ and $R^1 \neq H$. In the latter case it is generally a mixture, except when R^1 is very bulky, as for compound <u>66</u> where it is acetylenic (<u>67</u>). Concerning the reaction rates, these will depend on steric and electronic parameters. Regarding the former, the size of R^2 and R^3 on one hand and that of R^1 on the other hand will influence the ease of formation of the acetylenic and allenic alcohols, respectively, and the bulkier the carbonyl, the larger will be this influence. Among the latter parameters, the nature of the metal and of the solvent are of particular importance. This could explain why among the metals examined here, the lithium derivatives behave differently from the others.

Some workers assume nevertheless that an SE mechanism should not be excluded in those cases[113] where steric hindrance may completely prevent the SE2' process.

It may be noted also that Creary[51] has used a different approach to explain orientation in the reaction, namely through Pearson's "hard and soft acid and base" principle (HSAB).[111] In addition to the steric requirement, the choice of the reaction site would be governed by this principle. Thus "hard" aliphatic ketones bind preferentially to the allenic end, whereas more polarizable "soft" aromatic carbonyls prefer the propargylic. This hypothesis, which has already been proposed in similar cases,[45,112] appears to explain the observations, at least in the case of the lithio derivatives.

Not all the studies which have been published on the addition of allenic carbanions to carbonyls have been discussed here, but only those which are necessary for the understanding of mechanism. Among other examples which have been reported, steroidal propargylic alcohols which arise from ketosteroids are of pharmaceutical interest.[141] As well, the reaction with ketones containing an asymmetric carbon center, hindered ketones, for instance 4-t-butylcyclohexanone,[106] quinones[139] and glyoxal[140] has also been studied from a stereochemical point of view.

(b) Carbonation

Although not studied as widely as addition to carbonyls, carbonation of allenic carbanions has been investigated by a number of chemists, even in the early period. This reaction has been performed with magnesium and zinc derivatives as well as with lithio compounds.

As stated previously, the first pertinent publications deal with Grignards derived from tertiary propargylic bromides **4** and **5**[2-4,9b] which must possess the allenic structure **148**. The acids isolated were shown also to be allenic (**149**), no acetylenic acid **149a** being found.

$$MCR^1=C=CR^2R^3 \qquad HOOCCR^1=C=CR^2R^3 \qquad R^1C\equiv CCR^2R^3COOH \qquad HOOCC\equiv CCHR^2R^3$$

148	**149**	**149a**	**149b**

These results were confirmed many years later with the carbonation of the Grignards prepared from 1,1-dialkyl 3-bromo allenes[21a] and the lithio derivative of 1,1-dimethylallene,[51] which must possess the structure **148** (R^1 = H). However, in the case of the Grignard some α-acetylenic acid **149b** was also obtained.

With organometallics derived from primary and secondary propargylic bromides the results were quite different. From the former (**65**) a mixture of acids **150** and **151** was obtained.[5,7,9,89] The lithio derivative of but-2-yne behaved similarly.[37] However, whereas in the case of the Grignards equal proportions of the two acids were obtained, with the lithio derivatives much more of the allenic acid resulted.

$$HOOCCR=C=CH_2 \qquad RC\equiv CCH_2COOH \qquad HC\equiv CCHRCOOH \qquad R^1C\equiv CCHR^2COOH$$

150	**151**	**152**	**153**

Secondary propargylic bromides **64** give also rise to a mixture. From compounds **64** having a terminal acetylenic group acids **150** and **152** are formed,[89] whereas internal acetylenes **69** give rise to mixtures of acetylenes **153** and allenes **154**.[9b,89,69]

$$R^1CH=C=CR^2CO_2H \qquad R^1C\equiv CCR^2R^3M \qquad HO_2CCR^1=C=CR^2R^3$$

154	**155**	**156**

These examples show that primary propargylic bromides which afford a mixture of allenic and acetylenic organometallics yield acids arising from both forms, as in the case of carbonyls. On the other hand from secondary propargylic halides,

which give rise to allenic organometallics, the acids obtained have the same
skeleton whether they are acetylenic or allenic (compounds 150 and 152, 153 and
154) and the presence of isomers can be explained by a prototropic rearrangement
of the initially formed acetylenic acid.[89] In the case of propargyl bromide 10,
reaction also leads to a mixture of the respective allenic and acetylenic acids
157 and 158,[7,9] but this does not give any indication concerning the course of the
reaction.

An interesting case is that of dicyclopropylacetylene 159,[110] the monolithio

$HO_2CCH=C=CH_2$ $HC\equiv CCH_2CO_2H$ ▷CHC≡CCH◁   ▷C(CO_2H)C≡CC◁ $BrMgC\equiv CCHR^2R^3$

  157   158   159   160   161

derivative of which leads to the acetylenic acid 160 free from any allenic isomer.
The mechanism here can reasonably be also assumed as a SE2' process. From allenic
organometallics acetylenic acids are obtained, whereas the acetylenic organometallics yield allenic acids. When the propargylic carbon atom is too hindered
as in the case of tertiary compounds, allenic acids 156 might be formed from
the acetylenic organometallic 155 which should be also present. Regarding now the
α-acetylenic acids 149b, these may arise either from 156 (R^1 = H) through hydrogen
migration or via a rearrangement of the Grignard to the alkynide 161.[21a] Here also
Creary proposed to apply the HSAB principle, which with lithio derivatives can
be considered as valid since hard CO_2 should bind preferentially to the allenic
site.

(c) Addition to Schiff bases.

The first example of the reaction of an allenic carbanion with a Schiff
base was reported by Huet[114] who treated the lithio and the zinc derivative of
propargyl bromide (the former being prepared from the latter by treatment of the
zinc derivative with excess of phenyllithium) with isopropylidene isopropylamine
162. The product, obtained in poor yield, was a mixture of the acetylenic amine 163

i-PrCH=NPr-i $HC\equiv CCH_2$CH(Pr-i)NH(Pr-i)   i-PrNHCH(Pr-i)CH=C=CH_2

  162   163   164

with about 30% of the allenic isomer 164.
Moreau and Gaudemar[86,115,116] studied the regioselectivity of this reaction
under various conditions. They observed that unsubstituted allenylaluminium
bromide generally yields with Schiff bases practically pure acetylenic amines 165
in moderate yield.

$HC \equiv CCH_2CHR^2NHR^3$ $R^3NHCHR^2CH=C=CH_2$ $\emptyset CH=NR$ $i\text{-}PrCH=N\emptyset$

    <u>165</u>   <u>166</u>   <u>167</u>   <u>168</u>

With the corresponding magnesium and zinc derivatives, <u>165</u> is also obtained but is contaminated with the allenic amine <u>166</u> in proportions which increase from the former to the latter. The proportion of <u>166</u> also rises when HMPT is added to the reaction mixture.

From secondary bromides <u>64</u>,[86,116] via the magnesium and zinc derivatives, aldimines <u>167</u> yield pure acetylenic amines <u>169</u> whereas <u>168</u> gives rise to a mixture containing up to 13% of allenic amines <u>170</u>. Primary bromides <u>65</u> always give rise to a mixture of the acetylenic <u>171</u> and the allenic amines <u>172</u>, as shown in Table 27.

$HC \equiv CCHR^1CHR^2NHR^3$ $R^3NHR^2CH-CH=C=CHR^1$ $R^1C \equiv CCH_2CHR^2NHR^3$ $R^3NHCHR^2CR^1=C=CH_2$

    <u>169</u>   <u>170</u>   <u>171</u>   <u>172</u>

TABLE 27

Addition of the Grignard from $R^1C \equiv CCH_2Br$ to $R^2CH=NR^3$

R^1	R^2	R^3	<u>172</u> %	total yield %
Me	$\emptyset$	Me	78	45
Me	$\emptyset$	Et	35	32
Me	i-Pr	Me	45	45
Et	$\emptyset$	Me	65	58
Et	$\emptyset$	$\emptyset$	40	49
Et	i-Pr	Me	50	57
Et	Me	i-Pr	30	38
Bu	Et	Me	55	62
Bu	Me	Bu	60	43
C_5H_{11}	$\emptyset$	Me	58	38
C_5H_{11}	i-Pr	Me	70	33

On the other hand secondary propargylic bromides <u>64</u> give rise to an important stereoselectivity;[86,117] as was the case with carbonyl compounds (see Table 28), the threo isomer is always favoured. Zinc derivatives are more effective than the Grignards and a change in the reaction medium does not modify the result to an important extent.

TABLE 28

Addition of the organometallic derivatives of $HC{\equiv}CCHR^1Br$ to $ØCH{=}NR^2$

R^1	R^2	Mg derivative in ether		Zn derivative in THF		in THF + DMSO	
		threo %	total yield %	threo %	total yield %	threo %	total yield %
Me	Me	80	65	85	70	87	52
Me	Me			92	78*		
Me	Et	88	72	93	60	93	74
Me	i-Pr	66	43	83	54	80	42
Me	i-Pr	74	35*				
Me	i-Bu	90	87	95	64	89	39
Me	Ø	75	35	79	51	81	31
Pr	Me	77	47	88	48	90	42
Pr	Et	87	55	96	73		
Pr	i-Pr	74	33	84	39		
Pr	i-Bu	91	34	85	48		
Pr	Ø	75	40	80	62		

*One equivalent of LiBr added.

To conclude this topic, we note the addition of the zinc derivative of the propargylic ether 173 to benzalaniline 174, which gives regiospecifically (70% yield) the aminoether 175 together with the pyrrole 176 which arises from

$MeC{\equiv}CCH_2OMe$ $ØCH{=}NØ$ $MeC{\equiv}CCH(OMe)CH(NHØ)Ø$

173   174   175   176

its cyclization with elimination of methanol.[87] The stereochemistry of the reaction was not studied.

All these results seem to indicate that the addition of allenic carbanions to Schiff bases follows the same SE2' mechanism as with carbonyls.

Finally we want to mention in this section the closely related addition of allenic carbanions to iminium salts reported recently by Miginiac and co-workers.[11?] The study of the behaviour of the organometallic derivatives of primary and secondary propargylic bromides showed that the aluminium and magnesium derivatives of compounds 10 and 64 give with 177 the acetylenic amine 178, whereas the zinc derivatives lead to a mixture of 178 with the allenic amine 179. From primary propargylic bromides 65, on the other hand, mixtures of the acetylenic and allenic amines 180 and 181 are obtained with the magnesium and lithium derivatives, while the aluminium and zinc derivatives yield mainly the allenic amine.

$$i-PrCH=\overset{+}{N}(CH_2)_4, \ Cl^- \qquad HC\equiv CCHRCH(i-Pr)N(CH_2)_4 \qquad (CH_2)_4NCH(i-Pr)CH=C=CHR$$

177 178 R = H, Me 179

$$RC\equiv CCH_2CH(i-Pr)N(CH_2)_4 \qquad\qquad (CH_2)_4NCH(i-Pr)CR=C=CH_2$$

180 181

(d) Reaction with esters

Esters react with allenic organometallics 182 in two steps yielding as final products diacetylenic, acetylenic-allenic or diallenic alcohols, that is 183, 184 and 185 respectively.

$$MR^1C=C=CHR^2 \qquad (R^1C\equiv CCHR^2)_2CR^3OH \qquad R^1C\equiv CCHR^2CR^3(OH)R^1C=C=CHR^2$$

182 183 184

$$HOCR^3(R^1C=C=CHR^2)_2 \qquad CH_2=C=CHCOR \qquad HC\equiv CCH_2COR \qquad (HC\equiv CCR^1R^2)_2CHOH$$

185 186 187 188

With aluminium and lithium compounds no ketonic intermediate could be isolated. With the former[83a,85,119,125] (R^1 = H) almost pure acetylenic alcohols 183 are formed, whereas with the second only diallenic alcohols are obtained.[126] With magnesium reagents (R^1 = H) with moderate cooling diacetylenic alcohols are also formed but they are contaminated with some 184.[85,123,125] With Grignards derived from tertiary propargylic bromides 8 ($R^1 = R^2$ = Me; R^1 = Me, R^2 = Et) ethyl formate still yields a diacetylenic alcohol[127] 188 while ethyl acetate gives rise to a mixture of the acetylenic allenic alcohol 189a with its diallenic isomer 190a in the first case, and to the diallenic alcohol 190b solely in the second case.[21b]

$$HC\equiv CCRMeCMe(OH)CH=C=CRMe \qquad HOMeC(CH=C=CRMe)_2 \qquad\qquad \text{a) R = Me}$$
$$\text{b) R = t-Bu}$$

189 190

When the reaction is performed with 182 ($R^1 = R^2$ = H) at -80°[85,121,122,124] a mixture of acetylenic and allenic ketones 186 and 187 is obtained in satisfactory yield. The Grignard derivative of 3-deutero propargyl bromide 73 yielded with ethyl acetate at -80° a mixture of ketones 191, 192, 193 and 194 in the respective ratios of 32:56:4:8. This shows that the allenic ketones are formed by a dual process. The major amount (192) clearly arises from the acetylenic 191

$$MeCOCH_2C{\equiv}CD \qquad MeCOCH{=}C{=}CHD \qquad MeCOCD{=}C{=}CH_2 \qquad MeCOCH{=}C{=}CH_2$$

$$\underline{191} \qquad\qquad \underline{192} \qquad\qquad \underline{193} \qquad\qquad \underline{194}$$

through a protropic rearrangement, whereas the very small proportion of 193 is obtained directly via a SE mechanism; however this mechanism may be largely favoured by steric effects.[21b]

It appears then that the first step of the reaction is mainly a SE' process which yields the acetylenic ketone. This compound is not stable under the reaction conditions and rearranges partly to its allenic isomer. It has been shown that such an isomerisation can be easily achieved in a weekly basic medium.[122] The second step was shown to proceed via a SE2' process[85] in agreement with previous work on the addition of allenic Grignards on allenic ketones[120] which yields acetylenic allenic alcohols.

In this section should be mentioned also the reaction of allenic carbanions with carbonic and chloroformic esters. The latter reagent yields with the aluminium derivative of propargyl bromide, the triacetylenic alcohol 195,[83a,119] whereas the former gives with the corresponding Grignard a mixture of the same alcohol with about 1/3 of its diacetylenic monoallenic isomer 196.[127]

$$(HC{\equiv}CCH_2)_3COH \qquad\qquad (HC{\equiv}CCH_2)_2C(OH)CH{=}C{=}CH_2$$

$$\underline{195} \qquad\qquad\qquad\qquad \underline{196}$$

With the hindered tertiary propargylic bromide 5 ($R^1 = R^2 = $ t-Bu, $R^3 = \emptyset$), however, the allenic ester 197 was obtained[4] and, even more interestingly, the lithioderivative of the Schiff base 34 yielded the acetylenic ester 198.[131]

$$MeO_2CC(t{-}Bu){=}C{=}C\emptyset Bu{-}t \qquad\qquad Me_3SiC{\equiv}CCR(CO_2Me)N{=}CH\emptyset$$

$$\underline{197} \qquad\qquad\qquad\qquad \underline{198}$$

(e) Addition to amides

Very few studies have been effected on the reactions of allenic carbanions with disubstituted amides and most of these deal with Grignards.[19,81,128] From propargyl bromide 10 only hydration products 199 of the expected acetylenic or allenic ketones are formed. 2-Bromobut-3-yne 9 leads to a mixture of the allenic ketone 200 and the diketone 201. 1-Bromobut-2-yne 203 gives a mixture of the allenic and acetylenic ketones 202 and 204, whereas 2-bromopent-3-yne yields the

RCOCH$_2$COMe   RCOCH=C=CHMe   RCOCH$_2$COEt   RCOCMe=C=CH$_2$

<u>199</u>   <u>200</u>   <u>201</u>   <u>202</u>

MeC=CCH$_2$Br   MeC≡CCH$_2$COR   MeC≡CCHMeBr   RCOCMe=C=CHMe

<u>203</u>   <u>204</u>   <u>205</u>   <u>206</u>

MeC≡CCHMeCOR   CH$_2$=C=CMeOMe   R^{4}COCR^1=C=CR^{2}R^3 CH$_2$=C=CMeOMe

<u>207</u>   <u>208</u>   <u>209</u>   <u>210</u>

allenic ketone <u>206</u> contaminated with its acetylenic isomer <u>207</u>.

    This reaction has also been performed with lithio derivatives of allenes.[50] Thus various allenic carbonyl compounds <u>209</u> were obtained in good yields (R^1, R^2, R^3, R^4 = H, alkyl). From the allenic ether <u>210</u> the unstable allenic ketoethers <u>211</u> were formed in the same way and were hydrolyzed to the ethylenic diketones <u>211a</u>.[68]

        RCOCH=C=CMeOMe   RCOCH=CHCOMe

           <u>211</u>   <u>211a</u>

    From these results it is difficult to deduce information as to the reaction mechanism, but the fact that from a secondary propargylic bromide the formation of an acetylenic ketone is not favoured as it is in the case of ketones and esters, does not signify a concerted process. This is in agreement with Prevost's conclusions,[19,81] according to which owing to the basicity of the nitrogen atom the metal must coordinate strongly to it. The only way for the reaction to proceed further then lies in the formation of a free carbanion.

    (f) <u>Addition to carbon-sulfur double bonds</u>

    Allenic carbanions behave towards thioketones[129] and dithioesters[130] as carbophilic reagents. Allenyl magnesium bromide adds to 2,2,4,4-tetramethyl 3-pentanethione <u>212</u> yielding a mixture of the acetylenic thiol <u>212a</u> with its cyclization product <u>212b</u>. With methyl dithioacetate in the presence of methyl iodide the diacetylenic dithioketal <u>212c</u> is obtained in 80% yield.

t-BuCSBu-t   HC≡CCH$_2$C(t-Bu)$_2$SH   t-Bu$_2$ [ring structure with S]   (MeS)$_2$CMeCH$_2$C≡CH

<u>212</u>   <u>212a</u>   <u>212b</u>   <u>212c</u>

(2) Addition to carbon-carbon unsaturated bonds

In 1970 Moreau, Frangin and Gaudemar[132] reported the conjugated addition of allene organomagnesium, zinc and aluminium derivatives on alkylidene-malonic esters 213 which yielded the acetylenic esters 213a. This work was completed later[13] and some chemical properties of the reaction products were studied.

$RCH=C(CO_2Et)_2$ $HC\equiv CCH_2CHRCH(CO_2Et)_2$ $RCH=C(CN)CO_2Et$

  213   213a   214

With alkylidenecyanoacetates (214), Miginiac and co-workers[134] obtained a mixture of allenic (215) and acetylenic (216) cyanoacetates, whereas acrylamides 217 gave allenic amides 218 contaminated with about 5% of their acetylenic isomers.[135]

$CH_2=C=CHCHRCH(CN)CO_2Et$ $HC\equiv CCH_2CHRCH(CN)CO_2Et$ $RCH=CHCONEt_2$

    215   216   217

$CH_2=C=CHCHRCONEt_2$ $HC\equiv CCH_2CHRCONEt_2$ $Me_3SiC\equiv CCH(N=CH\emptyset)(CH_2)_2CO_2Me$

  218   219   220

The lithio derivative of the silylated acetylenic Schiff base 34 was added to methyl acrylate yielding the acetylenic compound 221.[64]

$Me_3SiC\equiv CCR(N=CH\emptyset)(CH_2)_2CO_2Me$

    221

  222   223   224

On the other hand, although ethylenic ketones are known to give 1,2-adducts with allenic carbanions,[136] the lithio derivative of the allene 222 reacted with the alkoxyethylenic ketone 223, yielding, after t-butanol elimination, the allenic ketone 224.[136]

Copper derivatives have also been used for conjugate addition to acetylenic esters. Thus dimethylallenyl lithium cuprate 225 gave with methyl propiolate the allenic ethylenic ester 226.[41] Similarly the cuprate derived from undeca-1,2 diene 227 led to the ester 228.[41] The acetylenic copper derivatives 229 add 1,6 to the diethylenic esters 230 yielding a mixture of allenic and acetylenic ethylenic esters 231 and 232.[79] Table 29 shows how the ratios of these products vary with the nature of R.

$(Me_2C=C=CH)_2CuLi$ $Me_2C=C=CHCH=CHCO_2Me$ $n-C_8H_{17}CH=C=CH_2$

225   226   227

$n-C_8H_{17}CH=C=CHCH=CHCO_2Me$ $R^1Me_2SiC\equiv CCH_2Cu$ $R^2CH=CR^3CR^4=CHCO_2Et$

228   229   230

$CH_2=C(R^1Me_2Si)CHR^2CR^3=CR^4CH_2CO_2Et$ $R^1Me_2SiC\equiv CCH_2CHR^2CR^3=CR^4CH_2CO_2Et$

231   232

TABLE 29

Addition of 229 to 230

R^1	R^2	R^3	R^4	$CH_2=C(R^1Me_2Si)CHR^2CR^3=CR^4CH_2CO_2Et$ 231 %	Total yield %
Me	H	H	H	80	42
t-Bu	H	H	H	95	36
Me	Me	H	H	18	32
t-Bu	Me	H	H	3	25
Me	H	Me	Me	85	35
t-Bu	H	Me	Me	85	20

Allenic carbanions have also be found to add to alkynyl carbanions. Such a reaction has been performed with the zinc derivatives of the primary bromides 65 which yield with terminal acetylenes' Grignards or zinc compounds, the vinyl-acetylenes 233.[142] The reaction can proceed further, giving rise to diacetylenes 234.

$R^1C\equiv CCH_2CR^2=CH_2$ $R^1C\equiv CCH_2CMeR^2C\equiv CR^1$

233   234

This type of addition seems to be limited to primary propargylic bromides with $R^1 \neq H$. Zinc derivatives of compounds 9 and 10 do not react.

There are also a few interesting cases to mention where the compound to be metalated plays at the same time the role of the electrophile. Thus the allenic sulfone 235, when treated with BuLi, undergoes a cyclodimerization according to Scheme 7, yielding the dithiaadamantane 236.[69]

$SO_2(CH=C=CH_2)_2$

235

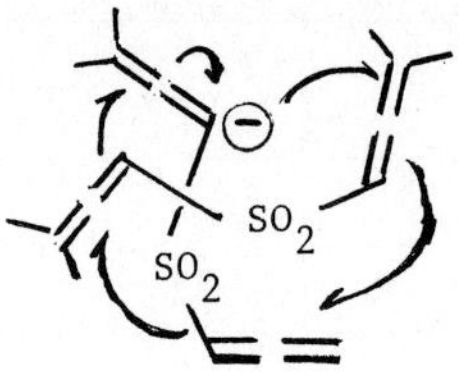

SCHEME 7

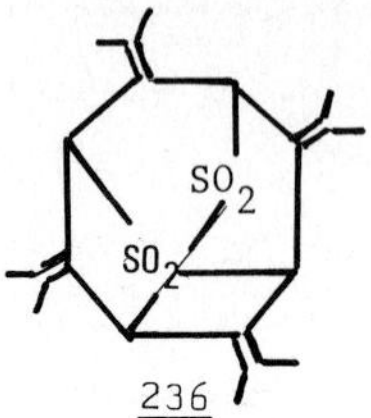

236

The diacetylenic diamine 237 (R=Me) gives with BuLi a carbanion which undergoes an intramolecular nucleophilic addition yielding instantaneously and quantitatively, via Scheme 8, the triene diamine 238.[137,138] When R is bulkier, the reaction can still take place but its rate is considerably decreased.

$R_2NCH_2C{\equiv}C(CH_2)_3C{\equiv}CCH_2NR_2$

237

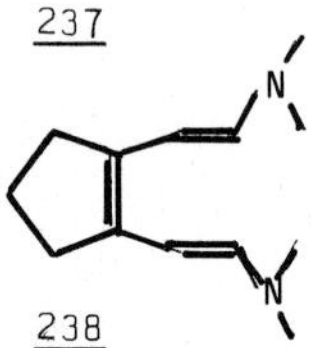

238

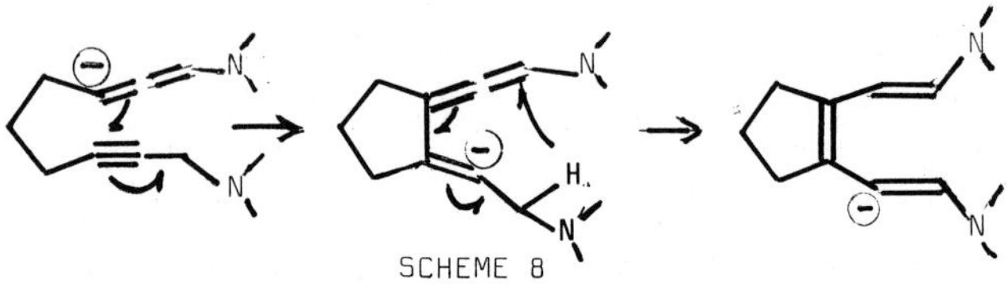

SCHEME 8

(B)   Substitution reactions

Allenic carbanions have been submitted to a large variety of substitution reactions with organic and inorganic electrophiles. Their classification has proven difficult and we have chosen to subdivide the present heading into two parts. The first part will be devoted to purely organic reagents, as a continuation of the previous section while the second will deal with metallic and non-metallic compounds. Finally we will discuss the mechanism of these reactions.

(1) Organic substitution reagents.

(a) Reaction with oxiranes

The first example of the reaction of allenic carbanions with an epoxide was reported by Burn and co-workers[145] who reacted allenylmagnesium bromide with the 5α-6α-epoxysteroid 239 and obtained the acetylenic alcohol 240.

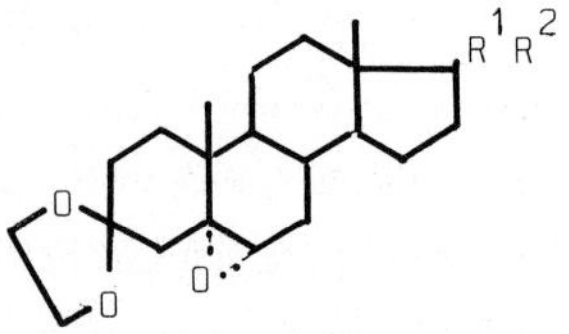

239

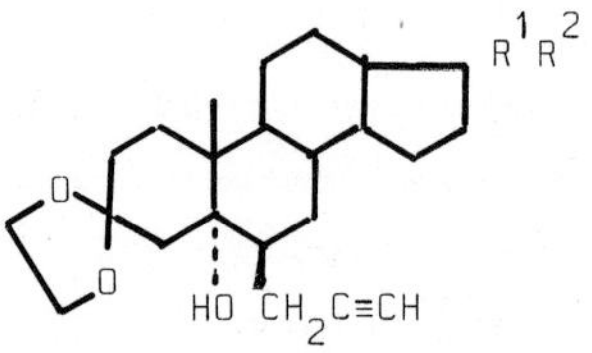

240

Similar behaviour was observed by Prevost and co-workers[19,139] who noted also that organoaluminium compounds are not reactive. Later, Vitali and Gardi[146] mentioned that the above Grignard yields, with 4α-5α epoxysteroids 241 carrying an alkoxy group on carbon 3, the allene 242 or the acetylene 243 according as it occupies the β or the α position. Vinylacetylenic chlorides 244 also led via their Grignards mainly to vinylacetylenic alcohols 245.[147]

RO$\cdots$

241

RO$\cdots$ OH

CH$_2$=C=CH

242

RO$\cdots$ OH

HC≡C-CH$_2$

243

$$CH_2=CHC\equiv CR^1R^2CCl$$

244

$$CH_2=CHC\equiv CR^1R^2C(CH_2)_2OH$$

245

With lithio compounds the results are quite different. Lithio-dimethyl-allene 84 yields with cyclohexene oxide in ether the allenic alcohol 246,[51]

$$(CH_2)_4 \begin{array}{c} CHOH \\ | \\ CHCH=C=CMe_2 \end{array}$$

246

$$CH_2=CHCH(OH)CH_2CH=C=CR^1R^2$$

247

a) R^1 = R^2 = Me

b) R^1 = R^2 = H

c) R^1 = H, R^2 = C$_7$H$_{15}$

whereas with butadiene oxide in THF a mixture of the allenic alcohol 247a and its acetylenic isomers 248a and 249a is formed.[52b] When HMPT is added in an amount equal to that of THF, only the allene 247a is formed. With allene

$$HC\equiv CCR^1R^2CH_2CH(OH)CH=CH_2$$

248

$$HC\equiv CCR^1R^2CH(CH=CH_2)CH_2OH$$

249

$$MeC\equiv CCH_2CH(OH)CH=CH_2$$

250

itself, in the absence of HMPT, the result is similar, leading to a mixture of 247b, 248b and 249b. With larger proportions of HMPT the only product is the acetylenic alcohol 250, whereas, with only one equivalent, a 50% mixture of 247b and 250 is formed. When smaller amounts of HMPT are used 248b and 249b are present again. However when the metalation time was shortened to 10 min, which was actually proven to be sufficient, practically pure 247b was formed, with only traces of 248b and 249b. Under the same conditions deca-1-2-diene affords 247c.

Metalated heterosubstituted allenes could also be reacted with oxiranes. Thus allenic ethers 251a led, via their lithio derivatives on reaction with ethylene oxide, to the allenic hydroxyethers 252a. With the thioethers 251b the reaction can be run in liquid ammonia using lithium amide or sodamide as metalating agent and affords the hydroxythioethers 252b.[56]

$$RCH=C=CHX \qquad RCH=C=CX(CH_2)_2OH$$

a) X = OMe

b) X = SMe

251   252

From this limited number of examples it is difficult to propose a mechanism for the behaviour of allenic carbanions towards oxiranes. It seems, however, to be similar to that with carbonyl compounds, as was already assumed in the case of Grignards.[19] With lithio derivatives in a non-polar solvent the reaction is probably assisted by the Li^+ cation. This, as previously stated, (see section IV A-1(a)(iii)) favours the formation of the allenic alcohol. HMPT will prevent such a process by solvating the cation.[52b]

The special case of the epoxysteroids 241 is worthy of further remarks. The normal reaction would take place when the RO group on carbon 3 occupies the α position and hence is remote from the reaction site. When in the β position, the attack of the heterocycle is assumed to be effected by a molecule of Grignard which is prior coordinated to OR. The transition state would then be as shown in Scheme 9 leading from a "propargylic" organometallic to the rearranged allene 242.[146] Such a hypothesis, which considers that the Grignard prepared from propargyl bromide is acetylenic, seems, however, to be in disagreement with other properties of this organometallic.

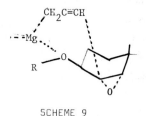

SCHEME 9

(b) Reaction with orthoesters

Very few studies have been published on the reaction of allenic carbanions with orthoesters. The aluminium derivative of propargyl bromide has been reported to afford with ethyl orthoformate the acetylenic acetal 253 in 62-76% yield.[139] On the other hand the allenic Grignard,[107] prepared either from the acetylenic or allenic bromides 5 (R^1 = H, R^2 = R^3 = Me) and 256, led to a mixture of the

$HC{\equiv}CCH_2CH(OEt)_2$ $HC{\equiv}CCMe_2CH(OEt)_2$ $(EtO)_2CHCH{=}C{=}CMe_2$ $Me_2C{=}C{=}CHBr$

    253   254   255   256

acetylenic and allenic acetals 254 and 255, with a large predominance of the former.[20b] These results might indicate that the reaction mechanism is similar to that with carbonyls.

      Quite recently this reaction was performed with phenyldiethyl orthoformate and found to proceed readily at room temperature. The results,[182] which are shown in Table 30, are in good agreement with the above conclusions.

TABLE 30

Reaction of organometallics derived from $R^1C{\equiv}CCHR^2Br$ with $\emptyset OCH(OEt)_2$

R^1	R^2	Metal	Solvent	Acetal	total yield %
H	H	Al	Ether	$HC{\equiv}CCH_2CH(OEt)_2$	80
H	H	Mg	Ether	$HC{\equiv}CCH_2CH(OEt)_2$	55
H	Me	Al	Ether	$HC{\equiv}CCHMeCH(OEt)_2$	50
H	Me	Mg	Ether	$HC{\equiv}CCHMeCH(OEt)_2$	50
Me	H	Zn	THF	$(EtO)_2CHCMe{=}CH_2$	70
Me	H	Mg	Ether	$(EtO)_2CHCMe{=}CH_2$ 77%	50
				+ $MeC{\equiv}CCH_2CH(OEt)_2$ 23%	

      (c)   Carbon-carbon coupling.

      During the earlier attempts to prepare Grignards from tertiary propargylic bromides 5, formation of dimeric material was observed.[4] Thus from 5 ($R^1 = R^2$ = t-Bu, $R^3 = \emptyset$) a mixture of dimers was obtained which was assumed to consist of the diallene 257 actually isolated and of a stereoisomer. A similar result was observed with the sodio derivative. Later, the formation of unidentified dimers

$t{-}BuC{=}C{=}C\emptyset Bu{-}t$ $(HC{\equiv}CCH_2)_2$ $(CH_2{=}C{=}CH)_2$ $HC{\equiv}CCH_2CH{=}C{=}CH_2$
     |
$t{-}BuC{=}C{=}C\emptyset Bu{-}t$

    257   258   259   260

from primary propargylic bromides was reported.[9a] With propargyl bromide itself dimerization in ether was found to be negligible,[7] though in THF solution it appeared to be complete[81,148] and gave rise to a mixture of 258, 259 and 260.[92,139a] A similar coupling was observed with Grignards prepared from tertiary propargylic chlorides[149] in the same solvent.

With Grignards prepared from allenic bromides 261, the ratio of dimerization products was shown to be noticeable even in ether solution.[21] Thus from 261a a

$$RMeC=C=CHBr \qquad\qquad RMeC=C=CHCRMeC\equiv CH \qquad\qquad CH_2=CR\text{-}CH=CHCRMeC\equiv CH$$

261   262   263

a) R = Et

b) R = Me $(HC\equiv CCRMe)_2$ $(RMeC=C=CH)_2$

c) R = t-Bu   264   265

mixture of 262a and 264a was obtained.[21a] 261b led in 20% yield to a mixture composed of 50% of 262b, 45% of 263b (which is an isomerization product of the former) and 5% of 264b.[21b] 261c yielded 45% of a mixture containing 65% of 262c and 35% of 265c. When allenic Grignards are treated with oxygen, the coupling products are formed in high yield. Thus 261a afforded 98.7% of the same mixture as above.

The fact that dimerization takes place suggests that the halide, which is used as a starting material to prepare the organometallic, can be brought to react with it under proper conditions. Actually very reactive alkylating agents such as sulfates attack easily allenic carbanions. Thus allenyl magnesium bromide yields with butyl sulfate a 1:1 mixture of hept-1-yne 266 and hepta-1-2-diene 267.[10,150] With less reactive compounds THF could be successfully used as an alkylation enhancing solvent[92,139a] as shown in Table 31. Instead of using THF as a reaction medium, adding cuprous chloride as a catalyst to the ether solution of the reactants also proved to be effective.[151] In most cases a mixture was obtained with predominance of the allene. However, with the dibromoolefin 268, allenyl-magnesium bromide gave, in 51% yield, the diacetylenic derivative 269.[152] In the same way 1-bromohept-2-yne 11 was said to afford with the same Grignards the diacetylene 270 containing a small amount of allenic material.[153] The allylic

$$HC\equiv CC_5H_{11} \qquad CH_2=C=CHC_4H_9 \qquad BrCH_2CMe\overset{E}{=}CMeCH_2Br$$

266   267   268

$$HC\equiv C(CH_2)_2CMe\overset{E}{=}CMe(CH_2)_2C\equiv CH \qquad BuC\equiv C(CH_2)_2C\equiv CH$$

269   270

phosphate 271 also reacted with the above Grignard in the presence of Cu_2Cl_2 leading to the pure allene 272.[154]

TABLE 31

Reaction of $R^1CH=C=CHMgBr$ with R^2X in THF

R^1	R^2	products	total yield %
H	$HC\equiv CCH_2$	$(HC\equiv CCH_2)_2$	
		$H_2C=C=CHCH_2C\equiv CH$	60
		$(H_2C=C=CH)_2$	
H	$CH_2=CHCH_2$	31% $CH_2=CH(CH_2)_2C\equiv CH$	75
		69% $CH_2=CHCH_2CH=C=CH_2$	
Et	$CH_2=CHCH_2$	33% $CH_2=CHCH_2CHEtC\equiv CH$	65
		66% $CH_2=CHCH_2CH=C=CHEt$	

$(EtO)_2P(O)OCH_2CH=CR^1R^2$ $CH_2=C=CHCH_2CH=CR^1R^2$ $R^1CH=C=CHSEt$

271   272   273

    In a few instances alkylation was performed on sodio-derivatives in liquid ammonia. This was the case with the allenic thioethers 273 and ethers 274 which led, respectively, to 275 and 276.[55] In general alkylation of acetylenes or allenes, (functionalized or not) is effected by lithioderivatives in THF solution. The

$H_2C=C=CHOR^1$ $R^1CH=C=CR^2SEt$ $CH_2=C=CR^2OR^1$ $RCH_2C\equiv CSiMe_3$

274   275   276   277

metalating agent is either butyllithium[31,40,41,52a,57,59,102] or a lithium dialkyl-amide.[40,51,68,131] Thus, Corey and coworkers obtained from the acetylenic silane 22 the acetylenic derivative 277,[31,155] while allene itself yielded undeca-1-2-diene 278 together with some undec-1-yne 279.[40] From terminal allenes 82 the internal allenes 280 were obtained.[41] Dimethylallenyllithium, 84, was alkylated via its cuprate 281 to 282.[41] With benzylic halides, however, 84 yielded mainly the acetylene 283 together with the allene 284, and the alkylidenecyclopropanes 285[51] formed presumably via the phenylcarbenoid intermediate.

$C_8H_{17}CH=C=CH_2$ $HC\equiv CC_9H_{19}$ $R^1CH=C=CHR^2$ $(Me_2C=C=CH)_2CuLi$

278   279   280   281

a) R^1 = Me, R^2 = H
b) R^1 = H, R^2 = Me

$Me_2C=C=CHC_8H_{17}$ $HC\equiv CCMe_2CH_2Ar$ $Me_2C=C=CHCH_2Ar$ $Ar\underset{}{\triangle}=CR^2_2$ (R^1_2)

282   283   284   285

From the vinylallenes 286 a mixture of acetylenes and allenes 287, 288 and 289, is formed in a ratio which varies with the solvent and with the metalation time;[156] in nonpolar medium 288 is the major component.

$R^1CH=CR^2CR^3=C=CH_2$

286

$R^1CH=CR^2CR^3R^4C\equiv CH$

287

$R^1R^4C=CR^2CHR^3C\equiv CH$

288

$R^1CH=CR^2CR^3=C=CHR^4$

289

$R^1C\equiv CCHR^2OMe$

290

$R^1R^3C=C=CR^2OMe$

291

From the lithio derivatives of the acetylenic ethers 290, the allenic ethers 291 are formed (R^3 = Me, Et, $(CH_2)_2NEt_2$).[57,112] The thioethers 292 led to the acetylenic substituted thioethers 293[102] as the sole products or if R^1 is bulky, together with the allenic thioether 294. From the thioether acetal 37 the allenic derivative 295 is obtained.[67]

The allenic ethers 296 lead to 297.[56a,68] However if R^1 is bulky and metalation effected by lithium dicyclohexylamide, 298 is formed, contaminated with only 5% of 297. From the substituted allenic ether 299, 300 is obtained[112] and from the disilylated ether 301, 302 is isolated.[157]

$Me_3SiC\equiv CCH_2SR^1$

292

$Me_3SiC\equiv CCHR^2SR^1$

293

$Me_3SiCR^2=C=CHSR^1$

294

$(EtO)_2C=C=CRSMe$

295

$CH_2=C=CHOR^1$

296

$CH_2=C=CR^2OR^1$

297

$R^2CH=C=CHOR^1$

298

$C_5H_{11}CR^1=C=CHOMe$

299

$C_5H_{11}CR^1=C=CR^2OMe$

300

$Me_3SiCH=C=C(OMe)SiMe_3$

301

$Me_3SiCR=C=C(OMe)SiMe_3$

302

With the lithioenolates 57 alkylation proceeds similarly, leading to the allenic ketones 303.[73]

Finally, alkylation of the lithio derivative of the Schiff base 34 and of the carbamate 35 yield only the acetylenic derivatives 304[64,131] and 305.[65]

$R^1COCH=C=CHCHXR^2$

303

$Me_3SiC\equiv CCHRN=CH\emptyset$

304

$Me_3SiC\equiv CCHRNHCO_2Bu-t$

305

The reaction of chloromethyl ethers with allenic carbanions is very similar to that of alkyl halides. The main difference lies in their higher reactivity which allows them to react even with organoaluminium compounds, yielding a mixture of allenic and acetylenic ethers 306 and 307,[83a] just as they do with Grignards.[19,150,158,159] However, 1,2-dibromoethyl ethyl ether affords only the acetylenic ether 308.[139a]

$MeOCH_2CH=C=CH_2$ $HC≡C(CH_2)_2OMe$ $HC≡CCH_2CH(OEt)CH_2Br$

  306   307   308

$ØC≡CCHR^1OR^2$ $MeOCH_2CØ=C=CR^1OR^2$

  309   310

Lithio compounds yield with chloromethyl ethers only allenic derivatives. Thus from the acetylenic ethers 309 the α-diethers 310 are formed[57] whereas from the allenic ethers 311 the α-diethers 312 are obtained.[59b]

$Me_3SiCR^1=C=CHOBu-t$ $Me_3SiCR^1=C=C(OBu-t)CH_2OMe$

    311   312

(2)   Metallic and non-metallic compounds

With the exception of water and other reagents which can be used to hydrolyse carbanions, most of the compounds included under this heading are metallic and non-metallic halides, or sometimes alkoxides or related molecules.

(a) Hydrolysis.

In most of the investigations on allenic carbanions a competing hydrolysis has been reported. It yields generally a mixture of allene and acetylene in variable ratios with the former predominating, though sometimes the allene is obtained practically pure.[20a,24,74,57] This offers a convenient method of converting the propargylic ethers 29 into ethylenic aldehydes via the allenic ethers 40.[57,60]

(b) Mercuration

Mercuric chloride and bromide react with organometallic derivatives of propargylic bromides yielding allenic or acetylenic mercury compounds. Thus 3-phenylpropargyl bromide 313 gives rise, via its zinc derivative (itself a mixture of allenic and acetylenic compounds), to the pure acetylenic organomercurial 314,[83a,143] whereas the secondary homologue 315 (the metallic derivative of which is allenic) affords the allenic mercury compound 316.

ØC≡CCH$_2$Br   ØC≡CCH$_2$HgX (X = Cl, Br)   ØC≡CCHMeBr   XHgCØ=C=CHMe

<u>313</u>   <u>314</u>   <u>315</u>   <u>316</u>

Me$_3$SnCH=C=CH$_2$ ClHgCH=C=CH$_2$

<u>317</u>   <u>318</u>

It is interesting to note that the allenic stannane <u>317</u> can also be used for the same reaction,[144] giving rise to the allene <u>318</u>.

(c) <u>Lead compounds.</u>

Grignards and organoaluminium derivatives of propargylic bromides react with triphenyllead halides, yielding mixtures of allenic and acetylenic lead derivatives.[160,161] Thus, from <u>10</u> a mixture of <u>319a</u> and <u>320a</u> is formed, containing

R^{1}C≡CCHR^{2}PbR^3$_3$ a) R^1 = R^2 = H$_2$, R^3 = Ø   R$^3_3$PbCR^1=C=CHR^2

<u>319</u>   b) R^1 = H, R^2 = Me, R^3 = Ø   <u>320</u>

        c) R^1 = Me, R^2 = H, R^3 = Ø

        d) R^1 = R^2 = H, R^3 = Me

        e) R^1 = R^2 = H, R^3 = Et

90% of the latter.[160,161] From the secondary bromide <u>9</u> the ratio of <u>320b</u> is 80%, whereas from <u>65</u> (R = Me) equivalent amounts of <u>319c</u> and <u>320c</u> are obtained.[161] With trimethyl and triethyllead halides, <u>10</u> gives rise to the allenes <u>320d</u> and <u>e</u>.[161]

(d) <u>Tin compounds.</u>

The Grignards derived from propargylic bromides react with triphenyl- or trimethylstannyl chloride yielding a mixture of the acetylenic stannane <u>321</u> and its allenic isomer <u>322</u>.[161,162] As can be seen in Table 32, the ratio of the above derivatives varies widely with the substituents of the stannyl grouping.

TABLE 32

Reaction of the Grignard prepared from R^{1}C≡CCHR^{2}Br with R^{3}SnCl

R^1	R^2	R^3	acetylenic stannane %	yield %
H	H	Ø	90	80
H	Me	Ø	80	70-75
Me	H	Ø	nearly pure	60
H	H	Me	30	30-40

With dichlorodimethyltin, allenylmagnesium bromide yields pure dimethyl-dipropargylstannane 323.[164]

In contrast, lithioallenes 324 yield only allenic trialkyl stannanes 325[163].

$R^1C \equiv CCHR^2SnR^3_3$　　　　$R^3_3SnCR^1=C=CHR^2$　　　　$(HC \equiv CCH_2)_2SnMe_2$

321　　　　　　　　　　322　　　　　　　　　　323

$R^1CLi=C=CHR^2$　　　　$R^3_3SnCR^1=C=CHR^2$

324　　　　　　　　325

(e) Silylation

A relatively important number of papers deal with the reaction of allenic carbanions with chlorosilanes. The Grignard prepared from propargyl bromide 10 and its higher homologues 9 and 21 (R^1 = H, R^2 = Me), known to be purely allenic for the first two and a mixture for the third, yield with trialkylsilyl chlorides the allenic silanes 326a and b, contaminated with acetylenic isomers 327.[161]

$R^3_3SiCR^1=C=CHR^2$　　$R^1C \equiv CCHR^2SiR^3_3$　　a) $R^1 = R^2$ = H　　　$R^1CLi=C=CHR^2$

326　　　　　　　　327　　　　　b) R^1 = H, R^2 = Me　　　328

　　　　　　　　　　　　　　　　c) R^1 = Me, R^2 = H

　　　　　　　　　　　　　　　　d) R^1, R^2 = alkyl

Regarding the lithio derivatives, from lithioallene itself, both 326a and 327a are obtained in significant amounts.[35] From "internal" lithioallenes 328, the allenic silanes 326d are formed with small amounts of the acetylenes 327d.[165,166] 1,1-Dimethylallene yields only the allenic silane 329.[51] However lithiated dicyclopropylacetylene 158 affords exclusively the acetylenic silane 330.[110]

$Me_2C=C=CHSiMe_3$　　▷$(SiMe_3)C \equiv C$◁　　$Me_3SiC(C_5H_{11})=C=CHOMe$

329　　　　　　　　330　　　　　　　　331

From lithiated derivatives of acetylenic and allenic ethers only allenic silanes are obtained. Thus the acetylenic ethers 29 ($R^1 = C_5H_{11}$, R^2 = Me) and ($R^1 = Me_3Si$, R^2 = t-Bu) yield the allenes 331[112] and 332.[59] From the allenic ethers 331, 332, 333 and 335, the allenic silanes 336,[112] 337,[59] 334[157] and 338[112] are respectively obtained. From the silylated acetylenic thioethers 339, however, the acetylenic silanes 340 are formed.[102]

$(Me_3Si)_2C=C=CHOBu-t$　　$CH_2=C=CHOMe$　　$CH_2=C=C(OMe)SiMe_3$　　$C_5H_{11}CMe=C=CHOMe$

332　　　　　　　　333　　　　　　　334　　　　　　　335

$Me_3SiC(C_5H_{11})=C=C(OMe)SiMe_3$ $(Me_3Si)_2C=C=C(OBu-t)SiMe_3$ $C_5H_{11}CMe=C=C(OMe)$
$|$
$SiMe_3$

336 337 338

$Me_3SiC\equiv CCH_2SR$ $Me_3SiC\equiv CCH(SR)SiMe_3$ $(HC\equiv CCH_2)_2SiR_2$ $(R = Me,\emptyset)$

339 340 341

It seems important to set off here the reaction of dichlorosilanes with allenylmagnesium chloride which, in contrast to most of the above results, affords only the diacetylenic silanes 341. Similarly p-phenylenebis(chlorodimethylsilane) 342 affords the diacetylenic derivative 343.[164]

$ClMe_2Si-\langle\bigcirc\rangle-SiMe_2Cl$ $HC\equiv CCH_2Me_2Si-\langle\bigcirc\rangle-SiMe_2CH_2C\equiv CH$

342 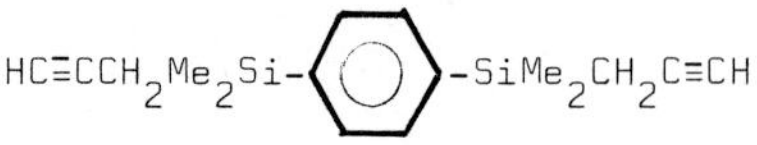 343

(f) Germanium halides

The reaction of allenic carbanions with germyl halides has been studied much less than with silyl halides. Contrasting with the latter, the former react with the Grignards derived from propargylic bromides to yield mainly the acetylenic germanes 344 together with 10-20% of the allene 345. Diethyl- and diphenyldichloro-germanes afford only the diacetylenic derivatives 346.[164] With the lithioallene 347, however, trimethylgermyl chloride leads to the allenic germane 348.[168]

$R^1C\equiv CCHR^2GeR_3$ $R_3GeCR^1=C=CHR^2$ a) $R^1 = R^2 = H$[161,167]
b) $R^1 = H$, $R^2 = Me$[161]
344 345 c) $R^1 = Me$, $R^2 = H$[161]

$(HC\equiv CCH_2)_2GeR_2$ $(R = Et,\emptyset)$ $EtCLi=C=CHPr$ $Me_3GeCEt=C=CHPr$

346 347 348

(g) Arsenic halides

Propargylic bromides 21 react via their Grignards with haloarsines 349, yielding mixtures of allenic and acetylenic arsines 350 and 351,[84] as shown in Table 33.

R^3_2AsX $R^3_2AsCR^1=C=CHR^2$ $R^1C\equiv CCHR^2AsR^3_2$

349 350 351

TABLE 33

Reaction of $\underline{349}$ with the Grignard derived from $R^1C\equiv CCHR^2Br$

R^1	R^2	R^3	Acetylenic arsine %	total yield %
H	H	Ø	89	75
H	H	Et	85	55
H	Me	Ø	65	65
H	Me	Et	27	45
Me	H	Ø	82.5	70
Me	H	Et	30	50
Me	Me	Ø	100	70
Me	Me	Et	80	60

(h) Boron compounds.

Grignards derived from propargylic bromides $\underline{21}$ react with boric acid esters yielding allenic, acetylenic boronates or a mixture of both. Thus from primary or secondary propargylic bromides, the allenic boronates $\underline{352}$ are formed whereas from aromatic $\underline{21}$'s (R^1 = Ø, R^2 = H, alkyl) only acetylenic derivatives $\underline{353}$ are obtained.[83b, 169,171] This reaction was extended later to the cyclic borates $\underline{354}$ which afforded

$$(R^3O)_2BCR^1=C=CHR^2 \qquad\qquad ØC\equiv CCHR^2B(OR^3)_2$$

$\underline{352}$ $\qquad\qquad\qquad\qquad\qquad$ $\underline{353}$

EtOB⟨O⟩(CH$_2$)$_3$

$\underline{354}$

the more stable (towards hydrolysis) allenic and acetylenic boronates, respectively $\underline{355}$ and $\underline{356}$.[172] In that case mainly allenic compounds $\underline{355a}$ were formed from aliphatic propargylic bromides whereas aromatic led to the acetylene $\underline{356b}$ as major product. From the silane $\underline{15}$ the main product was also the acetylene $\underline{356c}$ together with 24% of the allene $\underline{355c}$.

$(CH_2)_3$⟨O⟩$BCR^1=C=CHR^2$

a) R^1, R^2 = H,Me

b) R^1 = Ø, R^2 = H,Me

c) R^1 = Me$_3$Si, R^2 = H

d) R^1 = t-Bu, R^2 = H

$R^1C\equiv CCHR^2B$⟨O⟩$(CH_2)_3$

$\underline{355}$ $\qquad\qquad\qquad\qquad\qquad\qquad\qquad\qquad$ $\underline{356}$

From the lithioderivative of 2,2-dimethyl pent-3-yne $\underline{357}$, $\underline{356d}$ was obtained together with 10% of $\underline{355d}$, whereas 1-trimethylsilylpropyne $\underline{22}$ led to $\underline{356c}$ together with 30% of $\underline{355c}$.

Finally with the cyclic chloroborondiamide 358 pure acetylenic 359 was obtained from lithiated 22, whereas 357 led to the acetylene 360 containing 20% of the allene 361.

t-BuC≡CMe

$$ClB\underset{NMe}{\overset{NMe}{\diagdown}}(CH_2)_2$$

$$Me_3SiC≡CCH_2B\underset{NMe}{\overset{NMe}{\diagdown}}(CH_2)_2$$

357   358   359

$$t-BuC≡CCH_2B\underset{NMe}{\overset{NMe}{\diagdown}}(CH_2)_2$$

$$(CH_2)_2\underset{NMe}{\overset{NMe}{\diagup}}BC(Bu-t)=C=CH_2$$

360   361

(i) Sulfur and selenium compounds.

Disulfides 362 react with allenic lithio derivatives yielding allenic sulfides. Lithiodimethylallene 84 affords the compounds 363,[51] while the allenic ether 44 (R = H) similarly yields 364.[56b]

RSSR R = Ø,Me $Me_2C=C=CHSR$ $CH_2=C=C(OMe)SMe$ ØSeSeØ

362   363   364   365

An analogous reaction was performed recently with diphenyl diselenide 365 giving rise to the allenic selenoether 366.[173]

$R_2C=C=CHSeØ$ $RS(O)CMe=C=CHMe$ R = $Me-C_6H_4-$, $ØCH_2-$

366   367

Alkyl sulfinates react with allenic Grignards leading to the allenic sulfoxides 367.[174] If the sulfur atom is asymmetric, induction of chirality takes place stereospecifically in the allenic system.

(3) Mechanism

From the above results it can be seen that, among the substitution reagents, oxiranes and orthoesters occupy a special place since they behave similar to the carbonyl compounds. The other reagents differ in that there is apparently no longer a direct relationship between the structure of the organometallic and that of the product. Though lithium derivatives generally lead to allenes, exceptions to this are observed in the following circumstances :

     a) The acetylenic group carries a phenyl, vinyl, Si, t-Bu or a cyclo-propyl substituent; in these cases the major or even the sole product can be acetylenic.

b) The electrophile is the dibromoolefin <u>268</u> or a benzylic halide, which also afford predominantly an acetylene (allyl bromide follows the general rule however).

With the other organometallics a mixture is generally obtained which is often richer in the allenic component, with the following exceptions which also yield the acetylene as the major or exclusive product :

a) HgX_2 with <u>65</u> (R = Ø)

b) $Ø_3SnCl$ and Me_2SnCl_2

c) Me_2SiCl_2 and $ClMe_2Si$ Me_2SiCl

d) R_3GeCl

e) $Ø_2AsCl$ and Et_2AsCl (with un- and disubstituted propargylic bromides)

f) $(RO)_2BOEt$ with phenylated acetylenes.

For some of these reactions Gaudemar proposed a mechanism via free carbanions.[83a] In the absence of steric effects, polar factors (see section IIIB) control the orientation of the substitution leading preferentially to allenes. When both electronic and steric parameters are present, the latter prevail. The fact that, with derivatives of the more positive lithium, allenes are generally obtained as the major products is a strong argument in favour of this hypothesis. It applies perfectly well in the case of mercuric halides which attack preferably the allenic site in the absence of steric effects, and the propargylic when the former alone is hindered. With lead halides a similar tendency can be noted, although less pronounced since the proportion of allene is always significant. Interestingly, this proportion is much smaller when the metal is substituted with a phenyl than with an alkyl group. Steric effects do not seem to be important, however, since the proportion of allene is the same with trimethyl- and triethyllead halides.

In the case of tin compounds, the ratio of allene also depends largely on the metal substituents. When it is a phenyl group, the product is mainly acetylenic. A significant steric effect can also be noted since a substituent on the propargylic carbon atom reduces the proportion of acetylene.

Silylation leads mostly to allenes except with dicyclopropylacetylene, or when the silylating agent is dichlorodimethylsilane, or bis-(dichlorodimethyl-silyl)-benzene <u>342</u>.

In contrast, germanium halides yield, with the Grignards derived from propargylic halides, mainly acetylenes, regardless of the substituents. However lithium compounds again lead to allenes.

With haloarsines carrying a phenyl group, the acetylene is largely favoured in all cases. With ethyl substituents, when both sites of the Grignard are unsubstituted or when they carry the same substituent, the acetylene is still the major product. However, when only one of these centres is substituted, then the main product is the allene.

In the case of boron compounds, allenes are predominant except when the alkynyl group carries a phenyl, t-butyl or trimethylsilyl substituent.

Finally sulfur and selenium reagents lead only to allenic compounds. The asymmetric synthesis observed with sulfinates indicates that in this case the reaction follows a concerted mechanism.

From these results it seems difficult to draw a conclusion concerning the mechanism of these substitution reactions. Obviously a single process could not explain all these results. In a limited number of cases the mechanism proposed by Gaudemar[83a] could apply. The alternative hypothesis based on the HSAB principle which was proposed later[44,51,112] (See Section IV A-1,a,iii) might perhaps fit better. Hard electrophiles E_h (See Scheme 10) will react rather on carbon 1 which must display some hard character, thus leading to allenes, whereas soft electrophiles E_s will attack preferably the soft carbon 3 yielding acetylenes.

$$R^1-CE_h=C=CR^2R^3 \quad \longleftarrow \quad R^1-C\equiv\overset{\ominus}{C}-CR^2R^3 \quad \longrightarrow \quad R^1C\equiv CCR^2R^3E_s$$
$$\quad\;\; 1 \quad\;\; 3 \qquad\qquad\qquad 1$$

SCHEME 10

This concept should however be considered as relative. The degree of hardness and softness depends on several factors. The nature of the metallic counterion doubtless plays an important role. Strongly electropositive lithium will "harden" the carbanion, and a dissociating reaction medium will have the same effect. In contrast, substitution of the charge-carrying carbon with a phenyl, vinyl, cyclopropyl, or even with a silyl group which introduces a pπ-dπ conjugation, will "soften" it. Consequently less allenic product will be formed. On the other hand, whereas alkyl sulfates and halides can be considered as hard, as well as metallic and non-metallic halides or alkoxides, benzyl and allyl halides are generally regarded as soft. So are also metallic and non-metallic halides carrying a phenyl group. According as they are hard or soft, electrophiles will attack preferentially the harder or the softer site of the carbanion, in so far as steric effects allow it. Thus the presence of a t-butyl group on carbon 1 will also favour the formation of an acetylene.

Some of the above results are in agreement with such a hypothesis, but others appear to be in contradiction with it. For instance soft allyl bromide reacts at the hard carbon, leading with Grignards mainly to allenes. Aliphatic dichlorostannanes and silanes yield pure acetylenes. Disulfides, although being considered as soft, lead to allenes even when substituted with a phenyl group.

Also unexplained is the case of chlorodiethylarsine which yields an acetylene
with allenylmagnesium bromide or with its 1,3-dimethyl derivative, and an allene
with either of its monomethyl derivatives.

Finally it should be mentioned that the reaction of dimethylallenyllithium
84 with benzylic halides was assumed on the basis of NMR data observed during
reaction to follow an electron transfer mechanism.[51]

(C) Dicarbanions

The di-Grignard 19 was found to be a valuable starting material for the
synthesis of the acetylenic α,β'-diols 368.[8,10,17] With higher homologues of 19,
these derivatives do not seem however to have been studied.

$$HOCR^1R^2C{\equiv}CCH_2CR^1R^2OH \qquad R^1R^2CHC{\equiv}CH \qquad HOCR^3R^4C{\equiv}CCR^1R^2CR^3R^4OH$$

$$\underline{368} \qquad\qquad \underline{369} \; a)R^2 = H \qquad\qquad \underline{370}$$

In contrast, many investigations have been performed with the dilithio
derivatives of propyne, allene and their substituted derivatives, though most
of the studies deal with terminal acetylenes 369. A special interest has been
shown to these compounds which allow in many instances selective reactions, if
one excepts addition to carbonyls and carbon dioxide. With formaldehyde[178] or
acetaldehyde,[43] the former affords mainly the acetylenic diols 370 together with
some alcohols 371. However with cyclic ketones some alcohol 372 ($R^3,R^4 = (CH_2)_n$)
is also formed.[177,178] Carbonation of 369a yields the allenic acid 373, which
is probably the rearrangement product of the acetylenic diacid 374 initially
formed, together with a mixture of unidentified acetylenic monoacids.[177,178]
From dicyclopropylacetylene 159 however, the acetylenic diacid 375 was isolated.[110]

$$HOCR^3R^4C{\equiv}CCHR^1R^2 \qquad HC{\equiv}CCR^1R^2CR^3R^4OH \qquad HO_2CCR{=}C{=}CHCO_2H$$

$$\underline{371} \qquad\qquad\qquad \underline{372} \qquad\qquad\qquad \underline{373}$$

$$HO_2CCHRC{\equiv}CCO_2H \qquad \triangleright (CO_2H)C{\equiv}C(CO_2H) \triangleleft$$

$$\underline{374} \qquad\qquad\qquad\qquad \underline{375}$$

With most of the other electrophiles these dianions react selectively on
the propargylic carbon atom. With very reactive compounds such as trialkylsilyl chlorides,[33,35-39,43,44,46,58,110] trimethylgermyl chloride,[43] diethyl
sulfate,[43,44] dimethyl sulfate[58] and benzyl chloride when used in excess,
disubstitution takes place. However with equimolar amounts only compounds
monosubstituted on the propargylic site are formed.[179] With less reactive

TABLE 34

Selective monosubstitution of dilithioderivatives of acetylenes

Starting acetylene (allene)	Electrophile	Product	Reference
$RCH_2C{\equiv}CH$ (R = H, Me, Pr)	$\emptyset CH_2Cl$	$\emptyset CH_2CHRC{\equiv}CH$	179
$MeC{\equiv}CH$	MeI	$EtC{\equiv}CH$	176
$MeC{\equiv}CH$	BuBr	$C_5H_{11}C{\equiv}CH$	48,176
$MeC{\equiv}CH$	$Br(CH_2)_{10}Br$	$HC{\equiv}C(CH_2)_{12}C{\equiv}CH$	176
$C_5H_{11}C{\equiv}CH$	BuBr	$(C_4H_9)_2CHC{\equiv}CH$	42,49
$C_6H_{13}C{\equiv}CH$	PrBr	$C_5H_{11}CHPrC{\equiv}CH$	48
$\emptyset CH{=}C{=}CH_2$	MeI	$\emptyset CHMeC{\equiv}CH$	39
$\emptyset C{\equiv}CMe$	MeI	$\emptyset CHMeC{\equiv}CH$	175
$EtC{\equiv}CH$	BuBr	$BuCHMeC{\equiv}CH$	27
$MeCH{=}C{=}CH_2$	BuBr	$BuCHMeC{\equiv}CH$	27
$\emptyset SeCH_2C{\equiv}CH$	RX (R = Me, Et, **i-Pr**, $\emptyset CH_2$, $\emptyset(CH_2)_3$)	$\emptyset SeCHRC{\equiv}CH$	66
$RCH_2C{\equiv}CH$ (R = H, Me, Pr)	Me_3SiCl	$Me_3SiCHRC{\equiv}CH$	179
$RCH_2C{\equiv}CH$ (R = Et, Pr, Bu, C_5H_{11})	HMPT (as a source of $CH_2{=}NMe$)	$MeNHCH_2CHRC{\equiv}CH$	178
$RCH_2C{\equiv}CH$ (R = Et, Pr, Bu, C_5H_{11})	$\emptyset CH{=}N\emptyset$	$\emptyset NHCH\emptyset CHRC{\equiv}CH$	178
$R^1CH_2C{\equiv}CH$ (R^1 = H, Me, Pr)	$\underset{O}{\overset{}{CH_2{-}CHR^2}}$	$HOCHR^2CH_2CHR^1C{\equiv}CH$	179

electrophiles such as alkyl halides,[27,36,39,42,48,49,66,175,177,179] epoxides[179] and imines,[178] again monosubstituted derivatives on the same carbon atom are obtained. A similar result is reached with the selenoether 39.[66] These results are shown in table 34.

We shall also mention here the case of the dianion prepared from the acetylenic ether 30 in which the triple bond is internal. This anion, which is assumed to possess the allenic structure 376, yields with electrophiles the monosubstituted derivatives 377.[58]

$$\overset{\ominus}{\text{ØC}}=\text{C}=\overset{\ominus}{\text{C}}\text{ OMe}$$ ØCE=C=CHOMe   (E = alkyl, trimethylsilyl)

   376   377

These regiospecific substitutions were explained by Priester and West[44] on the basis of the HSAB principle.[111] Hard reagents should react on the hard alkynylic site and soft ones on the soft propargylic site. However, as hard trimethylsilyl chloride was shown to react on the latter carbon, Hommes, Verkruijsse and Brandsma[179] ruled out the idea and suggested that the stronger basicity of the propargylic carbanion is responsible for its higher reactivity. A similar explanation might apply to the anion 376 in which the site remote from the ether grouping should be the more basic.

(D) Rearrangements

We shall deal here only with the rearrangements in which strong covalent bonds are broken and formed. Prototropy will be omitted as it can be considered in general as a part of the overall reaction of the organometallic with the electrophile. However, caution is called for regarding such an assumption since it has been reported that metalation time may influence the composition of the final reaction mixture.[156]

The Wittig rearrangement has been used in a few instances for synthetic purposes. Thus the allylic propargylic ether 378 rearranges via its lithio derivative to the alcohol 379.[180] In an analogous way the dipropargylic ethers 380 yielded the acetylenic allenic alcohols 381.[181] Satisfactory results are obtained here only when $R^1 = R^4 = $ Me or $SiMe_3$ and $R^2 = R^3 = $ H. When $R^1 = R^2 = R^4$ and $R^3 = $ H a 31% yield of 382 is obtained together with 9% of 381.

It seems also interesting to report here the case of a silyl group shift which occured on a metalated species. Thus the silylated ether 383, on lithiation, rearranges to its isomer 384.[157]

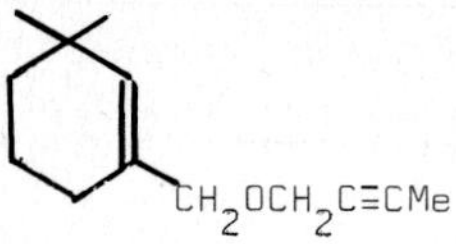

378

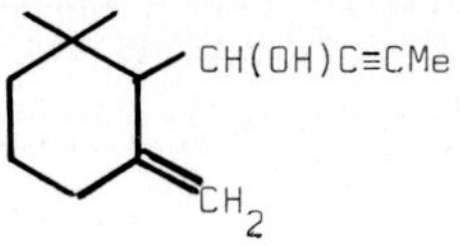

379

$$R^1C\equiv CCR^2R^3OCH_2C\equiv CR^4$$

380

$$R^2R^3C=C=CR^1CH(OH)C\equiv CR^4$$

381

$$R^1C\equiv CCR^2R^3CH(OH)C\equiv CR^4$$

382

$$RCH=C=C(OMe)SiMe_3$$

383

$$Me_3SiCR=C=CHOMe \qquad (R = n\text{-}C_6H_{13})$$

384

CONCLUSION

With this article we intended first to recall all the synthetic possibilities offered by the allenic-α-acetylenic carbanions. The abundance of studies published on that matter shows the importance of these reagents in the chemical field.

Our second and main aim, however, was an attempt to find among all the mentioned reactions which can be achieved with these entities, a global explanation of their mechanism. It must be concluded, however, that no single process can be put forward in order to account for all the above results. It appears actually that two different mechanisms can occur , each with a more or less important contribution, according to the nature of the reactants. The first of these processe is a concerted one while the second is of a free ion type. Thus it appears, in the case of organometallics derived from groups II and III metals, that with the following electrophiles : carbonyl compounds, imines, esters, oxiranes and ortho-esters, anSE2' mechanism would offer a satisfactory explanation for the regio and (or) the stereochemical course of the reaction. In the other cases, either when lithium derivatives are used or with other electrophiles than those mentioned above, the second process would occupy the main if not the exclusive place. The orientation of the substitution would then be governed by the HSAB principle.

REFERENCES

1(a). J.L. Moreau, "Organometallic Derivatives of Allenes and Ketenes", in
       "The Chemistry of Ketenes, Allenes and related compounds", Chemistry of
       Functional Groups, S. Patai Editor, J. Wiley and Sons, Chichester,
       New York (1980) 362.

 (b). J.Klein, "Propargylic Metalation" in "The Chemistry of carbon-carbon triple
       bond", ibid., (1978) 343.

 (c). L. Brandsma and H.D. Verkruijsse, Synthesis of Acetylenes, Allenes and
       Cumulenes, Studies in Organic Chemistry, Elsevier, Amsterdam, 8 (1981).

 (d). B.J. Wakefield, The Chemistry of Organic Lithium Compounds, Pergamon Press,
       Oxford, (1974).

 (e). J.H. Wotiz, Chemistry of Acetylenes, Marcel Dekker, New York, (1969) 366.

2.   P.L. Salzberg and C.S. Marvel, J.Am.Chem.Soc., 50 (1928) 1737.

3(a). S.S. Rossander and C.S. Marvel, ibid., 51 (1929) 932.

 (b). H.B. Gillespie and C.S. Marvel, ibid., 52 (1930) 3368.

 (c). B.W. Davis and C.S. Marvel, ibid., 53 (1931) 3840.

 (d). J.G. Stampfli and C.S. Marvel, ibid., 53 (1931) 4057.

 (e). H.E. Munro and C.S. Marvel, ibid., 54 (1932) 4445.

 (f). J. Harmon and C.S. Marvel, ibid., 55 (1933) 1716.

4.   J.H. Ford, C.D. Thompson and C.S. Marvel, ibid., 57 (1935) 2619.

5.   M.S. Newman and J.H. Wotiz, ibid., 71 (1949) 1292.

6.   G.R. Lappin, ibid., 71 (1949) 3966.

7.   C. Prevost, M. Gaudemar and J. Honigberg, C.R.Acad.Sci., 230 (1950) 1186.

8.   M. Gaudemar, ibid., 233 (1951) 64.

9(a). J.H. Wotiz, J.Am.Chem.Soc., 72 (1950) 1639, 73 (1951) 693.

 (b). J.H. Wotiz, J.S. Mathews and J.A. Lieb, ibid., 5503.

 (c). J.H. Wotiz and R.J. Palchak, ibid., 1971.

10.   M. Gaudemar, Ann.Chim., 1 (1956) 161.

11.   N.V. Komarov, M.F. Shostakovskii and L.N. Astaf'eva, J.Gen.Chem.USSR 31
       (1961) 1963.

12.   S.F. Karaev, Sh.O. Guseinov and E.A. Akhundov, ibid., 51 (1981) 1163.

13.   K. Eiter and co-workers, Lieb.Ann., (1978) 658.

14.   R.G. Daniels and L.A. Paquette, Tetrahedron Lett., (1981) 1579.

15.   J. Chenault and F. Tatibouet, C.R.Acad.Sci.C., 262 (1966) 499.

16.   J.L. Moreau and M. Gaudemar, Bull.Soc.Chim. France, (1970) 2171, 2175.

17.   R. Kuhn and H. Fischer, Chem.Ber., 92 (1959) 1849.

18.   M. Bertrand and co-workers, Tetrahedron Lett., (1979) 1845.

170

19.   M. Andrac, F. Gaudemar, M. Gaudemar, B. Gross, L. Miginiac, Ph. Miginiac
       and Ch. Prévost, Bull.Soc.Chim.France, (1963) 1385.

20(a). Y. Pasternak, Thèse de Doctorat Aix-Marseille (1964).

  (b). Y. Pasternak and J.C. Traynard, Bull.Soc.Chim.France, (1966) 356.

21(a). P.M. Greaves, S.R. Landor and M.M. Lwanga, Tetrahedron, 31 (1975) 3073.

  (b). R. Barlet, Bull.Soc.Chim.France, (1979) 132.

22.   G. Fontaine, C. André, Ch. Jolivet and P. Maitte, ibid., (1963) 1444.

23.   L. Miginiac-Groizeleau, Ph. Miginiac and C. Prévost, ibid., (1965) 3560.

24(a). J.P. Dulcère, J. Goré and M.L. Roumestant, ibid., (1974) 1119.

  (b). M.L. Roumestant, J.P. Dulcère and J. Goré, ibid., (1974) 1124.

25.   M. Bourguel, C.R.Acad.Sci., 177 (1923) 688. ibid., 179 (1924) 686.
       Ann.Chim. 3 (1925) 191, 325.

26.   T.H. Vaughn, J.Am.Chem.Soc., 55 (1933) 3453.

27.   K. Eberly, H.E. Adams, J.Organometal.Chem., 3 (1965) 165.

28.   R. West, P.A. Carney and I.C. Mineo, J.Am.Chem.Soc., 87 (1965) 3788.

29.   R. West and P.C. Jones, J.Am.Chem.Soc., 91 (1969) 6156.

30.   J.E. Mulvaney, T.L. Folk and D.J. Newton, J.Org.Chem., 32 (1967) 1674
       and references therein.

31.   E.J. Corey and H.A. Kirst, Tetrahedron Lett., (1968) 5041.

32.   J. Klein and S. Brenner, J.Am.Chem.Soc., 91 (1969) 3094.

33.   J. Klein and S. Brenner, J.Organometal.Chem., 18 (1969) 291.

34.   J. Klein and E. Gurfinkel, J.Org.Chem., 34 (1969) 3952.

35.   F. Jaffe, J.Organometal.Chem. 23 (1970) 53.

36.   J. Klein and S. Brenner, Tetrahedron, 26 (1970) 2345.

37.   J. Klein and J. Becker, Tetrahedron, 28 (1972) 5385.

38.   J. Klein and J. Becker, Chem.Com., (1973) 576.

39.   L. Brandsma, E. Mugge, Rec.Trav.Chim. Pays-Bas, 92 (1973) 628.

40.   G. Linstrumelle and D. Michelot, Chem.Com., (1975) 561.

41.   D. Michelot and G. Linstrumelle, Tetrahedron Lett.,(1976) 275.

42.   S. Bhanu, E.A. Khan and F. Scheinmann, J.Chem.Soc.Perkin I (1976) 1609.

43.   W. Priester, R. West and T. Ling Chwang, J.Am.Chem.Soc., 98 (1976) 8413.

44.   W. Priester and R. West, ibid., 8421.

45.   W. Priester and R. West, ibid., 8426.

46.   J.Y. Becker, J.Organometal.Chem., 118 (1976) 247.

47.   A.J. Quillinan, E.A. Khan and F. Scheinmann, J.Chem.Soc.,Chem.Comm.,
       (1974) 1030.

48.   S. Bhanu and F. Scheinmann, ibid., (1975) 817.

49.   A.J.G. Sagar and F. Scheinmann, Synthesis, (1976) 321.

50.   C. Clinet and G. Linstrumelle, N.J.Chim. 1 (1977) 373.

51.   X. Creary, J.Am.Chem.Soc., 99 (1977) 7632.

52(a). R. Baudouy, F. Delbecq and J. Goré, Tetrahedron Lett., (1979) 937.

  (b). G. Balme, A. Dutheau, J. Goré and M. Malacria, Synthesis, (1979) 508.

53.   Y. Yamakado, M. Ishiguro, N. Ikeda and H. Yamamoto, J.Am.Chem.Soc., 103
        (1981) 5568.

54.   L.I. Zakharkin et al., J.Gen.Chem. USSR, 51 (1981) 1337.

55.   L. Brandsma, H.E. Wijers and J.F. Arens, Rec.Trav.Chim., 82 (1963) 1040.

56(a). S. Hoff, L. Brandsma and J.F. Arens, ibid., 87 (1968) 916.

  (b). S. Hoff, L. Brandsma and J.F. Arens, ibid., 87 (1968) 1179.

  (c). S. Hoff, L. Brandsma and J.F. Arens, ibid., 88 (1969) 609.

57.   R. Mantione and A. Alvès, C.R.Acad.Sci,C, 268 (1969) 997.

58.   Y. Leroux and R. Mantione, Tetrahdron Lett., (1971) 591; J.Organometal.
        Chem. 30 (1971) 295.

59(a). R. Mantione and Y. Leroux, Tetrahedron Lett., (1971) 593.

  (b). R. Mantione and Y. Leroux, J.Organometal.Chem., 31 (1971) 5.

60.   E.J. Corey and S. Terashima, Tetrahedron Lett., (1972) 1815.

61.   F. Mercier, R. Epsztein and S. Holand, Bull.Soc.Chim. France, (1972) 690.

62.   R. Epsztein and F. Mercier, Synthesis, (1977) 183.

63.   M. Huché and P. Cresson, Tetrahedron Lett., (1975) 367.

64.   B.W. Metcalf and P. Casara, ibid., (1975) 3337.

65.   P. Casara and B.W. Metcalf, Tetrahedron Lett., (1978) 1581.
        B.W. Metcalf and P. Casara, J.Chem.Soc.,Chem.Comm., (1979) 119.

66.   H.J. Reich and S.K. Shah, J.Am.Chem.Soc., 99 (1977) 263.

67(a). R.M. Carlson and J.L. Isidor, Tetrahedron Lett., (1973) 4819.

  (b). R.M. Carlson, R.W. Jones and A.S. Hatcher, ibid., (1975) 1741.

68.   J.C. Clinet and G. Linstrumelle, ibid., (1978) 1137.

69.   A. Braverman, D. Reisman and M. Sprecher, ibid., (1979) 901.

70.   P.E. van Rijn, R.H. Everhardus and L. Brandsma, Rec.Trav.Chim., 99 (1980) 179.

71.   P.P. Montijn, H.M. Schmidt, J.H. van Boom, H.J.T. Bos, L. Brandsma and
        J.F. Arens, ibid., 84 (1965) 271.

72.   R. Mantione and A. Alvès, Tetrahedron Lett., (1969) 2483.

73.   K. Atsumi and I. Kuwajima, J.Am.Chem.Soc., 101 (1979) 2208.

74 (a). A.A. Petrov et al., J.Gen.Chem. USSR, 30 (1960) 231, 2221, 2226, 3846;
        35 (1965) 571, 618, 963.

  (b). Kh.V. Bal'yan et al., J.Org.Chem. USSR, 2 (1966) 1553, 1719, 1723, 1897,
        1902, 1906.

172

75(a). V.A. Kormer and A.A. Petrov, J.Gen.Chem. USSR, 30 (1960) 3846.

(b). A.A. Petrov, V.A. Kormer and M.D. Stadnichuk, ibid., 31 (1961) 1049.

(c). M.D. Stadnichuk, ibid., 36 (1966) 952.

(d). V.V. Markova, V.A. Kormer and A.A. Petrov, ibid., 37 (1967) 1886.

76.   L.N. Cherkassov et al., J.Org.Chem. USSR, 7 (1971) 1364, 2230; 8 (1972)
       2033; 9 (1973) 1380.

77.   T.B. Patrick, E.C. Haynie and W.J. Probst, J.Org.Chem., 37 (1972) 1553.

78.   G. Pourcelot, P. Cadiot and A. Willemart, C.R.Acad.Sci., 252 (1961)1630.

79.   B. Ganem, Tetrahedron Lett., (1974) 4467.

80.   M. Ishiguro, N. Ikeda and H. Yamamoto, J.Org.Chem., 47 (1982) 2225.

81.   Ch. Prévost, M. Gaudemar, L. Miginiac, F. Bardone-Gaudemar and M. Andrac,
       Bull.Soc.Chim. France, (1959) 679.

82.   M. Andrac, C.R.Acad.Sci., 248 (1959) 1356.

83(a). M. Gaudemar, Bull.Soc.Chim. France, (1962) 974; (1963) 1475.

(b). E. Favre and M. Gaudemar, ibid., (1968) 3724.

84.   J. Benaïm, C.R.Acad.Sci. C, 262 (1966) 937; Thèse de Doctorat, Paris (1968).

85(a). R. Couffignal, Thèse de Doctorat, Paris (1971); Bull.Soc.Chim.France,
       (1969) 3218.

(b). R.C. Couffignal and M. Gaudemar, ibid., 3550.

86.   J.L. Moreau and M. Gaudemar, ibid., (1973) 2549; 2729.

87.   F. Mercier, Thèse de Doctorat, Orsay, (1977), p. 52-58.

88.   Y. Pasternak, C.R. Acad.Sci., 255 (1962) 1750.

89.   J. Pansard and M. Gaudemar, Bull.Soc.Chim. France, (1968) 3332.

90.   M. Lequan and G. Guillerm, J.Organometal.Chem., 54 (1973) 153.

91.   E.R.H. Jones, J.H. Whitham and M.C. Whiting, J.Chem.Soc., (1957) 4628.

92.   L. Miginiac-Groizeleau, Bull.Soc.Chim. France, (1963) 1449.

93.   R. Epsztein et N. Le Goff, Tetrahedron Lett., (1970) 1897.

94(a). H. Chwastek, R. Epsztein and N. Le Goff, Tetrahedron, 29 (1973) 883.

(b). H. Chwastek, N. Le Goff, R. Epsztein and M. Baran-Marszak, ibid., 30
       (1974) 603.

95.   F. Mercier and R. Epsztein, J.Organometal.Chem., 108 (1976) 165.

96.   R. Epsztein and F. Mercier, Synthesis, (1977) 183.

97.   P.W. Collins, E.Z. Dajani, M.S. Bruhn, Ch. Brown, J.R. Palmer and
       R. Pappo, Tetrahedron Lett., (1975) 4217.

98(a). P. Läuger, M. Prost and R. Charlier, Helv.Chim.Acta, 42 (1959) 2379, 2394.

(b). M. Prost, M. Urbain and R. Charlier, ibid., 49 (1966) 2370.

(c). M. Prost, M. Urbain, A. Schumer, Ch. Houben and C. Van Meerbeck, ibid.,
       58 (1975) 40.

99. M. Andrac, Ann.Chim., 9 (1964) 287.

100. J.L. Moreau, Bull.Soc.Chim. France, (1975) 1248.

101. R. Mantione, Y. Leroux and H. Normant, C.R.Acad.Sci.C, 270 (1970) 1808.

102. R. Mantione and Y. Leroux, ibid., 272 (1971) 2201.

103. F. Mercier, Thèse de Doctorat, Orsay (1977), p. 31.

104. M. Karila, M.L. Capmau and W. Chodkiewicz, C.R.Acad.Sci.C, 269 (1969) 342.

105. M. Sanière-Karila, M.L. Capmau and W. Chodkiewicz, Bull.Soc.Chim. France, (1973) 3371.

106. W. Chodkiewicz et al., Bull.Soc.Chim.France, (1967) 2429; (1969) 976, 4023; (1971) 1824, 2248; (1972) 1417. Tetrahedron Lett., (1965) 1619; (1972) 37. C.R.Acad.Sci.C, 264 (1967) 921; 268 (1969) 1449; 269 (1969) 1556; 271 (1970) 1390; 272 (1971) 229, 486; 273 (1971) 759.

107. H. Feklin, Y. Gault and G. Roussi, Tetrahedron, 26 (1970) 3761.

108. M. Lequan and G. Guillerm, J.Organometal.Chem., 54 (1973) 153.

109. A. Horeau and H. Kagan, Tetrahedron, 20 (1964) 2431.

110. G. Köbrich and D. Merkel, J.Chem.Soc., Chem.Comm., (1970) 1452.

111. R.G. Pearson and J. Songstad, J.Am.Chem.Soc., 89 (1967) 1827.

112. Y. Leroux and C. Roman, Tetrahedron Lett., (1973) 2585.

113. R. Gélin, S. Gélin and M. Albrand, Bull.Soc.Chim. France, (1971) 4546.

114. J. Huet, ibid., (1964) 952.

115. J.L. Moreau and M. Gaudemar, ibid., (1971) 3071.

116(a). C. Nivert and L. Miginiac, C.R.Acad.Sci. C, 272 (1971) 1996.

  (b). J.L. Moreau and M. Gaudemar, C.R.Acad.Sci. C, 274 (1972) 2015.

117. J.L. Moreau and M. Gaudemar, Bull.Soc.Chim. France, (1975) 1211.

118. G. Courtois, M. Harama and Ph. Miginiac, J.Organometal.Chem., 218 (1981) 1.

119. M. Gaudemar, C.R.Acad.Sci., 239 (1954) 1303.

120. M. Bertand and J. Le Gras, Bull.Soc.Chim. France, (1963) 2136.

121. M. Gaudemar and R. Couffignal, C.R.Acad.Sci. C, 265 (1963) 42.

122. R. Couffignal and M. Gaudemar, ibid., 266 (1968) 224.

123. H. Driguez and J.C. Traynard, ibid., 267 (1968) 497.

124. R. Couffignal and M. Gaudemar, Bull.Soc.Chim. France, (1969) 898.

125. D. Plouin and R. Glénat, ibid., (1973) 737.

126. O.V. Perepelkin, Izv.Vyssh.Uchebn,Zaved.Khim.Tekhnol., 14 (1971) 561; C.A. 75 (1971) 63063s.

127. G. Pfeiffer and H. Driguez, C.R.Acad.Sci. C, 267 (1968) 773.

128. F. Bardone-Gaudemar, ibid., 243 (1956) 1895; Ann.Chim., 3 (1958) 52.

129. S. Masson, M. Saquet and A. Thuillier, Tetrahedron, 33 (1977) 2949.

130. D. Paquer and M. Vazeux, J.Organometal.Chem., 140 (1977) 257.

131.   D. Taub and A.A. Patchett, Tetrahedron Lett., (1977) 2745.

132.   J.L. Moreau, Y. Frangin and M. Gaudemar, Bull.Soc.Chim.France, (1970) 4511.

133.   M. Bellassoued, Y. Frangin and M. Gaudemar, Synthesis, (1978) 150.

134.   G. Daviaud, M. Massy-Barbot and Ph. Miginiac, C.R.Acad.Sci., C, 272 (1971)
        969.

135.   G. Daviaud and Ph. Miginiac, Tetrahedron Lett., (1971) 3251.

136.   P. Condran, Jr, L. Hammond, A. Mourino and W.H. Okamura, J.Am.Chem.Soc.,
        102 (1980) 6260.

137.   R. Epsztein and B. Herman, J.Chem.Soc.Chem.Comm., (1980) 1250.

138.   R. Epsztein and N. Le Goff, Tetrahedron Lett., (1981) 1965.

139(a). L. Miginiac-Groizeleau, Ann.Chim., 6 (1961) 1071.

   (b). R.S. Skowronski and W. Chodkiewicz, C.R.Acad.Sci., 251 (1960) 547.

   (c). W. Chodkiewicz, P. Cadiot and A. Willemart, ibid., 253 (1961) 954.

   (d). M. Montebruno, F. Fournier, J.P. Battioni and W. Chodkiewicz, Bull.Soc.
        Chim. France, (1974) 283.

140.   R. Epsztein and N. Le Goff, Tetrahedron Lett., (1970) 1897.

141(a). J.A. Cella, E.A. Brown and R.R. Burtner, J.Org.Chem., 24 (1959) 743.

   (b). Zh.S. Sydykov and G.M. Segal, Bioorg.Khim., 2 (1976) 1531.

142.   Y. Frangin, E. Favre and M. Gaudemar, C.R.Acad.Sci., C, 282 (1976) 277.

143.   M. Gaudemar, ibid., 254 (1962) 1100.

144.   A. Jean, G. Guillerm and M. Lequan, J.Organometal.Chem., 21 (1970) P1.

145.   D. Burn, G. Cooley, V. Petrow and G.O. Werton, J.Chem.Soc., (1959) 3808.

146.   R. Vitali and R. Gardi, Tetrahedron Lett., (1972) 1651.

147.   M. G. Voskonyan, A.A. Pashayan and Sh.O. Badanyan, Arm.Khim.Zh. 28 (1975) 7

148.   L. Miginiac, C.R.Acad.Sci., 247 (1958) 2156.

149.   G.F. Hennion and C.V. Digiovanna, J.Org.Chem., 31 (1966) 970.

150.   M. Gaudemar, C.R.Acad.Sci., 243 (1956) 1216.

151(a). G. Pfeiffer, Bull.Soc.Chim.France, (1962) 776.

   (b). P. Perriot and M. Gaudemar, C.R.Acad.Sci., C, 272 (1971) 698.

152.   R.E. Ireland, N.I. Dawson and Ch.A. Lipinski, Tetrahedron Lett., (1970) 224

153.   K. Mori, M. Tominaga and M. Matsui, Tetrahedron, 31 (1975) 1846.

154.   M. Bourgain-Commerçon, J.F. Normant and J. Villieras, J.Chem.Res. (1977) (S
        183, (M) 2101.

155.   E.J. Corey, J.A. Katzenellenbogen, N.W. Gilman, S.A. Roman and B.W. Erickso
        J.Am.Chem.Soc., 90 (1968) 5618.

156.   R. Baudouy, F. Delbecq and J. Goré, Tetrahedron Lett.,(1979) 937.

157.   J.C. Clinet and G. Linstrumelle, ibid., (1980) 3987.

158.   M. Bertrand, C.R.Acad.Sci., 244 (1957) 619.

159. M.G. Voskanyan, A.A. Pashayan, and Sh.O. Badanyan, Arm.Khim.Zh. 28 (1975) 791.

160. J.C. Masson, M. Lequan, W. Chodkiewicz and P. Cadiot, C.R.Acad.Sci., C, 257 (1963) 1111.

161. J.C. Masson, M. Lequan, W. Chodkiewicz and P. Cadiot, Bull.Soc.Chim. France, (1967) 777.

162. M. Lequan and P. Cadiot, ibid. (1965) 45.

163. L.N. Cherkasov and V.S. Zavgorodnii, J.Gen.Chim. USSR, 38 (1968) 2713.

164. A.M. Sladkov and L.K. Luneva, ibid. 36 (1966) 570.

165. L.N. Cherkasov, L.M. Zubritskii and K.V. Balyan, ibid., 38 (1968) 2046.

166. L.M. Zubritskii, K.V. Balyan and L.N. Cherkasov, ibid., 39 (1969) 2636.

167. P. Mazerolles, Bull.Soc.Chim. France, (1960) 856.

168. L.N. Cherkasov and K.V. Balyan, J.Gen.Chem. USSR, 39 (1969) 1139.

169. E. Favre and M. Gaudemar, C.R.Acad.Sci., C, 262 (1966) 1332.

170. J. Blais, J. Soulié and P. Cadiot, ibid., 271 (1970) 589.

171. E. Favre and M. Gaudemar, ibid., 272 (1971) 111.

172. A. L'Honoré, J. Soulié and P. Cadiot, ibid., 275 (1972) 229.

173. A. Haces, E.M.G.A. van Kruchten and W.H. Okamura, Tetrahedron Lett., 23 (1982) 2707.

174. M. Cinquini, S. Colonna and Ch.J.M. Stirling, J.Chem.Soc.Chem.Comm., (1975) 256.

175. J. Klein, and S. Brenner, J.Org.Chem., 36 (1971) 1319.

176. S. Bhanu and F. Scheinmann, J.Chem.Soc.Perkin I, (1978) 1218.

177. G.R. Khan, K.A. Pover and F. Scheinmann, J.Chem.Soc.Chem.Comm., (1979) 215.

178. K.A. Pover and F. Scheinmann, J.Chem.Soc. Perkin I, (1980) 2338.

179. H. Hommes, H.D. Verkruijsse and L. Brandsma, Rec.Trav.Chim.Pays Bas, 99 (1980) 113.

180. K.H. Schulte-Elke, V. Rautenstrauch and G. Ohloff, Helv.Chim.Acta, 54 (1971) 1805.

181. M. Huché and P. Cresson, Tetrahedron Lett., (1975) 367.

182. F. Barbot, L. Poncini, B. Randriancelina and P. Miginiac, J.Chem.Res., (1981) (S) 343, (M) 4016.

CHAPTER 4

Stereoselective Aldol Condensations

Clayton H. Heathcock

Department of Chemistry, University of California
Berkeley, California 94720

Contents

> *"Nature, it seems, is an organic chemist having some predilection for the aldol and related condensations ..."*
>
> J.W. Cornforth

I. INTRODUCTION

The carbonyl group occupies a central place in organic synthesis. For the purpose of building up larger molecules from smaller ones, the two most fundamental processes are the reactions of enols or enolates with electrophiles (eqn.1) and the reactions of carbonyl compounds with nucleophiles (eqn. 2). These two important processes are combined in two organic reactions, the aldol and Claisen condensations (eqn. 3).

$$\underset{\text{C=C}}{\overset{\text{OH}}{|}} \text{ or } \underset{\text{C=C}}{\overset{\text{O}^-}{|}} + \text{E} \longrightarrow \underset{\text{C--C--E}}{\overset{^+\text{OH}}{|}} \text{ or } \underset{\text{C--C--E}}{\overset{\text{O}}{||}} \qquad (1)$$

$$\underset{\text{--C--}}{\overset{^+\text{OH}}{||}} \text{ or } \underset{\text{--C--}}{\overset{\text{O}}{||}} + \text{N:} \longrightarrow \underset{\text{--C--N}}{\overset{\text{OH}}{|}} \text{ or } \underset{\text{--C--N}}{\overset{\text{O}^-}{|}} \qquad (2)$$

$$\underset{\text{--C--}}{\overset{\text{O}}{||}} + \underset{\text{C=C}}{\overset{\text{O}^-}{|}} \longrightarrow \underset{\text{--C--C--C--}}{\overset{^-\text{O} \qquad \text{O}}{| \qquad ||}}$$

$$\underset{\text{--C--}}{\overset{^+\text{OH}}{||}} + \underset{\text{C=C}}{\overset{\text{OH}}{|}} \longrightarrow \underset{\text{--C--C--C--}}{\overset{\text{OH} \quad ^+\text{OH}}{| \qquad ||}} \qquad (3)$$

The aldol condensation[*] is a venerable reaction, having been discovered in 1838[1] almost thirty years before the structural theory was developed. Of all reactions, it is the only one to which an entire volume of *Organic Reactions* has been devoted.[2] As suggested by the quote which opens this chapter[3] the aldol condensation is important not only to the experimental scientist, but generally as a method whereby many of the important compounds of nature are synthesized. Thus, a vast array of natural products are built up from acetate, propionate, and succinate units by routes involving some variant of this important reaction.

In this chapter, we shall review one aspect of the aldol condensation-- stereochemistry. Even though this important reaction is almost a hundred and fifty years old, systematic stereochemical investigations were not carried out until relatively recently. There were two reasons for this lag. First, before the advent of nuclear magnetic resonance spectroscopy it was almost impossible to analyze accurately mixtures of diastereomers. Second, the reaction itself is of a nature that stereochemical investigations are difficult. If the aldol condensation is carried out in the traditional manner, in protic solvents under reversible conditions, one must contend with equilibration of diastereomeric products and

---

[*] In this article, we shall define aldol condensation somewhat more liberally than the term has been used. For our purposes, an aldol condensation is a reaction in which an enol or enolate of any type (aldehyde, ketone, ester, amide, thioamide) adds to an aldehyde or ketone carbonyl to give a β-hydroxycarbonyl compound. By this definition, we eliminate *acylation* reactions of enolates, which are then left to be known as Claisen condensations. Note that this definition includes within the general class of "aldol condensations" many reactions commonly known by other names: Reformatsky, Perkin, Knoevenagel, etc.

ith dehydration of the initial β-hydroxy carbonyl compounds. The solution to
his problem was the introduction of hindered, non-nucleophilic bases which
low carbonyl compounds to be converted into their enolates without self-
ondensation.

Of these important bases, lithium diisopropylamide (LDA) has turned out
o be the most useful. Frostick and Hauser introduced diisopropylamino-
aagnesium bromide as a catalyst for the Claisen condensation in 1949.[4] In
950, Hamell and Levine reported the first use of LDA for the same purpose.[5]
owever, it was some time before Wittig and co-workers reported that LDA
an be used to deprotonate aldimines and that the resulting N-lithioenamines
ondense with aldehydes and ketones to give β-hydroxy imines (eqn. 4).[6,*]

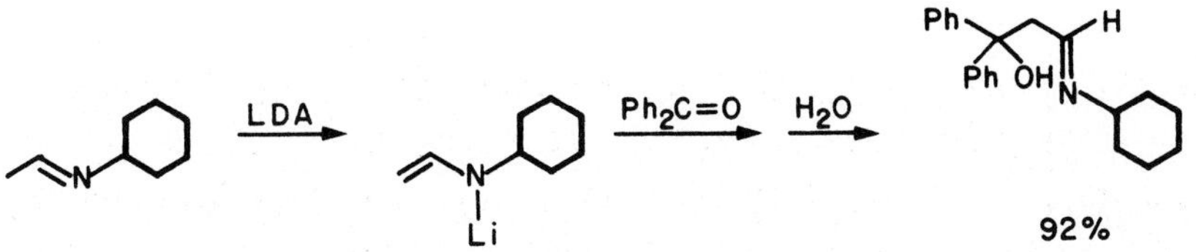

$$(4)$$

The first use of a preformed ester enolate in synthesis appears to have
een by Dunnavant and Hauser, who prepared the enolate of ethyl acetate
sing lithium amide in ammonia.[8] The resulting enolate was found to react with
ldehydes and ketones giving β-hydroxy esters in low yield. Ethyl acetate
ad been prepared earlier by treating a THF solution of ethyl acetate with
odium bis(trimethylsilyl)amide.[9] However, Rathke showed that the sodium
nolate is unstable even at -78 °C and introduced the lithium enolate, which he
repared with lithium bis(trimethylsilyl)amide.[10] The lithium enolate was found
o be stable indefinitely in THF solution at -78 °C and to react smoothly with
ldehydes and ketones (e.g., eqn. 5).

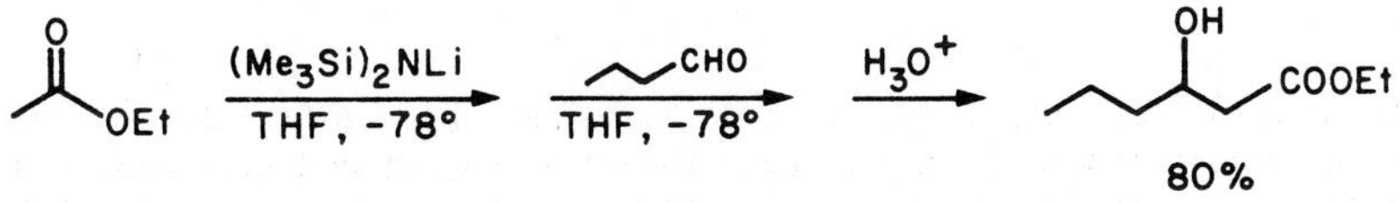

$$(5)$$

---

* In the course of his investigations of the directed aldol condensation, Wittig also discovered that
thium dialkylamides can function as reducing agents,[6b,c] a phenomenon recently rediscovered by
ther investigators.[7]

The first careful investigations of aldol stereoselection were carried out by the French chemist J.E. Dubois. In a series of papers in 1967-69, Dubois and his co-worker M. Dubois established that enolates of cyclic ketones show a kinetic preference for formation of *threo* aldols.[11,*] These investigations were carried out in the traditional manner for aldol condensations, using systems such as KOH in methanol. By working at low temperature and analyzing the product composition at low conversion, Dubois was able to establish that the kinetic preference for these (E) enolates to produce the *threo* aldol can be quite high (e.g., eqn. 6).[11b]

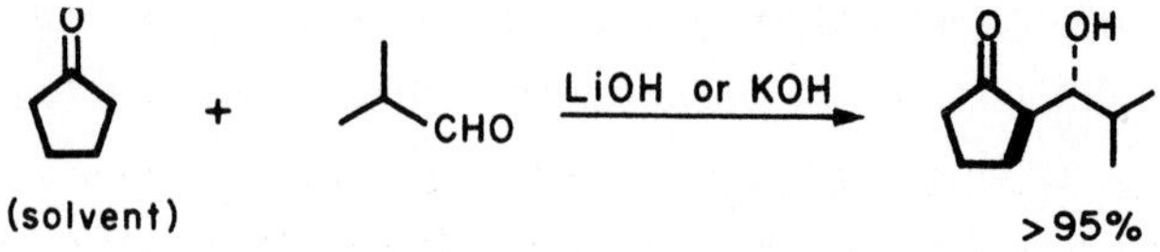

(solvent)   >95%   (6)

In 1972, Dubois and Fellman reported an important result, one which stimulated much of the subsequent work on aldol stereoselection. It was found that the bromomagnesium enolate of ethyl *t*-butyl ketone reacts with benzaldehyde to give solely the *erythro* aldol (eqn. 7).[12] The importance of this result

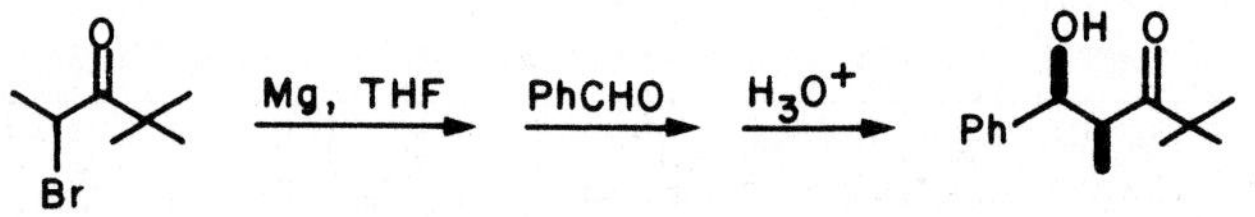

(7)

was not that a highly stereoselective condensation had been found, but that the intermediate enolate was shown to have the (Z) configuration. Thus, a potentially useful generalization had been established: (E) enolates show *threo* kinetic stereoselectivity and (Z) enolates show *erythro* kinetic stereoselectivity (eqns. 8 and 9).

---

* In this chapter, we shall use the stereochemical descriptors *erythro* and *threo* in the following sense: When the backbone of the aldol is written in an extended (zig-zag) manner, if the α-alkyl substituent and the β-hydroxy substituent both extend toward the viewer (bold bonds) or both extend away from the viewer (dashed bonds), this is the *erythro* diastereomer.

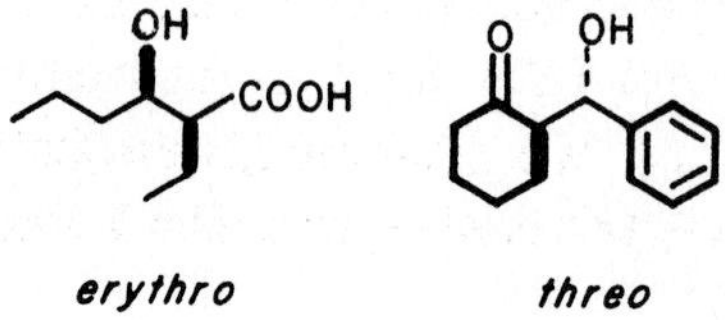

erythro   *threo*

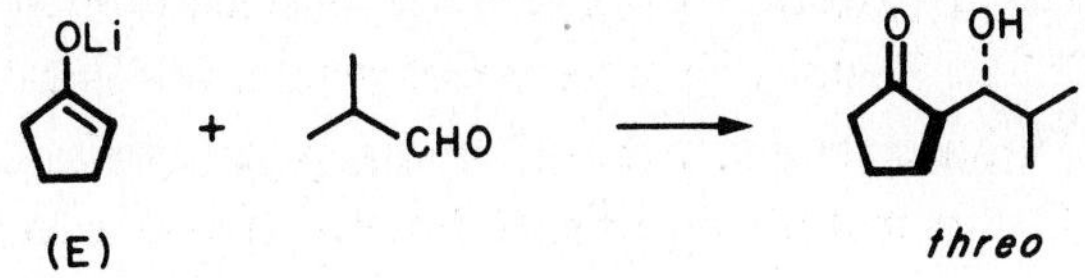

$$\text{(E)} \quad + \quad \text{CHO} \quad \longrightarrow \quad \textit{threo} \qquad (8)$$

$$\text{(Z)} \quad + \quad \text{PhCHO} \quad \longrightarrow \quad \textit{erythro} \qquad (9)$$

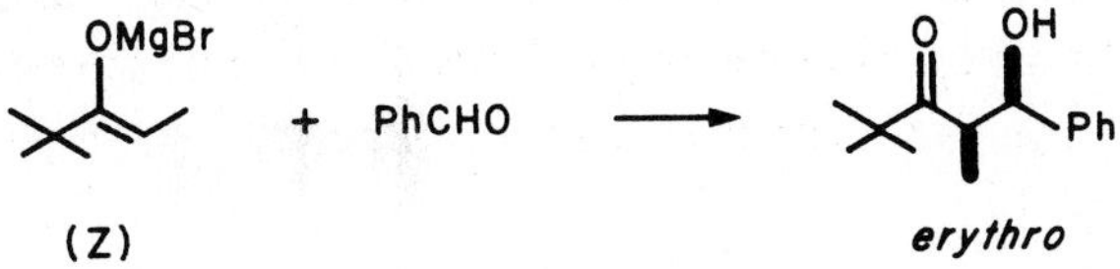

However, a subsequent report by Dubois and Fellman cast doubt on the scope of this generalization. It was found that the two stereoisomeric enolates of diethyl ketone show differing degrees of kinetic stereoselectivity in their reactions with pivaldehyde.[13] The (Z) enolate was found to be fairly *erythro*-selective, giving an *erythro:threo* ratio of 88:12 (eqn. 10). However, the (E) enolate was found to be indiscriminate, giving an *erythro:threo* ratio of 48:52 (eqn. 11).

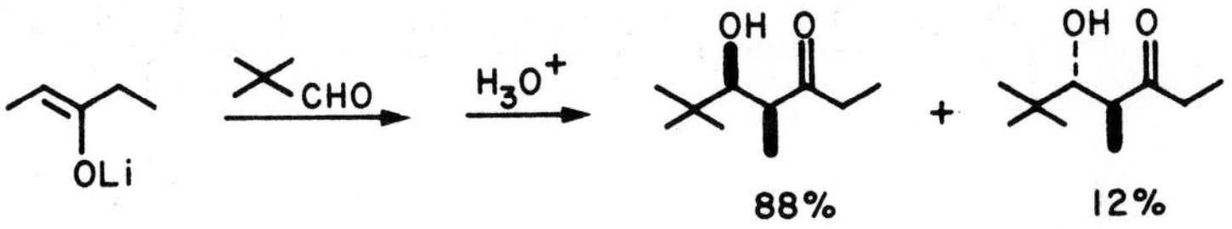

$$\qquad (10)$$

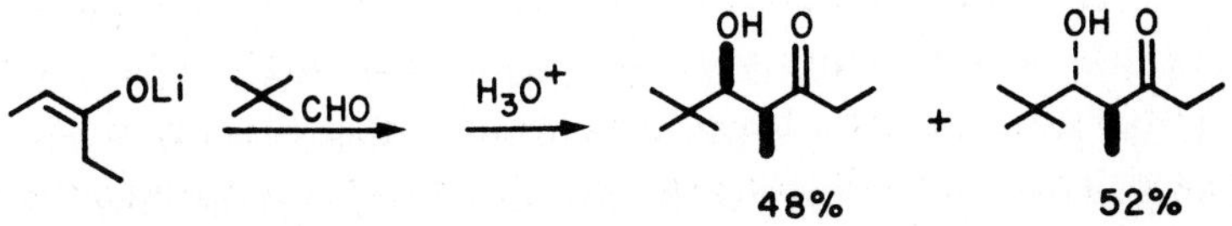

88%   12%

$$\qquad (11)$$

48%   52%

Concurrently with Dubois' investigations, House carried out a study of the condensation of ketone enolates with aldehydes.[14] House and his co-workers used preformed lithium enolates to which they added magnesium bromide or zinc chloride prior to addition of the aldehyde. In most cases, these reaction conditions favor formation of the *threo* diastereomer, presumably as a result of thermodynamic control (*vide infra*) . For example, condensation of phenylacetone with benzaldehyde was found to give a similar *threo:erythro* ratio regardless of whether the pure (Z) enolate or a 60:40 mixture of (Z) and (E) enolates is employed (eqn. 12).

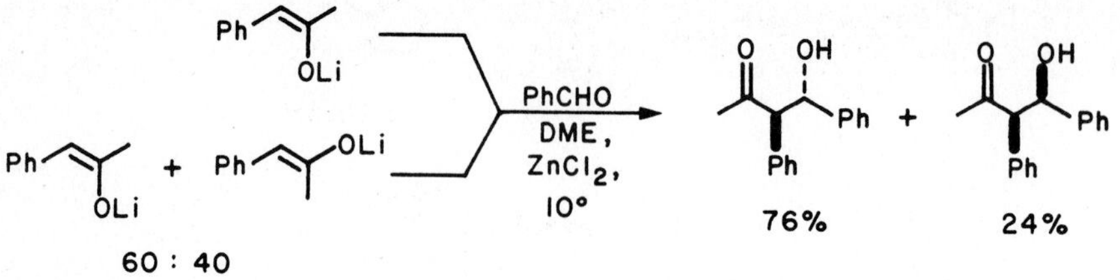

$$\text{(12)}$$

60 : 40   76%   24%

One other important paper helped pave the way for the extensive investigations of the past four years. In 1975 Fenzl and Koster, at the Max Planck Institute in Mülheim, reported a study of the reactions of diethylboron enolates with aldehydes.[15] It was found that propiophenone, which forms the (Z) diethylboron enolate, reacts with propionaldehyde to give *only* the *erythro* β-ketoboronate. Furthermore, Fenzl and Koster showed that diethyl ketone, which gives a 9:1 mixture of (Z) and (E) diethylboron enolates, reacts with benzaldehyde to produce a 9:1 mixture of *erythro* and *threo* β-ketoboronates (eqn. 13).

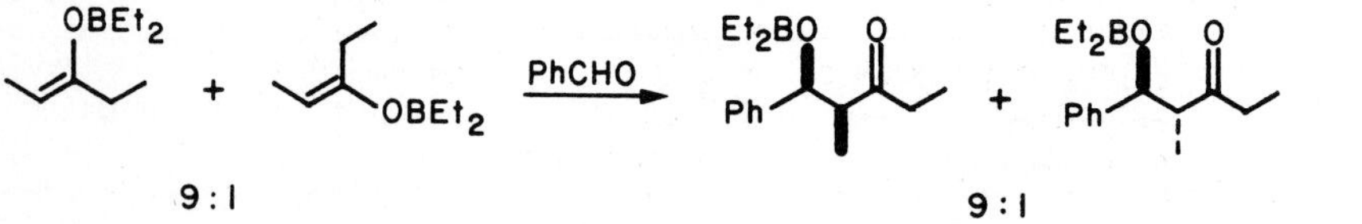

$$\text{(13)}$$

9 : 1   9 : 1

It was from these roots that the recent extensive work on aldol stereoselection has grown. In the remainder of this article, we will briefly summarize this burgeoning field. The review will be selective rather than exhaustive and will concentrate more on preparative than on mechanistic aspects. It will also be rather personal, since it will dwell more on contributions from our group at Berkeley than on the contributions of other workers.

II. Kinetic Stereoselection with Lithium Enolates

The work of Dubois and of Fenzl and Koster clearly established that (Z) enolates are *erythro-*selective. However, the situation with regard to (E) enolates was somewhat confusing. We began our investigations at Berkeley by surveying a series of condensations of preformed lithium enolates with benzaldehyde under conditions of strict kinetic control (THF, -78°C, 10 seconds reaction time).[16,17] The compounds studied are shown in Scheme I; data are collected in Table I.

Scheme I*

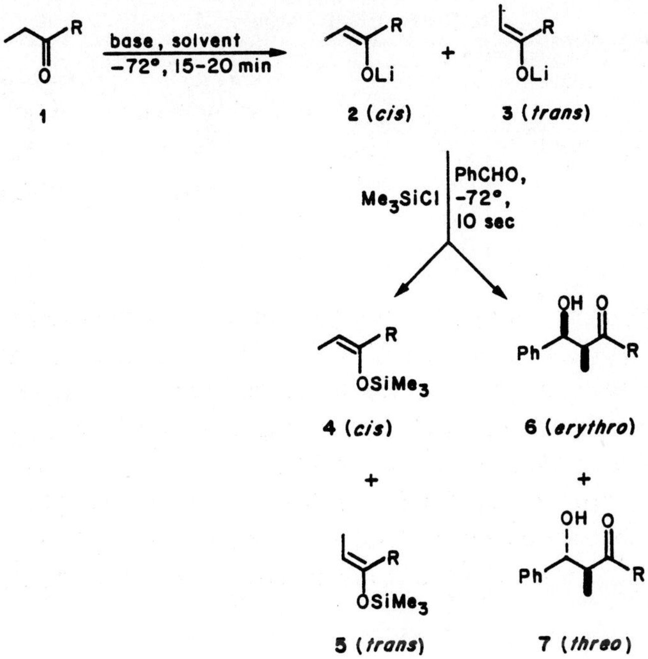

$R = $
a: OH(OLi)
b: CH$_3$O
c: CH$_3$OCH$_2$CH$_2$OCH$_2$O
d: t-C$_4$H$_9$O
e: (i-C$_3$H$_7$)$_2$N

f: H
g: C$_2$H$_5$
h: i-C$_3$H$_7$
i: t-C$_4$H$_9$
j: 1-adamantyl

k: Me$_3$Si
l: C$_6$H$_5$
m: mesityl

---

* In Scheme I and Table I, and for the remainder of this Chapter, we shall use the stereochemical descriptors *cis* and *trans* for enolates. The *cis* enolate is one having the O⁻ or OH on the same side of the double bond as the substituent attached to the α-carbon. For most ketones, *cis* is the same as (Z) and *trans* is the same as (E). However, for certain important ketones, such as acylsilanes, the *cis* enolate has the (E) configuration. Another reason for using the *cis-trans*, rather than the

cis, (Z)   *cis*, (E)

(E)-(Z) terminology is that ester enolates of a given configuration always have the same name. Thus, there is an ambiguity in naming the following three *cis* enolates by the (E)-(Z) nomenclature:

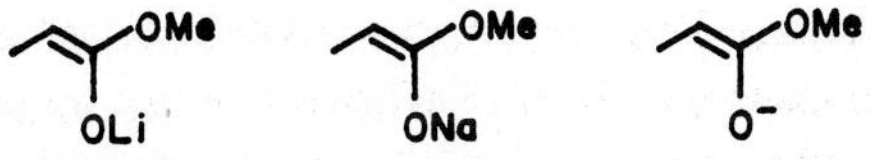

cis, (E)   *cis*, (Z)   *cis*, (?)

Table I. Reaction of Preformed Lithium Enolates of Compounds 1 with Benzaldehyde (Scheme I)

R	base[a,b]	cis/trans[c] (2/3=4/5)	erythro/threo (6/7)
LiO	LDA	—	45:55
CH_3O	LDA	5:95	62:38
$CH_3OCH_2CH_2OCH_2O$	LDA	d	23:77
$t\text{-}C_4H_9O$	LDA	5:95	49:51
$t\text{-}C_4H_9O$	LTMP	d	35:65
$(i\text{-}C_3H_7)_2N$	LDA	81:19	63:37
$(i\text{-}C_3H_7)_2N$	LTMP	52:48	68:32
H	e	100:0	50:50
H	e	0:100	65:35
C_2H_5	LDA	30:70	64:36
C_2H_5	LCPA	35:65	62:38
C_2H_5	LBTMSA	66:34	77:23
C_2H_5	LTMP	20:80	66:34
$i\text{-}C_3H_7$	LDA	60:40[h]	82:18
$i\text{-}C_3H_7$	LCPA	59:41[h]	75:25
$i\text{-}C_3H_7$	LBTMSA	>98:2	90:10
$i\text{-}C_3H_7$	LTMP	32:68	58:42
$i\text{-}C_3H_7$	e	0:100	45:55
$t\text{-}C_4H_9$	LDA	>98:2	>98:2
$t\text{-}C_4H_9$	LBTMSA	>98:2	>98:2
1-adamantyl	LDA	>98:2	>98:2
$(CH_3)_3Si$	LDA	38:62	58:42
C_6H_5	LDA	>98:2	88:12
C_6H_5	LCPA	>98:2	87:13
C_6H_5	LBTMSA	>98:2	88:12
C_6H_5	LTMP	>98:2	83:17
$2,4,6\text{-}(CH_3)_3C_6H_2$	LDA[f]	8:92	8:92
$2,4,6\text{-}(CH_3)_3C_6H_2$	LDA	5:95	8:92
$2,4,6\text{-}(CH_3)_3C_6H_2$	LCPA[g]	4:96	9:91
$2,4,6\text{-}(CH_3)_3C_6H_2$	LBTMSA[g]	87:13	88:12
$2,4,6\text{-}(CH_3)_3C_6H_2$	LTMP[g]	d	4:96

a. Unless otherwise noted, reactions were carried out in THF containing ca. 15% hexane from the n-BuLi used to generate the amide base. b. LDA=lithium diisopropylamide; LCPA=lithium cyclohexylisopropylamide; LBTMSA=lithium bistrimethylsilylamide; LTMP=lithium 2,2,6,6-tetramethylpiperidide. c. The prefixes cis and trans refer to CH_3 and OLi. d. Not determined. e. Enolate generated by treating the trimethylsilyl enol ether with methyllithium. f. In ether. g. Enolate insoluble. h. About 10% of the tetrasubstituted enolate is formed.

Several trends emerge from the data in Table I. First, the nature of the amide base used to accomplish deprotonation can have a pronounced effect of the *cis/trans* ratio in the resulting enolates. The similar lithium bases, lithium diisopropylamide (LDA) and lithium cyclohexylisopropylamide (LCPA), give nearly identical *cis/trans* ratios in all cases. However, lithium 2,2,6,6-tetra-methylpiperidide (LTMP) usually gives more *trans* enolate than LDA while lithium bistrimethylsilylamide (LBTMSA) gives more *cis* enolate. Data for a series of compounds are summarized in Table II. In addition to the effect of base on the *cis/trans* ratio, Table II shows that, with LDA and LTMP, the amount of *cis* enolate smoothly increases as the size of R increases.

Table II. Percent *Cis* Enolate (2) as a Function of Base (Scheme I)				
R	LTMP	LDA	LBTMSA	[A]/[B] (Scheme II)
CH_3O		5	a	13
$2,4,6\text{-}(CH_3)_3C_6H_2$	4	8	87	
C_2H_5	16	30	66	51
$i\text{-}C_3H_7$	32	56	100	82
$(i\text{-}C_3H_7)_2N$	52	81	a	
C_6H_5	100	100	100	221
$t\text{-}C_4H_9$	100	100	100	>4000

a. Esters and amides do not react with this base.

The trends in Table II may be explained in the following manner. It is assumed that the base approaches the C-H bond not along its axis, but more over the face of the carbonyl group (Scheme II). This results in a steric interaction between the base and the methyl group which raises the energy of transition state $A^{\ddagger}$ relative to that of $B^{\ddagger}$. Thus, all other things being equal, there is a bias in favor of formation of the *trans* enolate, and this bias is more pronounced with a more bulky base (LTMP) than with a less bulky one (LDA). However, conformation A is favored relative to conformation B, even when R=H. As R becomes larger, the [A]/[B] ratio is expected to increase. The ground state conformer ratios shown in Table II are obtained by adding to 800 cal mole^{-1} (the experimental conformational free energy difference between conformations A and B for propionaldehyde) one-half the conformational free energy difference of an R-substituted cyclohexane (an estimate of the gauche

Scheme II. Transition State for Formation of *cis* and *trans* Enolates

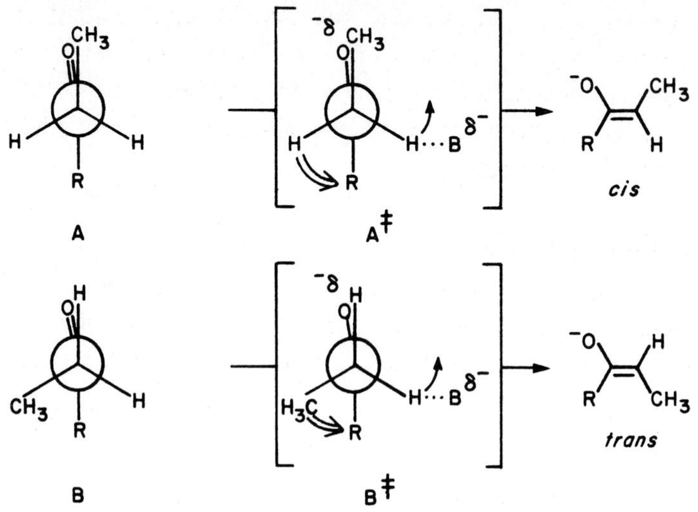

R:CH$_3$ interaction in conformation B). Thus, it is clear that as R becomes larger, the amount of *trans* enolate will decrease as a consequence of the increasing importance of the R:CH$_3$ interaction in B‡.

The second important trend which may be seen from the data in Table I is that *cis* and *trans* enolates do not generally show the same degree of kinetic stereoselection. As had previously been found by Dubois and Fellman,[13] the *cis*

R	*cis* enolate 2	*trans* enolate 3
CH_3O		1.5
$t\text{-}C_4H_9O$		1.0
H	1	1.5
C_2H_5	9	1.5
$i\text{-}C_3H_7$	9	1.0
C_6H_5	7	
$t\text{-}C_4H_9$	>50	
1-adamantyl	>50	
$2,4,6\text{-}(CH_3)_3C_6H_2$	>50	<0.02

Table III. Kinetic Stereoselectivity for the Reaction of Various Enolates with Benzaldehyde (Scheme I)

enolate of diethyl ketone (**2g**) shows kinetic stereoselectivity* of 9 while the *trans* isomer **3g** shows no stereoselectivity. In fact, the only nonselective *cis* enolate studied is the one derived from propionaldehyde and the only highly selective *trans* enolate studied is the one derived from ethyl mesityl ketone. When the R group attached to the carbonyl is bulky, both *cis* and *trans* enolates are highly selective, with the *cis* isomer giving the *erythro* aldol and the *trans* isomer giving the *threo* aldol.

III. Transition State Hypotheses

In 1957, Zimmerman and Traxler proposed that the preferred formation of *threo* adducts in the Ivanov condensation of phenylacetic acid with benzaldehyde can be understood in terms of a six-center transition state (Figure I).[18]

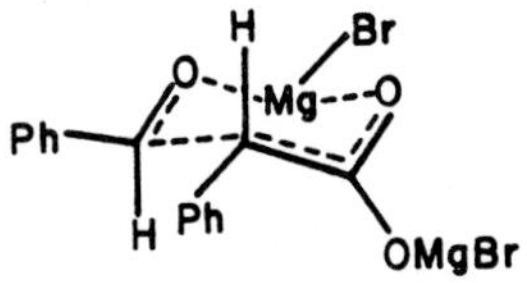

Figure I. Zimmerman Transition State

Since evidence had accrued that links the structure of the enolate to the stereostructure of the aldol, Dubois and co-workers adopted the Zimmerman transition state hypothesis to explain the observed stereochemistry of the reaction.[11,13] Thus, the two transition states depicted in Figure II would lead from a *cis* enolate to an *erythro* and a *threo* aldol, respectively. The two transition states are presumed to differ mainly in the R:R' interaction, which is worse in the transition state leading to the *threo* diastereomer than is the one leading to the *erythro* diastereomer. A similar pair of transition states are possible with the *trans* enolate, also leading to *erythro* and *threo* aldols (Figure III). The Zimmerman-Dubois transition state explains the observed *cis*-enolate:*erythro*-aldol relationship and also nicely explains why the stereoselectivity increases as the size of R' increases. It does not explain why *cis* lithium enolates seem to be inherently more stereoselective than their *trans* counterparts.

The suggestion has been made[17] that this difference arises from an asymmetry in the transition states depicted in Figures II and III. That is, if one looks down the newly-forming C-C bond, the transition state for the aldol condensation might be as depicted in Figure IV. In this model, there are two potentially serious interactions—between R_1 and R_4 and between R_2 and R_5.

---

* Kinetic stereoselectivity is defined as the ratio of *erythro* and *threo* aldols which is produced by the reaction of the enolate with a given aldehyde.

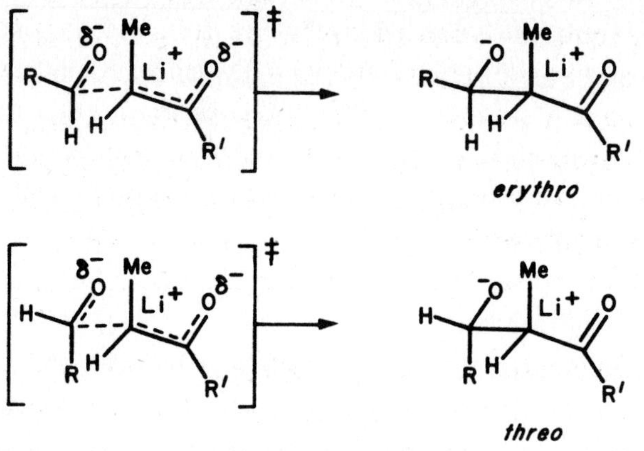

188

Figure II. Proposed Transition States Leading from a
cis-Enolate to *Erythro* and *Threo* Aldols

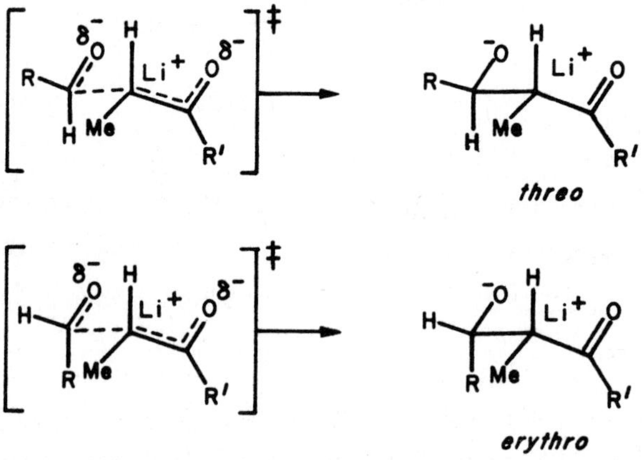

Figure III. Proposed Transition States Leading from a
trans-Enolate to *Erythro* and *Threo* Aldols

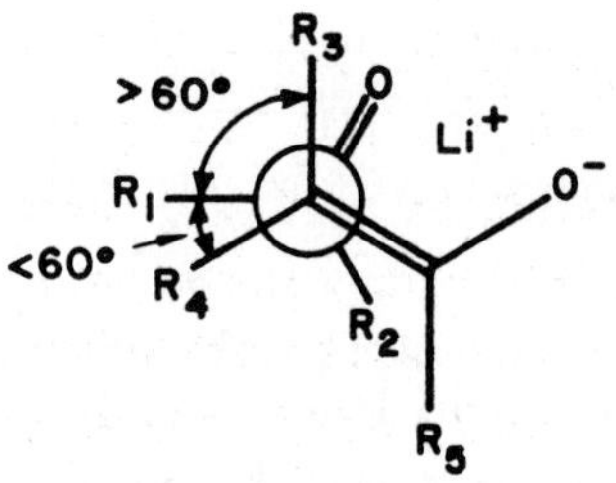

Figure IV. Transition State for the Aldol
Condensation

For a *cis* enolate, R_4 is hydrogen and the dominant interaction is between R_2 and R_5. Even when R_5 is relatively small, *cis* enolates are *erythro* selective. As R_5 becomes larger, the *erythro* selectivity becomes even larger. However, for *trans* enolates, both the $R_1:R_4$ and $R_2:R_5$ interactions must be considered. With groups of moderate size (C_2H_5, i-C_3H_7) the $R_1:R_4$ interaction apparently just balances the $R_2:R_5$ interaction. Only when R_5 becomes very large do *trans* enolates become highly *threo* selective. This model is also consistent with the observation that *cis* enolate **2g** reacts seven to eight times faster than *trans* enolate **3g**.[13]

The chelated transition state also serves to explain the stereochemistry observed in reactions of boron enolates. For example, Masamune and co-workers have reported the condensations depicted in eqns. 14 and 15.[19] Note

$$\text{(14)} \qquad 75\% \qquad 25\%$$

$$\text{(15)} \qquad > 5\% \qquad > 95\%$$

that the same trends are manifest by these reactions; the *cis* enolate is highly *erythro* selective and the *trans* enolate is less selective. However, an important difference is also seen. The *cis* enolate in eqn. 15 is highly *erythro* selective even though it does not have a bulky group attached to the carbonyl. Similar observations have been made by Evans and co-workers.[20]

The observation that boron enolates are more stereoselective than are comparable lithium enolates suggests that the boron chelate is "tighter" than that for lithium, thus maximizing steric effects. Indeed, the hypothesized six-center transition state in the case of lithium may not involve covalent bonds at all. Evidence that this may be so comes from several experiments carried out at Berkeley. First, it was found that the same kinetic stereoselectivity is observed in condensations carried out in the presence of four equivalents of HMPT as is seen in pure THF.[17] Second, potassium enolates have been found to display the same kinetic stereoselectivity as their lithium counterparts (although the resulting potassium aldolates equilibrate *much* faster).[17] Finally, even tetraalkylammonium enolates show the same kinetic stereoselectivity.[17] These observations clearly indicate that, if the postulated six-center transition state intervenes in the aldol condensation, ligand association of the two oxygens with the cation is not required. However, even a Coulombic attraction between these oxygens (both of which bear a fractional negative charge in the transition state) and the cation is probably sufficient to organize the transition state so that both oxygens point more or less in the same direction, as indicated in Figures II-IV.

The model put forth in Figure IV appears to be applicable only when the group attached to the enolate double bond (R_3 or R_4) is relatively small. Jeffery, Meisters and Mole have examined the reactions of the *cis* and *trans* dimethylaluminum enolates of methyl neopentyl ketone with acetaldehyde and benzaldehyde.[21] With acetaldehyde, the *cis* enolate gives a *threo* aldol (eqn. 16) and the *trans* enolate gives an *erythro* aldol (eqn. 17), just the opposite of what

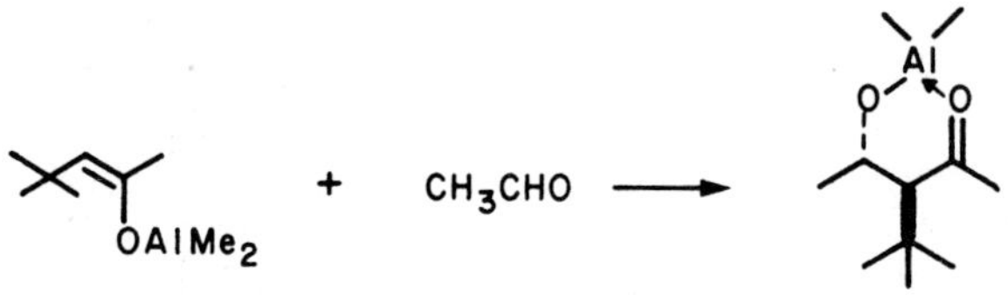

$$\text{(16)}$$

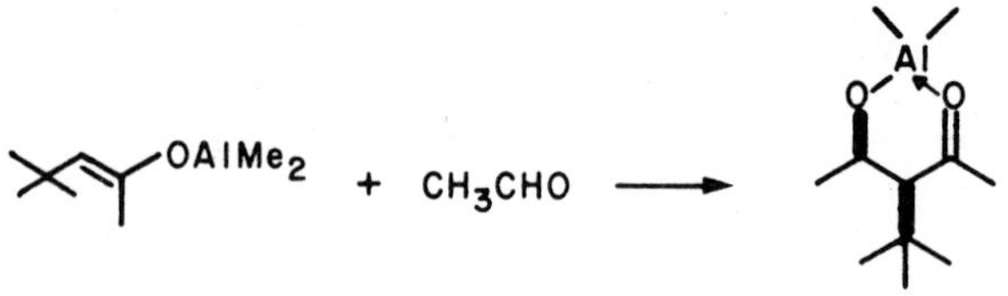

$$\text{(17)}$$

we have seen with other enolates. Even more interesting is the observation that with benzaldehyde, *both* the *cis* and *trans* enolates give a *threo* aldol (eqn. 18). This reversal of the normal pattern must have its origin in the *t*-butyl

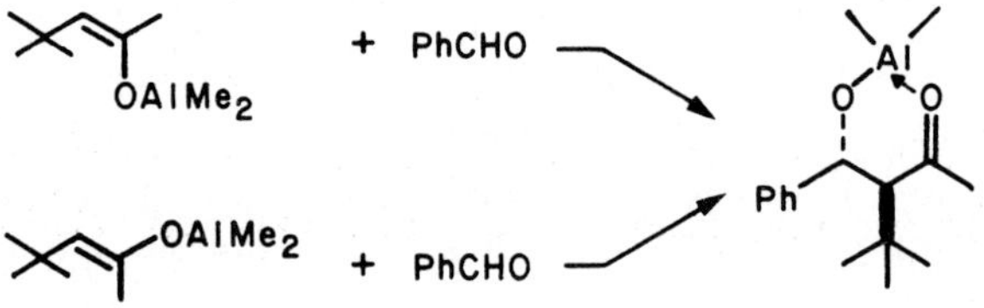

$$\text{(18)}$$

group. Jeffrey and co-workers invoke the transition states depicted in Figure V. It is argued that with benzaldehyde the *t*-butyl:phenyl interaction dominates and that both diastereomers of the enolate lead to the *threo* aldol. With acetaldehyde, it is postulated that *trans→erythro* is not so bad, and that the *erythro* aldol is formed through this transition state. However, it was not rigorously established that the products of these reactions are formed under kinetic control. These reactions are further complicated by the fact that the aluminum enolates are known to exist as dimers.

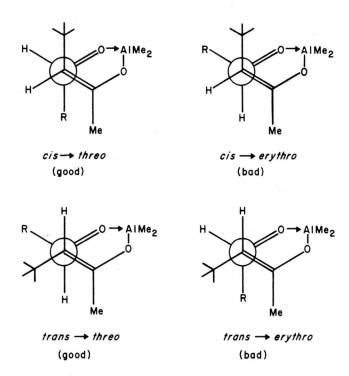

Figure V. Possible Transition States for Reaction of the Diethylaluminum Enolates of Methyl Neopentyl Ketone With Aldehydes

Heng and Smith have examined the reactions of the corresponding zinc enolates.[22,*] In this investigation, the enolate was generated by conjugate addi-

---

[*] Note that the stereostructure of the aldol obtained from eqn. 16 has been confirmed by single-crystal x-ray analysis.[23] Thus, the structures assigned to compounds **4-8** in ref. 22 should probably be reversed.

tion of a dialkylcuprate to mesityl oxide. One can make the reasonable assumption that this process gives the *cis* enolate. Aldol condensations were carried out by the House procedure (addition of anhydrous zinc chloride, followed by the aldehyde.)[14] In most cases, both *erythro* and *threo* aldols were obtained (eqn.19) with the *erythro:threo* ratio being different at -78° and at 0°. Some of

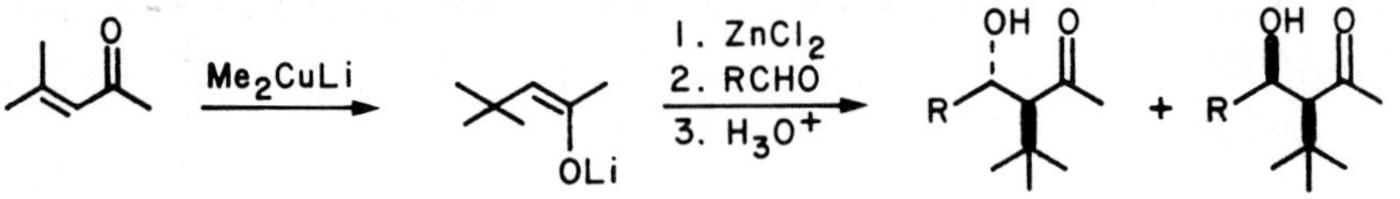

Smith's data are summarized in Table IV. It is not clear what to make of these results. The House procedure usually leads to thermodynamic control (*vide infra*). However, it would seem that Smith's data at -78° would reflect at least partial kinetic control, since substantially different ratios are obtained at 0°. However, more experiments would have to be carried out to establish this point. Nevertheless, a clear caveat emerges from the Jeffrey and Smith results; the Zimmerman-Dubois transition state hypothesis should be applied with caution in systems having bulky substituents on the enolate double bond.

Table IV. Condensation of the *Cis* Enolate of Methyl Neopentyl Ketone with Various Aldehydes (Eqn. 19)

R	Yield of Aldol, %	*erythro/threo* ratio, -78°	*erythro/threo* ratio, 0°
CH_3	96	—	only *threo*
CH_3CH_2	50	2.8	7.4
C_6H_5	72	2.0	1.0
p-$CH_3OC_6H_4$	46	1.8	0.2

Table I shows that both methyl propionate and *t*-butyl propionate give entirely the *trans* enolates, but that these enolates show poor kinetic stereoselectivity. However, the enolate derived from (2-methoxyethoxy)methyl propionate (which is presumably the *trans* isomer) is moderately *threo* selective. Meyers and Reider[24] observed this same phenomenon (eqn. 20, Table V) and

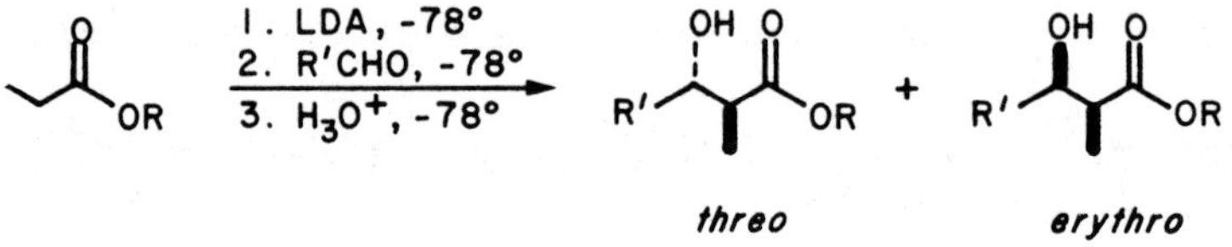

Table V. Kinetic Stereoselectivity of Aldol Condensations of Alkoxyalkyl Ester Enolates (Eqn. 20)

R	R'CHO	ratio	%threo	%erythro
Me	i-PrCHO	1.2:1	55	45
CH_2OMe	i-PrCHO	8.5:1	90	10
$CH_2O(CH_2)_2OMe$	i-PrCHO	10:1	91	9
Me	CH_3CHO	1.3:1	57	43
CH_2OMe	CH_3CHO	2:1	67	33
Me	PhCHO	1.2:1	55	45
$CH_2O(CH_2)_2OMe$	PhCHO	3:1	75	25

proposed a mechanism (Figure VI) in which the oxygens of the ester group chelate with the lithium, thus disrupting the six-centered transition state which has usually been presumed to be responsible for the stereoselection of the reac-

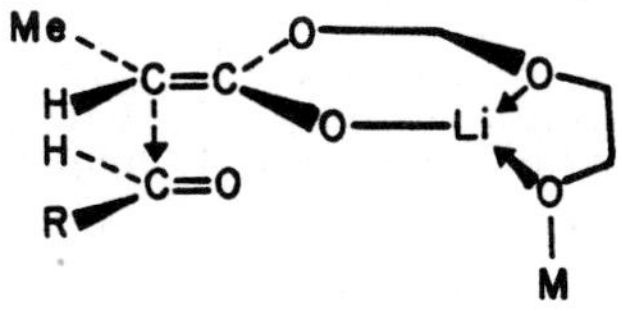

Figure VI. Meyers–Reider Proposed Transition State for the Aldol Condensation With (2–Methoxyethoxy)methyl Esters

tion. The transition state depicted in Figure VI would explain the observation that aldehydes with more bulky R groups show higher stereoselectivity than those with smaller R groups. However, it also predicts that esters of higher acids should show higher *threo* selectivity than the propionate ester. This prediction is not borne out, since the isovalerate ester shown in eqn. 21 reacts with benzaldehyde to give a 1:1 ratio of *erythro* and *threo* aldols.[17]

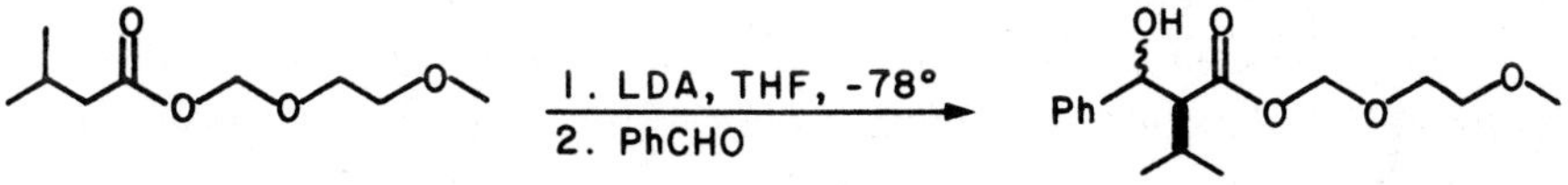

1. LDA, THF, -78°

2. PhCHO

(21)

8

erythro:threo = 1:1

In order to account for the effect of changing the metal cation on the stereoselectivity of addition of carboxylic acid dianions to aldehydes (Ivanov reaction), Mulzer has postulated a totally different mechanism.[25] The reaction upon which the Mulzer postulate is based is the condensation of phenylacetic dianions with pivaldehyde (eqn. 22). Relevant data are summarized in Table

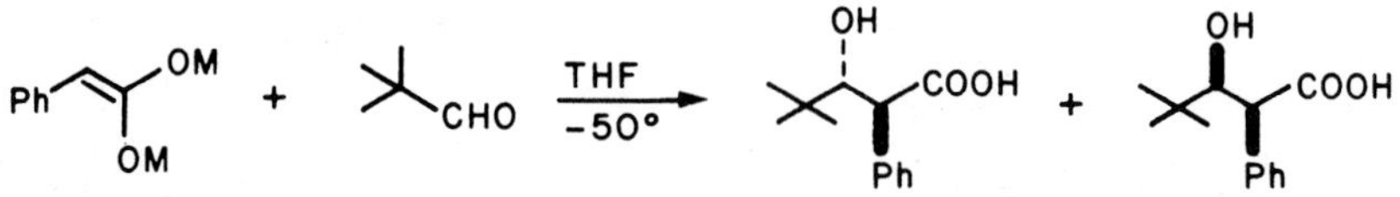

$$Ph\overset{OM}{\underset{OM}{\diagdown}} \;+\; \underset{CHO}{\diagup} \;\xrightarrow[-50°]{THF}\; \underset{Ph}{\overset{OH}{\diagup}}COOH \;+\; \underset{Ph}{\overset{OH}{\diagup}}COOH \qquad (22)$$

VI. The metals in the Table are arranged in order of decreasing charge:radius ratio. Mulzer argues that, if a six-center chelate is responsible for the stereoselectivity of the reaction, that the tighter transition state should be that involving metals with high charge:radius ratios and that these reactions should give rise to greater *threo/erythro* ratios. However, the opposite trend seems to emerge from the data in Table VI. To account for this dilemma, Mulzer pro-

Table VI. Stereochemistry of the Condensation of Phenylacetic Acid Dianions with Pivaldehyde (Eqn. 22)	
M	*threo:erythro*
$Mg_{1/2}$	58:42
$Zn_{1/2}$	55:45
Li	70:30
Li/cryptofix 2.1.1	70:30
Na	79:21
Na/crytpofix 2.2.1	82:18
K	79:21
$n\text{-}Bu_4N$	85:15
K/cryptofix 2.2.2	90:10
K/18-crown-6	>97:3

posed that approach of the aldehyde and enolate are dominated by a HOMO (enolate):LUMO(aldehyde) interaction, as shown in Figure VII. This would require that the reactants approach one another face-to-face, with the metal ions being in the plane of the enolate system. The transition state is postulated to be like that of a cycloaddition, since the HOMO of the enolate is like that of

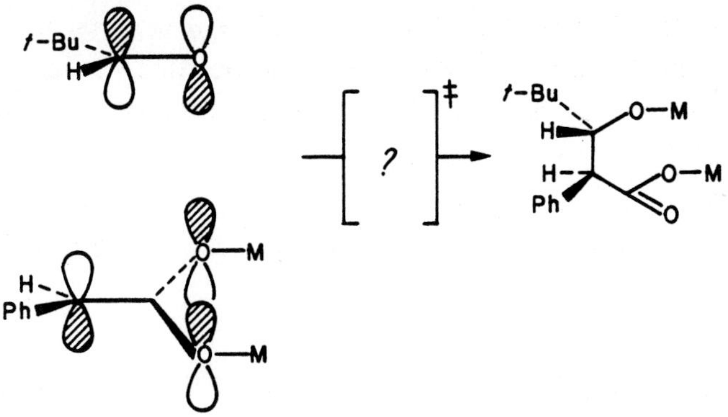

Figure **XII**. Mulzer Mechanism

allyl anion. However, cycloaddition would give rise to a highly unstable product (eqn. 23), which is estimated to be over 68 kcal mole^{-1} less stable than the

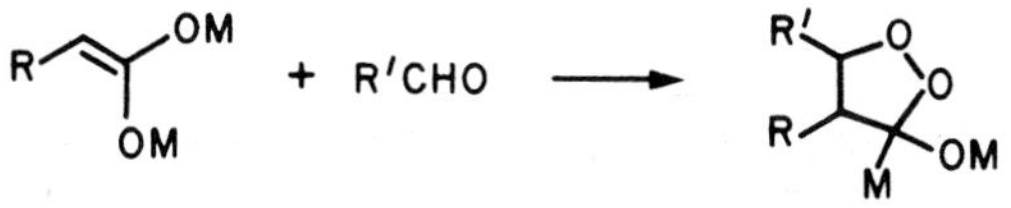

$$\text{(23)}$$

actual product of the condensation. Hence, it is speculated that the
HOMO:LUMO interaction of Figure VII only serves to orient the reactants in
their approach to one another and that, at least in the later stages of the reaction, a conventional metal-chelated arrangement obtains. The exact nature of
the transition state in this proposal is unclear.

One additional mechanism has been suggested to the author by Professor
Phillip Stotter of the University of Texas at San Antonio. Professor Stotter
notes that enolates often react with electrophiles first on oxygen. He suggests
that a possible mechanism is addition of the enolate oxygen to the carbonyl
group to give a hemiacetal anion which then undergoes an "alkoxide-promoted
[1,3]sigmatropic rearrangement" to give the aldolate. An appealing aspect of
the Stotter hypothesis is that it would appear to explain why *cis* enolates are
more stereoselective than *trans* enolates.

Noyori and coworkers, at Nagoya University, have recently reported
interesting results which bear on the question of whether the aldol condensation occurs through a chelated transition state, as in the Zimmerman-Dubois
model, or an "open" one. It was found that treatment of a trimethylsilyl enol

ether with tris(diethylamino)sulfonium (TAS) diflurotrimethylsiliconate generates the TAS enolate (e.g. **11**).[26] It was first established that the aldol conden-

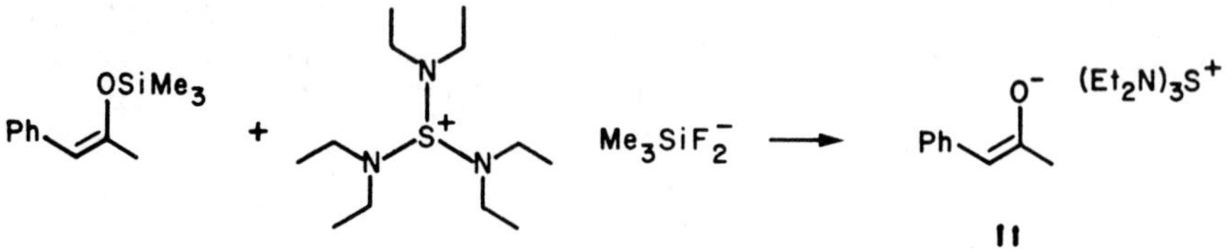

sation is thermodynamically unfavorable in the absence of a cation which can be chelated by the product aldoloxide. Thus, the equilibrium **11** + benzaldehyde ↔ **12** lies far to the left. However, addition of trimethylsilyl fluoride captures the aldolate and drives the condensation to completion.

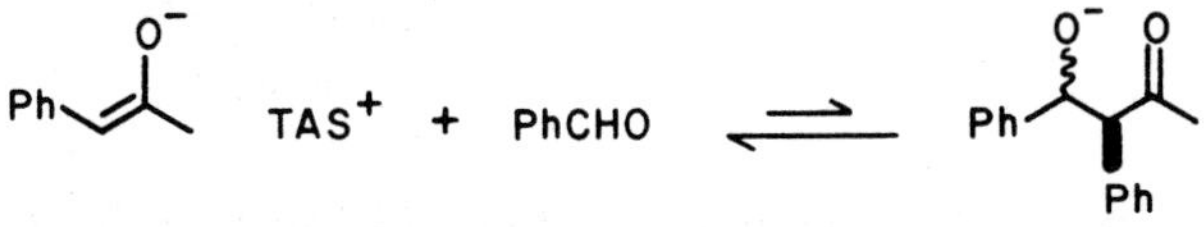

Furthermore, it was found that condensation of TAS enolates with various aldehydes gives predominantly *erythro* aldols, regardless of the geometry of the enolate.[27] Thus the diastereomeric silyl enol ethers derived from ethyl mesityl ketone both react with benzaldehyde to give the *erythro* aldol (eqn. 24).

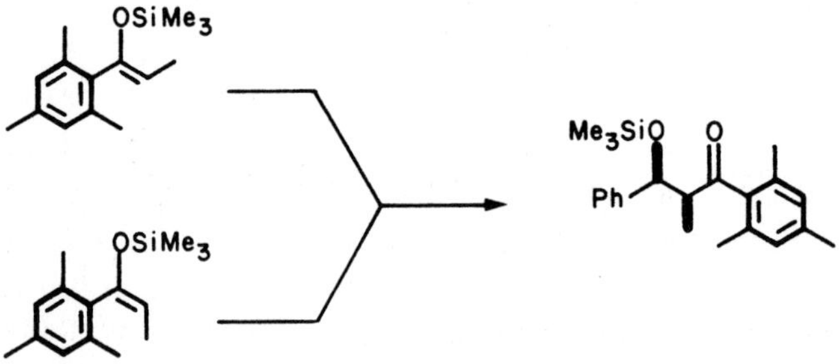

(24)

The TAS enolate results are interpreted by Noyori and coworkers in terms of an "open" transition state (Figure VIII) in which the two partially negative

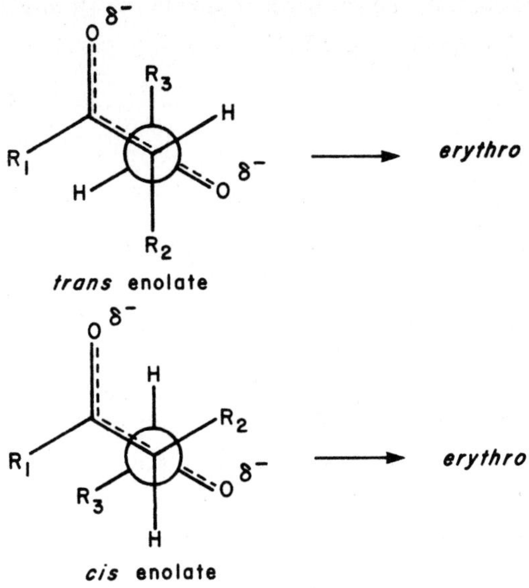

Figure VIII. Noyori Open Transition States

oxygens are as far apart as possible. As suggested in Figure VIII, the important interaction is considered to be that between R_2 and R_3. Thus, both the *trans* and *cis* enolates lead mainly to the *erythro* aldol.

IV. Equilibration: Thermodynamic Stereoselection

Aldols and their metal salts may undergo *erythro-threo* equilibration by reverse aldolization or by enolization. When the aldol condensation is carried out by addition of an aldehyde to a preformed enolate, the principal mechanism for *erythro-threo* equilibration seems to be by reverse aldolization (eqn. 25).

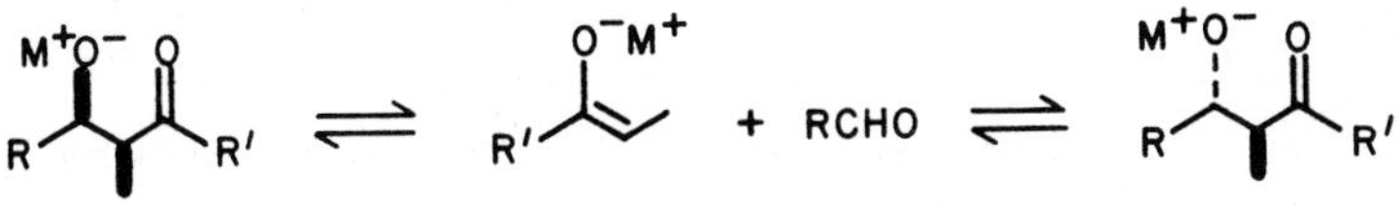

The literature permits a number of generalizations regarding this mechanism for *erythro-threo* equilibrations. First, *erythro-threo* equilibration may be *much slower* than reverse aldolization, if the enolate is highly stereoselective. For example, the *cis* enolate of ethyl *t*-butyl ketone has been found to show *erythro-threo* selectivity of 80:1 with benzaldehyde.[17] Therefore, reversal of the *erythro* aldol must occur, on the average, 80 times before one *erythro* molecule is converted into one *threo* molecule. This is demonstrated by the equilibrations shown in eqns. 26 and 27.[17]

$$p\text{-MeOC}_6\text{H}_4 \underset{\text{Li}^+\text{O}^-}{\overset{\text{O}}{\bigwedge\!\!\bigvee}} + \text{PhCHO} \;\underset{}{\overset{\text{THF, }0°}{\rightleftharpoons}}\; Ph \underset{\text{Li}^+\text{O}^-}{\overset{\text{O}}{\bigwedge\!\!\bigvee}} + p\text{-MeOC}_6\text{H}_4\text{CHO} \qquad (26)$$

$$t_{1/2} = 15 \text{ min}$$

$$Ph \underset{\text{Li}^+\text{O}^-}{\overset{\text{O}}{\bigwedge\!\!\bigvee}} \;\underset{}{\overset{\text{ether, }25°}{\rightleftharpoons}}\; Ph \underset{\text{Li}^+\text{O}^-}{\overset{\text{O}}{\bigwedge\!\!\bigvee}} \qquad (27)$$

$$t_{1/2} = 8 \text{ hrs}$$

A second generalization which may be drawn is that the rate of equilibration is highly solvent dependent. For example, it has been found that *erythro-threo* equilibration of lithium aldolates is much faster in pentane than in THF.[28] Thus, the *erythro* lithium aldolates depicted in eqn. 28 all equilibrate to their *threo* counterparts in less than two hours at 25° in pentane. Equilibrations are

$$Ph \underset{\underset{R}{\text{Li}^+\text{O}^-}}{\overset{\text{O}}{\bigwedge\!\!\bigvee}} \;\rightleftharpoons\; Ph \underset{\underset{R}{\text{Li}^+\text{O}^-}}{\overset{\text{O}}{\bigwedge\!\!\bigvee}} \qquad (28)$$

$$R = \text{Et, } n\text{-Pr, } n\text{-Bu}$$

much slower in THF or ether as shown by eqn. 27. This behavior is understandable in terms of the simple energy diagram depicted in Figure IX. In both the reactants and products of the aldol condensation, the negative charge is localized on one oxygen, but in the transition state it is shared between two oxygens. A polar solvent is expected to stabilize the reactants and products more than the transition state, and hence increase the activation energy for reaction.* Thus, both aldol condensation and aldol reversal should be more rapid in a nonpolar solvent than in a polar one.

---

* In Figure IX the aldol condensation is shown to be endothermic. This is based on the results of recent experiments by Noyori and co-workers which were discussed in the previous section.

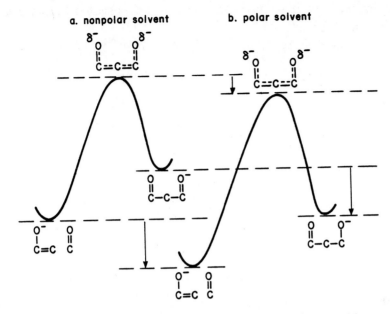

Figure IX. The Effect of Solvent Polarity on the Rate of the Aldol Condensation

Another generalization which may be drawn is that the rate of reverse aldolization (and hence of *erythro-threo* equilibration) depends upon the cation which is associated with the aldolate. Boron aldolates are very stable and do not undergo equilibration even at elevated temperatures. For example, compound **13** is unchanged by refluxing in ether for several hours, even in the presence of DBU. For the alkali metal aldolates, there is evidence that lithium

13   14

aldolates equilibrate most slowly and potassium aldolates most rapidly. For example, the *threo* potassium aldolate derived from the reaction of ethyl mesityl ketone and benzaldehyde undergoes equilibration at a substantial rate even at -78° while the corresponding lithium aldolate isomerizes with a half life of 5 min at 0° (eqn. 29).[17] The rate of equilibration is further accelerated by the addition of 18-crown-6.[29]

$$\underset{\text{Ph}}{\overset{M^+O^-\quad O}{\diagdown}} \quad \rightleftharpoons \quad \underset{\text{Ph}}{\overset{M^+O^-\quad O}{\diagdown}} \tag{29}$$

Aldol condensations may be carried out by treating a mixture of the trimethylsilyl enol ether of a ketone and an aldehyde with a tetraalkylammonium fluoride (eqn. 30).[30,17] The intermediate in this case is presumably a

$$\underset{\text{OSiMe}_3}{\diagdown} \quad + \quad \text{PhCHO} \quad \xrightarrow[\text{THF, } -78°]{(n-\text{Bu})_4\text{N}^+\text{F}^-} \quad \underset{\text{Ph}}{\overset{\text{Me}_3\text{SiO}\quad O}{\diagdown}} \tag{30}$$

15

"naked enolate", in which the counterion is a tetraalkylammonium ion. Under these conditions, *erythro-threo* equilibration has been found to be more rapid than with a lithium cation. For example, the silylated aldol **15** rearranges to its *threo* counterpart almost instantaneously at room temperature in the presence of tetrabutylammonium fluoride in THF.[17]

$$\underset{\text{Ph}}{\overset{\text{Me}_3\text{SiO}\quad O}{\diagdown}} \quad \xrightarrow[\text{THF, } -25°]{(n-\text{Bu})_4\text{N}^+\text{F}^-} \quad \underset{\text{Ph}}{\overset{\text{Me}_3\text{SiO}\quad O}{\diagdown}}$$

Erythro-threo equilibration has also been demonstrated with magnesium[12,14] and zinc[14,17] aldolates. The situation with regard to *rate* of equilibration is somewhat confusing. In some cases zinc aldolates equilibrate more rapidly than their lithium counterparts and in some cases less rapidly. For example, in the ethyl mesityl ketone-benzaldehyde system (eqn. 29) the lithium aldolate equilibrates with a half life at 0° of 5 min while the zinc aldolate equilibrates with a half life of about 30 sec at -78°.[17] On the other hand, for the propiophenone-benzaldehyde system (eqn. 31) just the reverse behavior is seen. The lithium aldolates equilibrate with a half life of about 1 min at -60° while the zinc aldolates equilibrate with a half life of approximately 4 min at -10°.

$$\underset{\text{Ph}}{\overset{M^+O^-\quad O}{\diagdown}}\text{Ph} \quad \rightleftharpoons \quad \underset{\text{Ph}}{\overset{M^+O^-\quad O}{\diagdown}}\text{Ph} \tag{31}$$

A final generalization regarding the rate of *erythro-threo* equilibration by reverse aldolization may be made. The rate increases with decreasing basicity of the enolate and with increasing steric repulsion in the aldolate. For example, aldolates derived from ketone enolates are more prone to undergo equilibration than aldolates derived from esters, amides, or carboxylate salts. The reason for this observation is intuitive, since reverse aldolization is more exothermic when a less basic enolate is released. This is probably why aldolates derived from propiophenone are so much more prone to *erythro-threo* equilibration than are aldolates derived from alkyl ketones.[17] Aldols which are substituted by sterically-demanding groups at the α-carbon seem to equilibrate more rapidly than normal, presumably because of relief of steric repulsion. For example, Mulzer found that aldolates **17** and **18** undergo equilibration in THF at 25° but that aldolates **19** and **20** do not equilibrate under these conditions.[31]

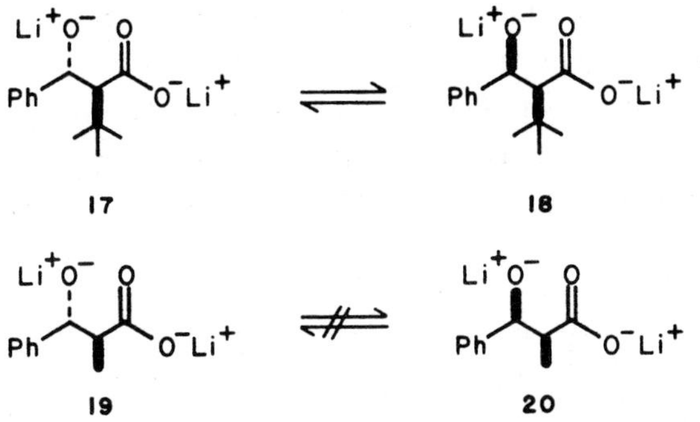

It has often been assumed that *threo* aldols predominate at equilibrium. This generalization stems from a report by Dubois and Fellman, who found that the magnesium aldolate derived from the magnesium enolate of ethyl *t*-butyl ketone and benzaldehye equilibrates to the *threo* diastereomer when kept for several hours in ether at room temperature (eqn. 32).[12] Similar results have

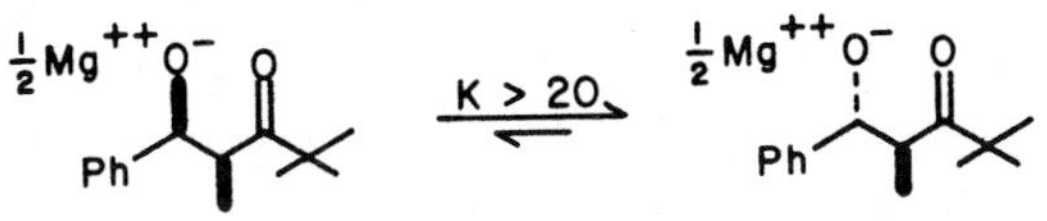

$$(32)$$

been found in the lithium and zinc salts of this aldolate.[17] However, the result is not general. For example, the lithium aldolates derived from benzaldehyde and either diethyl ketone or ethyl isopropyl ketone give a *threo:erythro* of only 56:44 at equilibrium (eqn. 33).[17] Furthermore, the lithium aldolates derived from benzaldehyde and ethyl mesityl ketone (eqn. 29) give an *erythro:threo* ratio of 77:23 at equilibrium.[17]

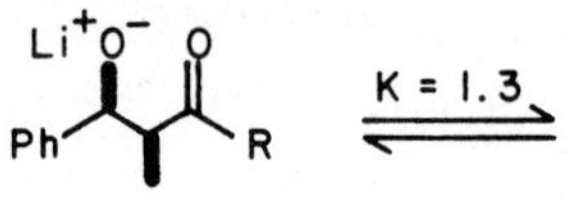

$$\text{(33)}$$

R = Et, *i*-Pr

The selectivity observed in the condensation of zinc enolates (House method)[14] is usually ascribed to thermodynamic control. Indeed, experiments performed at Berkeley indicate that these reactions *are* often under thermodynamic control, and that the presence of a zinc rather than a lithium cation often changes the equilibrium to favor the *threo* diastereomer.[17] For example, the ratio for the equilibrium depicted in eqn. 29 is *threo:erythro*=10:1 when $M^+=\frac{1}{2}Zn^{++}$ and *erythro:threo*=3:1 when $M^+=Li^+$.[17] The equilibrium ratios for the reaction shown in eqn. 31 are *threo:erythro*=1.1 for lithium and 3 for zinc.[17] It has been proposed that thermodynamic *threo*-selectivity results because there are fewer non-bonded interactions present in the chelated *threo* aldolate than in the chelated *erythro* aldolate (Figure X).[15,17] This is a reasonable argument and would explain why zinc aldolates display greater *threo* thermodynamic selectivity than do lithium aldolates.

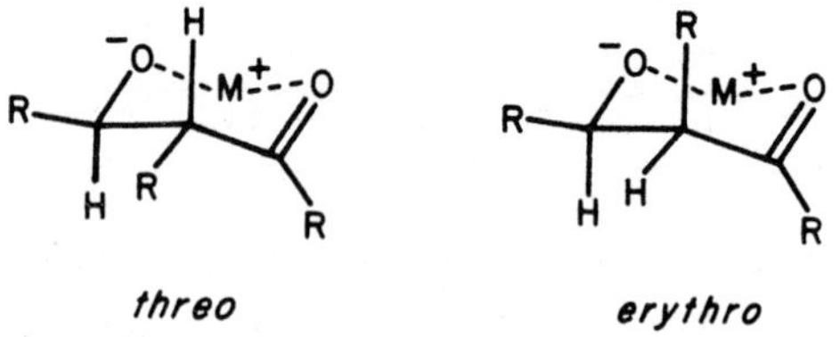

threo   *erythro*

Figure X. Chelated *Threo* and *Erythro* Aldolates

Mulzer and co-workers have studied the condensation of carboxylic acid dianions with aldehydes (eqn. 34).[31] It was found for a number of cases that the kinetic products (THF, -50°, 10 min) change to an equilibrium ratio with time (THF, 25°, 1-3 days). Selected data are summarized in Table VII.

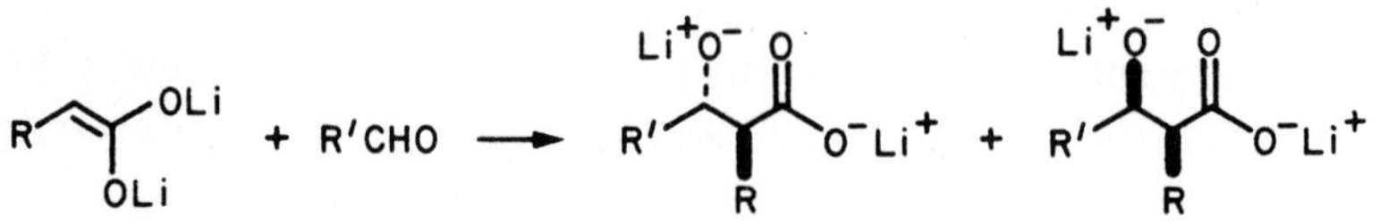

$$\text{(34)}$$

Table VII. Reported Stereoselectivity for Reaction of Carboxylic Acid Dianions with Aldehydes (Eqn. 34)

R	R	threo/erythro (kinetic)	threo/erythro (thermodynamic)
t-Bu	i-Pr	3.8	9.0
t-Bu	t-Bu	4.0	49.0
t-Bu	Ph	1.5	7.3
Ph	Ph	1.5	7.3
Ph	t-Bu	1.9	49.0
Ph	i-Pr	1.9	4.0

E. Cations other than Lithium

The majority of the work which has been done on the use of preformed enolates has involved lithium enolates. However, considerable effort has been expended on enolates associated with other cations, particularly boron and zinc. In addition, there have been studies involving aluminum, zirconium, and tetraalkylammonium enolates. Finally, some stereochemical data has emerged from a study of the titanium tetrachloride promoted condensation of O-trimethylsilyl ketone acetals with aldehydes.

As has been mentioned previously in this article, boron enolates offer an attractive advantage relative to lithium enolates, since both *cis* and *trans* enolates exhibit higher kinetic stereoselectivity than their lithium counterparts. Representative data are tabulated in Table VIII.[20b]

The major problem with the use of boron enolates has been in finding methods for the stereoselective generation of *cis* and *trans* enolates so that advantage can be taken of this high stereoselectivity. This problem has been partly solved, principally through the efforts of Masamune at MIT[19,32] and Evans at CalTech.[20] Masamune found that the boron enolates produced in the Hooz reaction[33] are almost exclusively the *trans* isomers and that these can be quantitatively isomerized to the *cis* isomers by treatment with lithium phenoxide or pyridine in benzene (e.g., eqn. 35).[19] A disadvantage of this method is

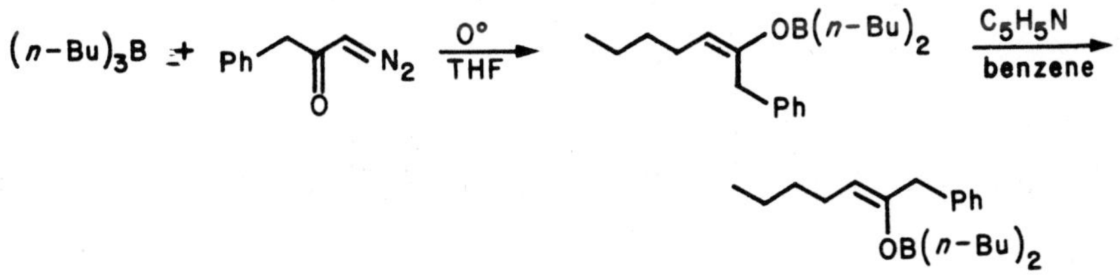

(35)

Table VIII. Influence of Cation of Kinetic Aldol Reactions with Benzaldehyde

Enolate	Metal (M)	*Erythro:Threo*
Me_3C — OM (enolate)	Li	>98:2
	MgBr	>97:3
	$B(C_{14}H_9)_2$	>97:3
Ph — OM (enolate)	Li	88:12
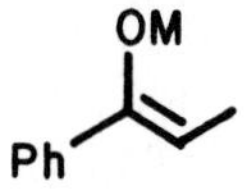	$B(C_4H_9)_2$	>97:3
Et — OM (enolate)	Li	80:20
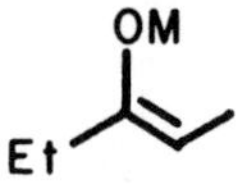	$B(C_4H_9)_2$	>97:3
t-BuS — OM (enolate)	Li	60:40
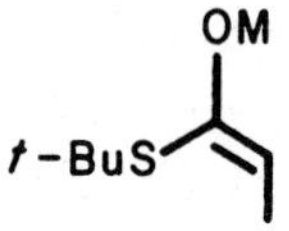	$B(C_4H_9)_2$	5:95
cyclohexenyl OM (enolate)	Li	48:52
	$Al(Et)_2$	50:50
	$B(C_5H_9)C_6H_{13}$	4:96

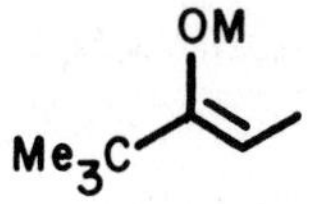

the necessity of the requisite borane and the fact that only one of the three boron alkyl groups is utilized. However, the use of other boranes (thexyl, 9-BBN, etc.) would presumably solve these problems.

Both Masamune[32] and Evans[20] have shown that Mukaiyama's method for preparing boron enolates[34] may be used for stereoselective formation of *cis* enolates. For example, treatment of S-*t*-butyl propanethioate with dicyclo-pentylboron triflate in the presence of diisopropylethylamine gives cleanly the *trans*-enolate (eqn. 36).[32,20] Evans has carried out an extensive study of this

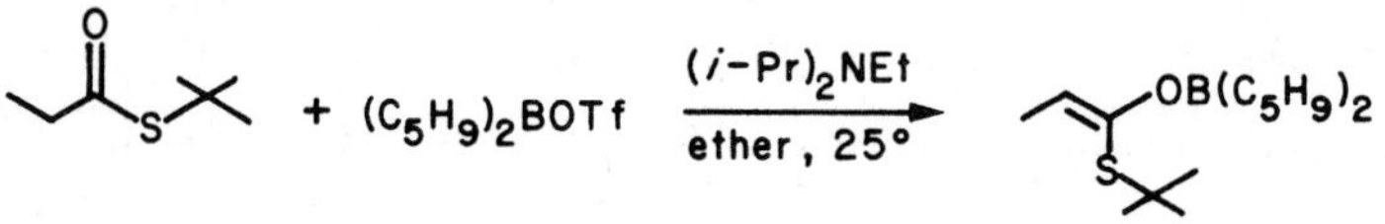

$$\text{(propanethioate)} \quad + \quad (C_5H_9)_2BOTf \quad \xrightarrow[\text{ether, 25°}]{(i\text{-Pr})_2NEt} \quad \text{(enolate)}—OB(C_5H_9)_2 \qquad (36)$$

21

procedure, studying the effect of boron ligand, base, solvent, and the group attached to the carbonyl.[20b] A number of generalizations may be drawn from this study. First, it is important to use a hindered amine; diisopropylethylamine gives much better stereoselectivity than lutidine. Second, the size of the boron ligand is important. For example, di-*n*-butylboron triflate reacts with diethyl ketones to give solely the *cis* enolates whereas dicyclopentylboron triflate gives a 4:1 mixture of *cis* and *trans* isomers (eqn. 37). Interpretation of this trend is

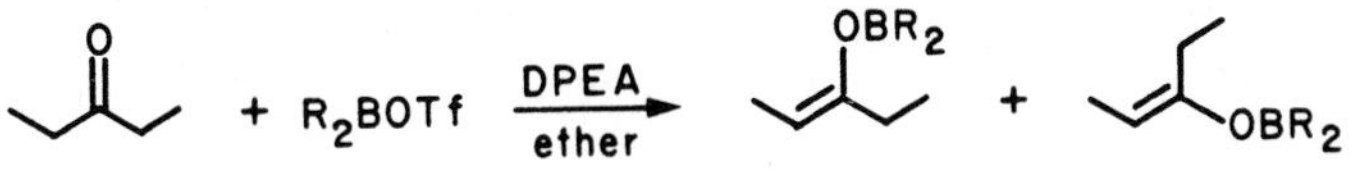

$$
\text{(image of eqn. 37)} \tag{37}
$$

R = *n*-C₄H₉: > 97   > 3
R = C₅H₉: 82   18

complicated by the fact that the more hindered triflate $(C_5H_9)_2BOTf$ is less reactive and must be used at 25°. Finally, with either triflate, the ketones studied exhibit a *decrease* in *cis* enolate formation with *increasing* steric bulk of the group attached to the carbonyl. Data for enolization of various carbonyl compounds with di-*n*-butylboron triflate are collected in Table IX.[20b] As the Table

Table IX. Kinetic Enolate Formation with $(n\text{-Bu})_2BOTf^a$		
Substrate	Conditions	*cis/trans*
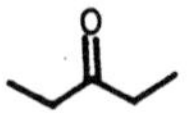 (structure)	-78°, 30 min	>97/3
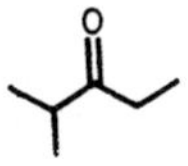 (structure)	-78°, 30 min	45/55
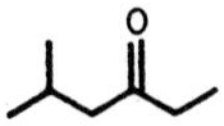 (structure)	-78°, 30 min	>97/3
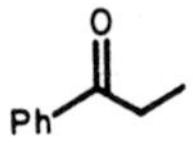 (structure)	+25°, 1 hr	>97/3
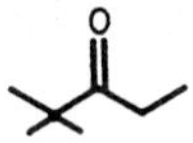 (structure)	+35°, 2 hr	>97/3
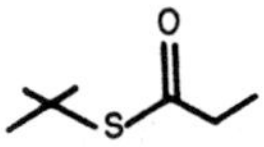 (structure)	0°, 30 min	<5/95

a. In each case the less substituted enolate is formed.

indicates, the method is most useful for the formation of *cis* enolates; the only substrate which gives the *trans* enolate selectively is S-*t*-butyl propanethioate. A major disadvantage with this procedure is that it fails with esters and amides.

Masamune also explored the addition of *t*-butyl di-*n*-butylthioborinate to methyl ketene and found that the *cis* boron enolate is produced with good stereochemical purity (eqn. 38).[32a] However, a more convenient method for

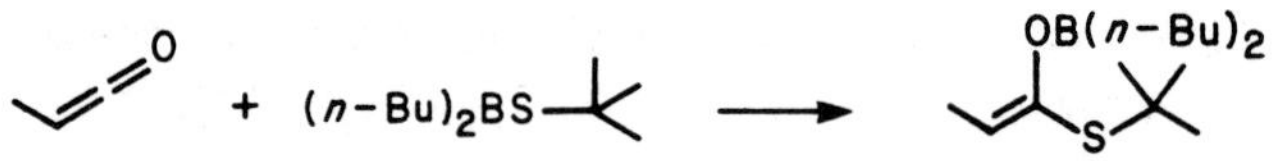

$$\text{(equation 38)} \tag{38}$$

generating the *cis* enolate turns out to be reaction of S-phenyl propanethioate with 9-borabicyclo[3.3.1]non-9-yl trifluoromethanesulfonate (9-BBN triflate) and diisopropylethylamine (eqn. 39).[32b] Reagents **21** (eqn. 36) and **22** (eqn. 39) are useful for *threo*-selective and *erythro*-selective condensations, respectively.

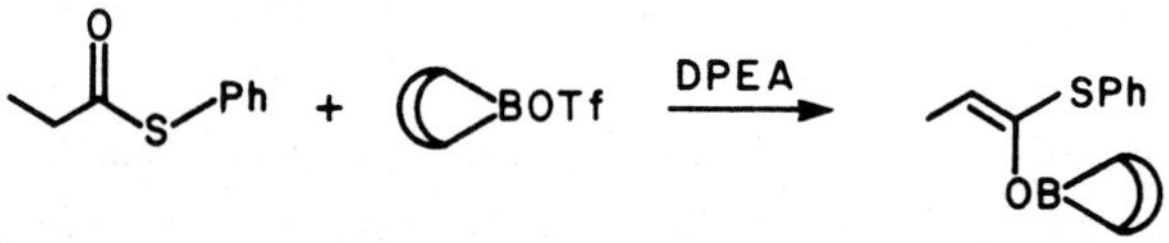

$$\text{(equation 39)} \tag{39}$$

Not much work has been done on the stereoselectivity of aldol condensations involving aluminum enolates. The only reports are the previously discussed work of Jefferey, Meisters, and Mole,[21] and a brief report by Yamamoto and co-workers.[35] The latter work involves *in situ* generation of an aluminum enolate by treatment of an α-bromoketone or ester with a mixture of diethylaluminum chloride and zinc. Several examples are shown in eqns. 40-42. In fact, the stereoselectivity shown by these reactions is about what one

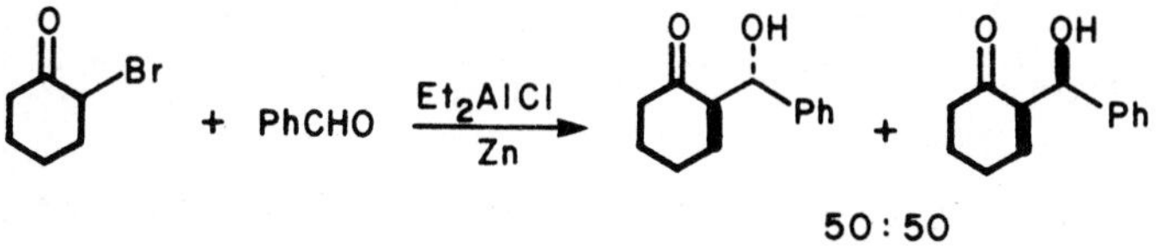

$$\text{(equation 40)} \tag{40}$$

50 : 50

$$\text{(equation 41)} \tag{41}$$

44 : 56

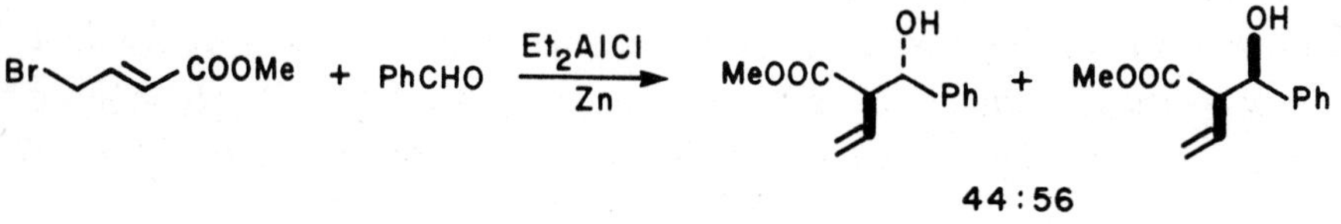

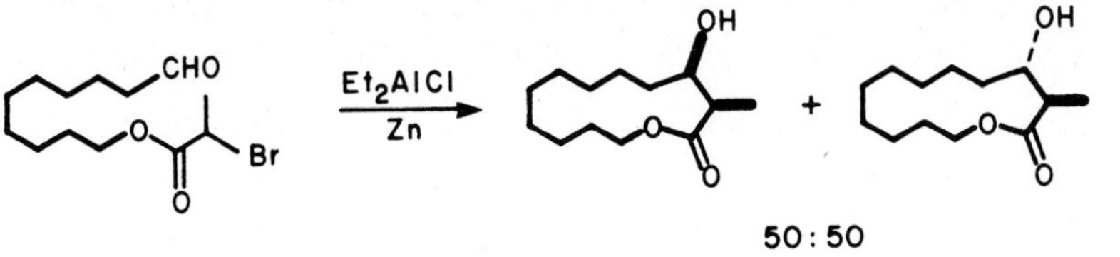

$$(42)$$

50 : 50

would expect of lithium enolates and certainly not what one would expect of boron enolates, especially in the case of eqn. 40 (see Table VIII).

The definitive work on zinc and magnesium enolates has been done by House.[14] As has been mentioned several times in this review, the zinc enolates seem to operate under thermodynamic control. Furthermore, it appears that in many cases the *threo* zinc aldolates are considerably more stable than the corresponding *erythro* isomers. Thus, the House method is an attractive method for preparation of *threo* aldols. However, when using this method, one is at the mercy of thermodynamics and stereoselectivity is often not very high.

Evans[36] and Yamamoto[37] have investigated zirconium enolates, which are prepared by addition of dicyclopentadienylzirconium dichloride (Cp_2ZrCl_2) to a solution of the lithium enolate in THF at -78°. Surprisingly, both *cis* and *trans* enolates are *erythro* selective. Data for the condensation of diethyl ketone with benzaldehyde (eqn. 43) are tabulated in Table X, along with results for the

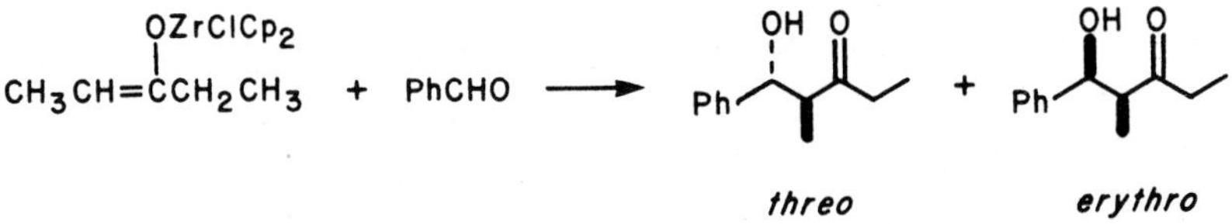

$$(43)$$

cyclohexanone and cyclopentanone enolates.[37] Note that the *trans* enolate of diethyl ketone appears to be highly *erythro* selective, but that the *cis* enolate is considerably less selective.

Evans and McGee have examined the reactions of the zirconium enolates of a number of compounds with benzaldehyde[36] and compared the results with those obtained from the corresponding lithium enolates (Table XI).[17] As is seen from the Table, the *erythro*-selectivity of the amide enolates is excellent, and it is with these compounds that the zirconium method will probably be most useful. Similar high *erythro* selectivity is seen in condensation of the zirconium amide enolates with several other aldehydes.[36]

Table X. Condensation of Zirconium Enolates with Benzaldehyde	
Enolate	*Erythro:Threo*
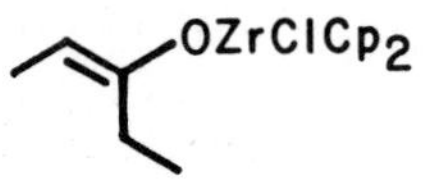 $OZrClCp_2$ (*trans:cis*=86:14)	88:12
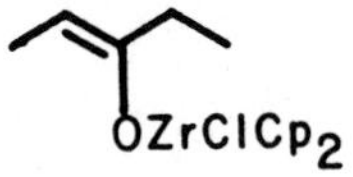 $OZrClCp_2$ (*trans:cis*=8:92)	67:33
$OZrClCp_2$	72:28
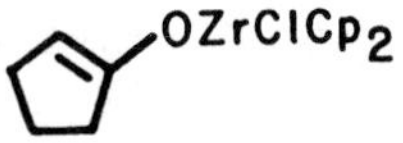 $OZrClCp_2$	74:26

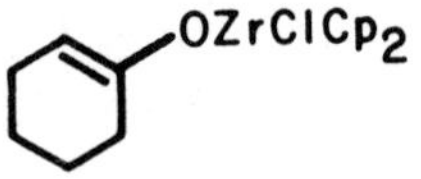

It is clear that different factors are at work in determining the stereo-chemistry of the zirconium enolate condensations than operate in the lithium and boron enolate reactions. Yamamoto[37] has speculated that the zirconium reaction may proceed through an "open" (non-chelated) transition state. How-ever, Evans has found that the bis(pentamethylcyclopentadienyl)zirconium enolate does not undergo the aldol condensation.[36b] This result is interpreted to mean that ligation of the aldehyde carbonyl onto the metal is an essential step in the process. Evans speculates that the coordinatively unsaturated zirconium enolate may add the aldehyde ligand to generate the reactive intermediate (Figure XI). There is reason to believe that the O-Zr-O bond angle in such a com-plex would be rather small (60-80°). Thus, the normal chair-like geometry of the Zimmerman-Dubois transition state would be greatly perturbed.

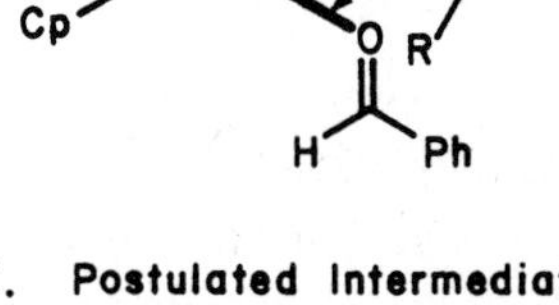

Figure XI. Postulated Intermediate in the Zirconium Enolate Condensation

Table XI. Condensation of Zirconium and Lithium Enolates with Benzaldehyde

Substrate	Enolate Ratio *cis:trans*	Product Ratios (M)	
		erythro:threo (lithium)	*erythro:threo* (zirconium)
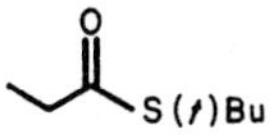 (image) S(*t*)Bu	10:90	63:37	93:7
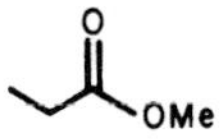 (image) OMe	5:95	52:48 (62:38)	87:13
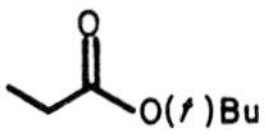 (image) O(*t*)Bu	5:95	37:63 (49:51)	72:28
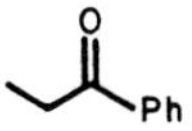 (image) Ph	>98:2	82:18 (88:12)	90:10
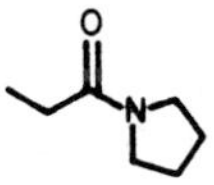 (image) pyrrolidine	>95:5	60:40	95:5
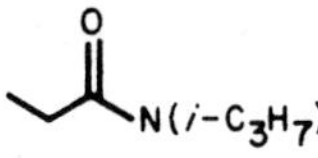 (image) N(*i*-C$_3$H$_7$)$_2$	81:19	61:39 (63:37)	>98:2

Chan and co-workers have investigated the stereochemistry of the titanium tetrachloride promoted condensation of silyl enol ethers with aldehydes.[38,39] The two diastereomeric trimethylsilyl enol ethers (ketene acetals) derived from ethyl propionate may be prepared as depicted in eqn. 44.

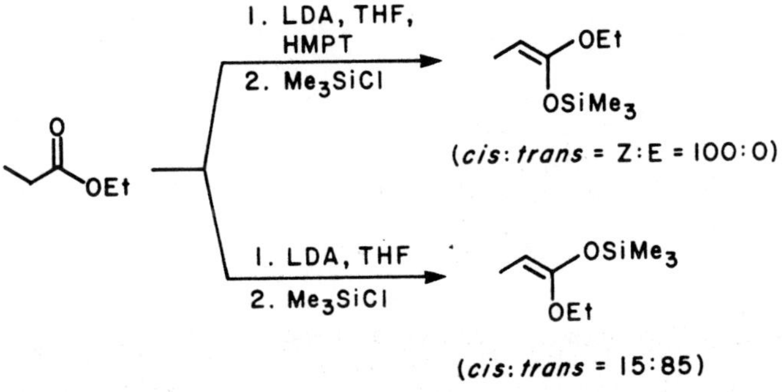

$$
\text{1. LDA, THF, HMPT} \quad \text{2. Me}_3\text{SiCl} \longrightarrow \text{OEt / OSiMe}_3 \quad (cis:trans = Z:E = 100:0)
$$

$$
\text{1. LDA, THF} \quad \text{2. Me}_3\text{SiCl} \longrightarrow \text{OSiMe}_3 / \text{OEt} \quad (cis:trans = 15:85)
$$

(44)

Condensation of the (E) ketene acetal with various aldehydes gives the *threo* aldol with good selectivity (eqn. 45). On the other hand, the (Z)-ketene acetal is non-selective (eqn. 46). The results have been explained by the conventional six-centered transition state.[39]

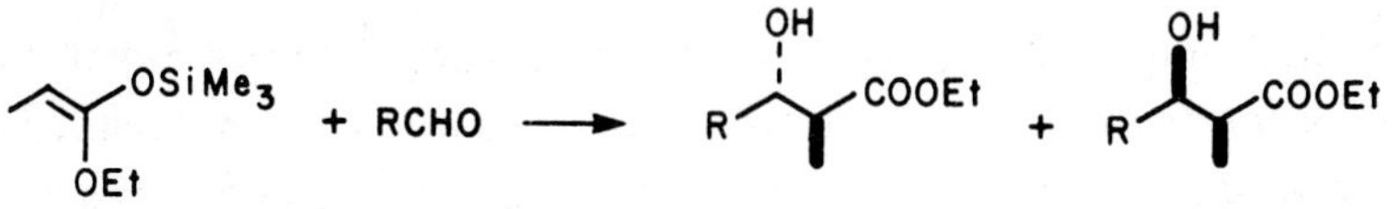

$$\text{(45)}$$

threo
(74-100%)

erythro
(0-26%)

$$\text{(46)}$$

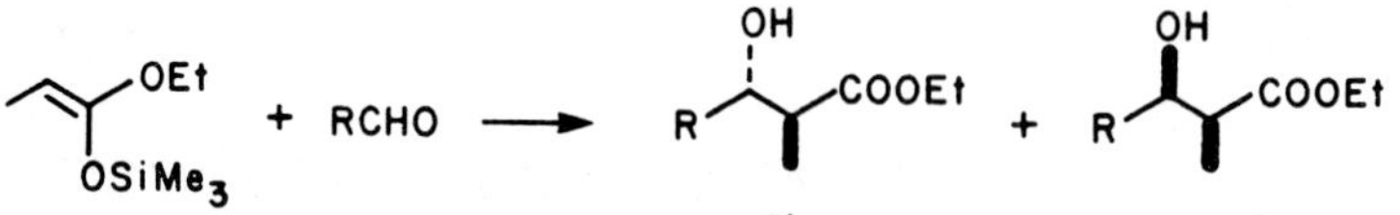

threo
(50-67%)

erythro
(33-50%)

VI. *Erythro*-Selective Synthetic Reagents

A useful *erythro*-selective propionic acid or propionaldehyde equivalent (23) has been developed at Berkeley.[17] Like ethyl *t*-butyl ketone, compound 23 gives solely the *cis* enolate, which reacts with a variety of aldehydes to give *erythro* aldols (e.g., eqn. 47). The resulting aldols may be cleaved by treatment

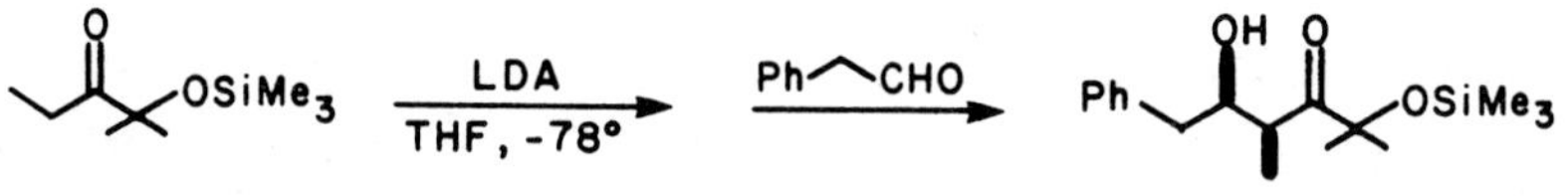

$$\text{(47)}$$

23

(74%)

with periodic acid in methanol to give the corresponding β-hydroxy acids (eqn. 48). Alternatively, the aldol may first be reduced ($LiAlH_4$ or $NaBH_4$) and the

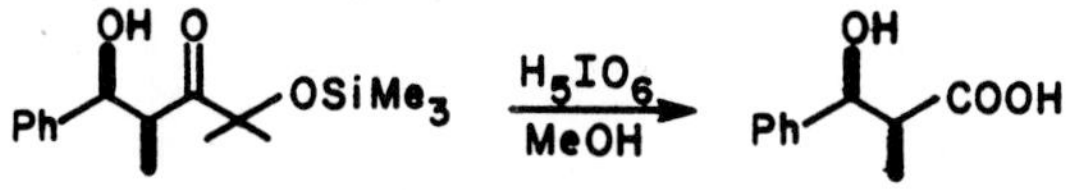

$$\text{(48)}$$

(77%)

resulting diol cleaved by buffered sodium periodate to obtain the β-hydroxy aldehyde (e.g., eqn. 49).[40] Finally, the aldols resulting from reagent 23 can be

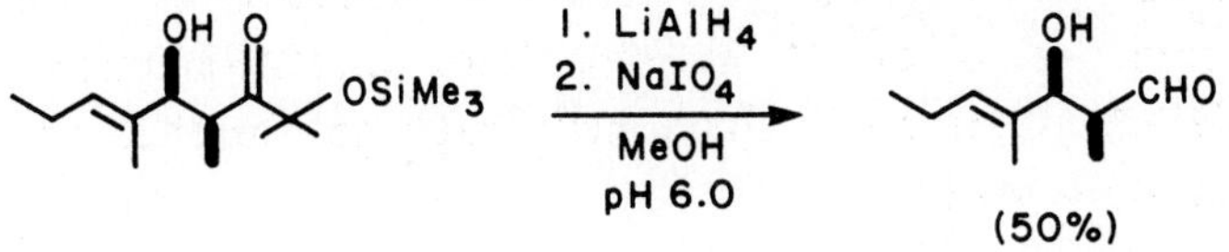

$$\text{(49)}$$

converted into aldols formally derived from other ketones.[41] Protection of the hydroxy group as the tetrahydropyranyl ether, followed by addition of an aryl- or alkyllithium reagent gives a product which can be cleaved by periodic acid to obtain the new aldol (eqn. 50).

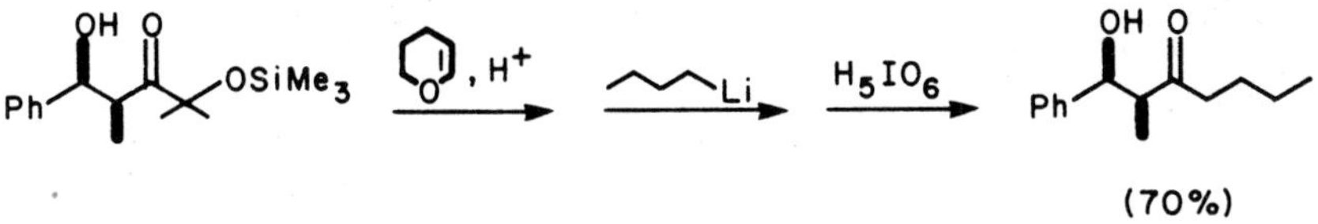

$$\text{(50)}$$

Analogs of compound **23** have also proven to be useful as *erythro*-selective reagents. An example is compound **24**, which has been used in the synthesis of **25**, an isomer of the naturally-occurring corynomycolic acid (eqn. 51).[28]

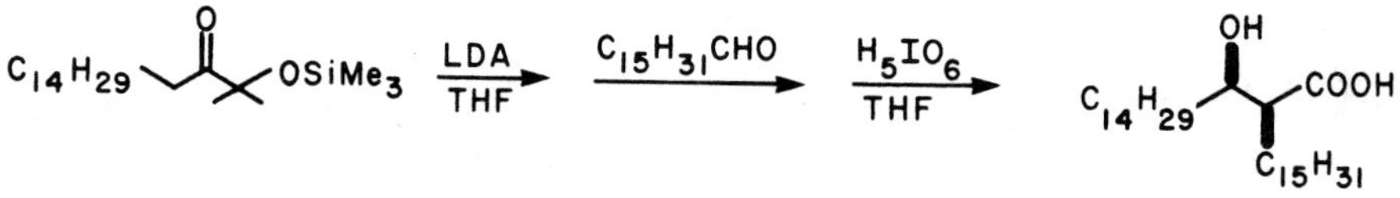

$$\text{(51)}$$

24   **25**

The only other generally useful type of reagent which is highly *erythro*-selective is the boron enolate of S-phenyl propanethioate (**22**) introduced by Masamune.[32b] This compound has been found to react with a variety of aldehydes to give *erythro* aldols of high (>97:3) stereochemical purity (e.g., eqn. 52).

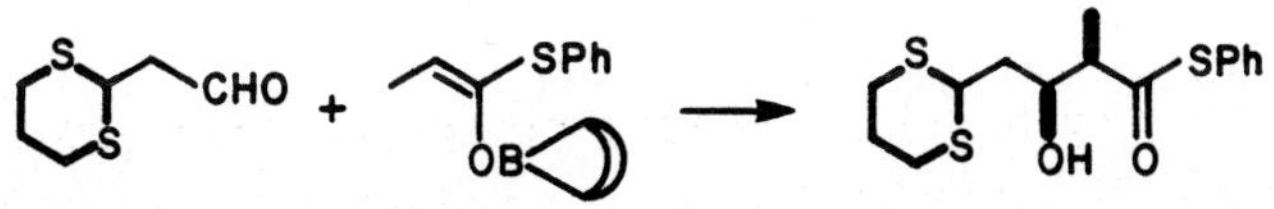

$$\text{(52)}$$

22

VII. *Threo*-Selective Synthetic Reagents

Several methods are available to achieve *threo*-selectivity, all involving enolates of esters or thiol esters. As shown in Tables I and II, esters give almost entirely the *trans* enolates. However, with normal esters, the kinetic stereoselectivity is poor. Investigations at Berkeley indicated that phenyl propionates are significantly more *threo* selective than simple alkyl propionates. For example, *p*-methoxyphenyl propionate condenses with benzaldehyde to give the *threo* and *erythro* aldols in a ratio of 80:20 (eqn. 53).[42] Reasoning that a

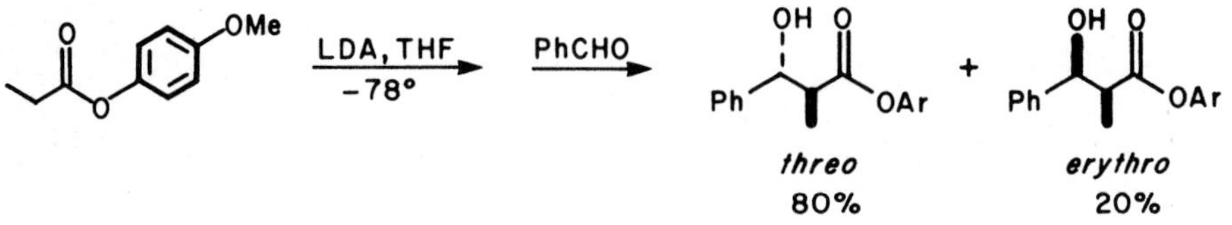

more bulky aryl group should result in even higher *threo* selectivity, the aryl propionates **26-28** were prepared and investigated.[43] The enolate of ester **26** reacts with various aldehydes to give predominantly the *threo* aldol (eqn. 54).

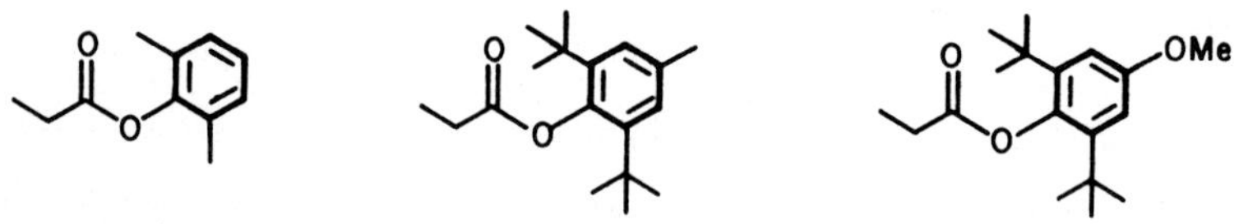

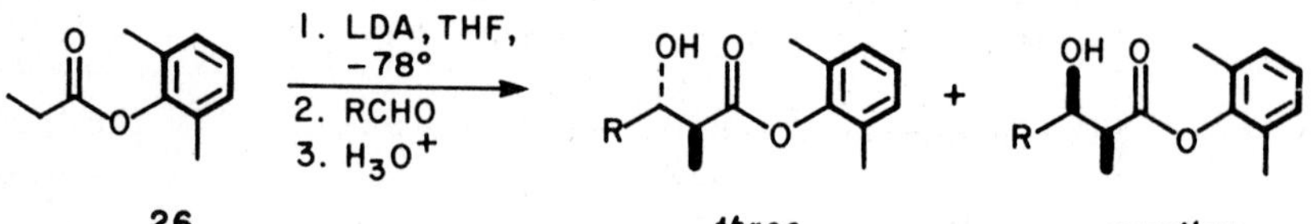

Data are summarized in Table XII. With benzaldehyde and hexanal ester **26** shows only modest selectivity. However, its selectivity is excellent with α-branched aldehydes. Recent investigations have shown that other 2,6-dimethylphenyl (DMP) esters are also stereoselective, even with simple aldehydes. Thus 2,6-dimethylphenyl 4-pentenoate reacts with several aldehydes to give only *threo* adducts (eqn. 55).[44]

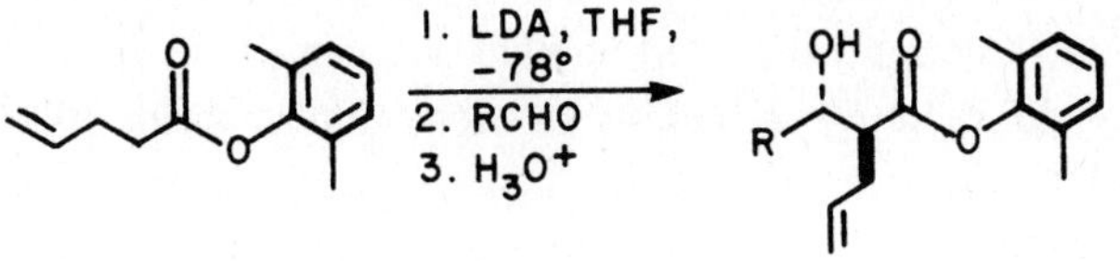

(55)

The more hindered esters **27** and **28** give only *threo* aldols with all aldehydes studied. For example, ester **27**, prepared from butylated hydroxytoluene (BHT), condenses with benzaldehyde to give only the *threo* product (eqn. 56), [44] In a similar fashion, ester **28**, prepared from dibutylated hydroxyanisole (DBHA), condenses with hexanal to give only the *threo* aldol (eqn. 57).[44]

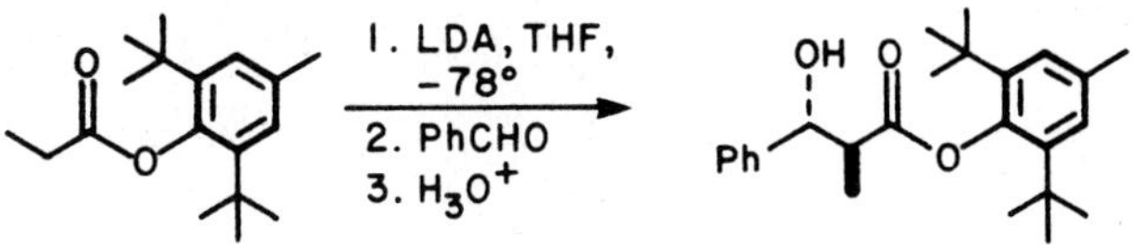

(56)

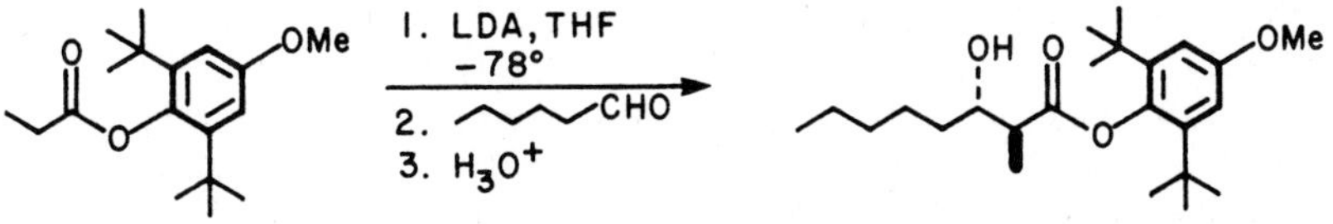

27

(57)

28

Although the BHT esters are highly *threo*-selective, the resulting aldols are so hindered that the ester function may not be hydrolyzed without aldol

Table XII. Reactions of Ester 26 with Various Aldehydes (Eqn. 54)		
R	aldol yield, %	*threo/erythro*
C_6H_5	72	86/14
$n\text{-}C_5H_{11}$	70	86/14
$i\text{-}C_3H_7$	78	>98/2
$t\text{-}C_4H_9$	82	>98/2
$C_6H_5(CH_3)CH$	81	>98/2

equilibration. However, they may be reduced cleanly to give 1,3-diols (eqn. 58).[44] If the hydroxy group is first protected, the BHT aldol may be converted into a protected β-hydroxy aldehyde by reduction and reoxidation (eqn. 59).[45]

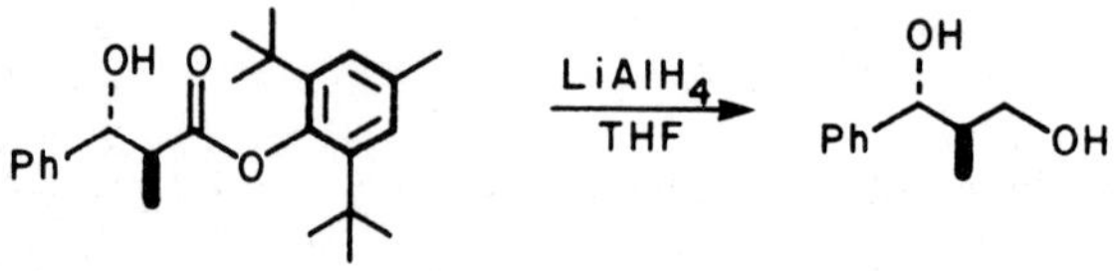

$$\text{(58)}$$

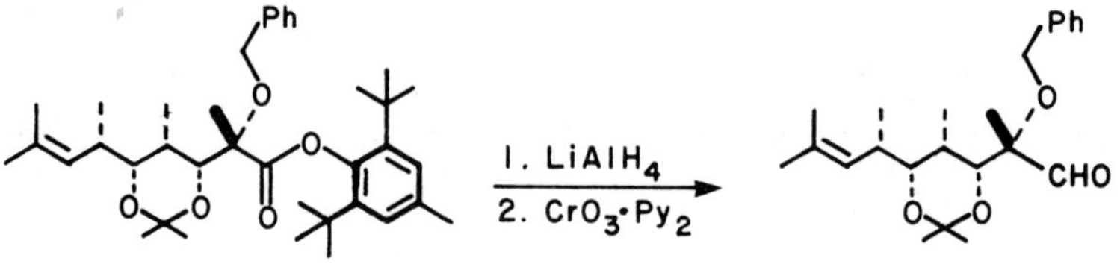

$$\text{(59)}$$

The DBHA esters were introduced because the ester group can be removed by a mechanism not involving nucleophilic attack at the ester carbonyl. Thus, treatment of the acetylated aldol with ceric ammonium nitrate (CAN) or silver(II) oxide readily affords the β-hydroxy acid (eqn. 60).[44]

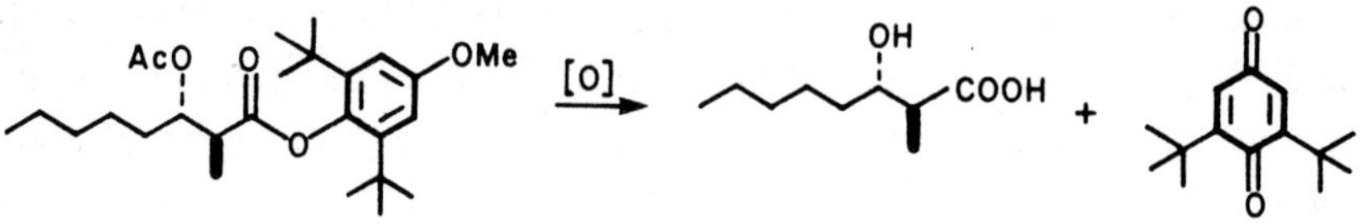

$$\text{(60)}$$

The BHT and DBHA esters of O-benzyllactic acid also show good *threo* selectivity (e.g., eqn. 61).[45,*] Although the enolate geometry has not been established, it is presumably that in which the benzyloxy and aryloxy groups are *trans*.

---

* As with the simple aldols, we have adopted the *erythro-threo* terminology to describe aldols derived from α-alkoxy esters. For simplicity, we utilize the same convention, ignoring the α-alkoxy group. Thus, when the aldol is written with its backbone in an extended (zig-zag) manner, if the α-alkyl substituent and the β-hydroxy group both project toward the viewer (bold bonds) or away from the viewer (dashed bonds), this is the *erythro* diastereomer.

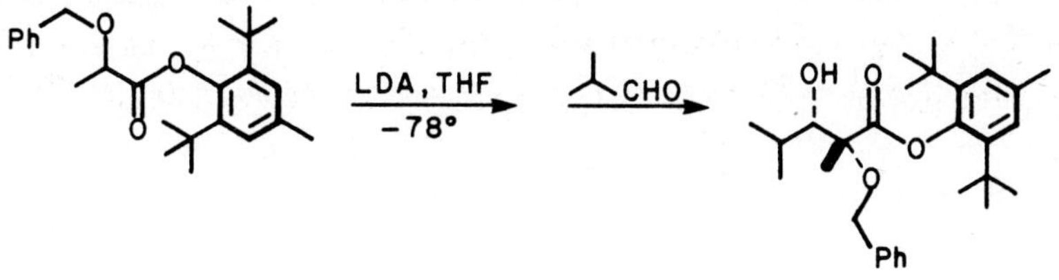

(61)

28

The other effective *threo*-selective reagent is the S-*t*-butyl propanethioate (**21**) developed independently by Masamune and Evans.[20,32] Masamune has used this reagent to prepare several *threo* aldols (e.g., eqn. 62).[32a]

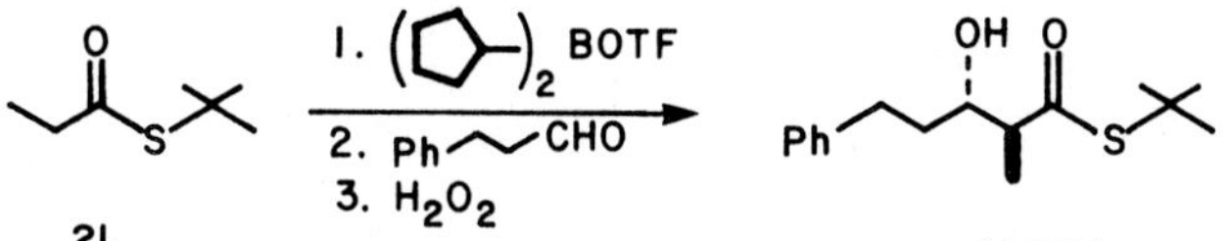

(62)

(68%)

VIII. Diastereoface Selection: The "Crams's Rule Problem"

The aldol condensation can show two distinct kinds of stereoselection. If two new asymmetric carbons are formed, as is the case in all the reactions we have discussed thus far, they may have either the *erythro* or the *threo* relative configuration. This kind of stereoselection may be termed *simple diastereoselection*. A different kind of stereoselection is possible when either reactant is chiral. For example, addition of an enolate to a chiral aldehyde can give two *erythro* aldols, resulting from attack on either of the diastereotopic faces of the aldehyde (eqn. 63). Similarly, if the enolate is chiral, and the aldehyde is achiral there are also two *erythro* aldols resulting from attack at the diastereotopic faces of the enolate (eqn. 64) We term this kind of stereoselection *diastereoface selection*.

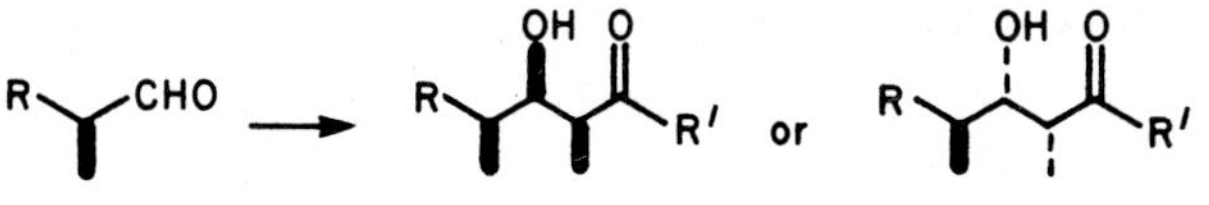

(63)

(S)   (S,R,S)   (S,S,R)

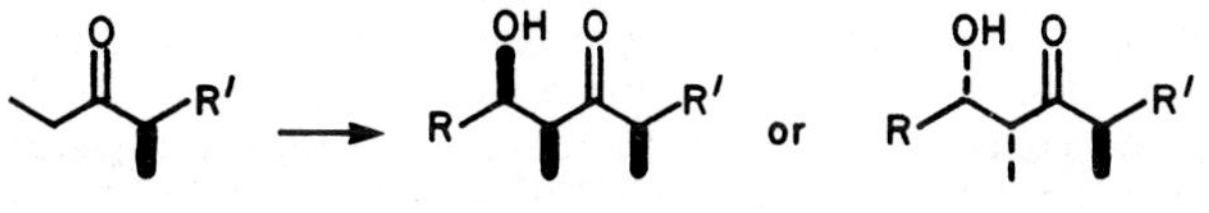

(64)

(R)   (R,S,R)   (S,R,R)

As we have seen, several reagents are available which give good simple diastereoselection in the aldol condensation. It is not so easy to achieve high diastereoface selection in additions to chiral aldehydes. For example, the enolate of ketone **23** reacts with α-phenylpropionaldehyde to give the two *erythro* aldols **29** and **30** in a ratio of 4:1 (eqn. 65).[17] The major product in eqn. 65 is

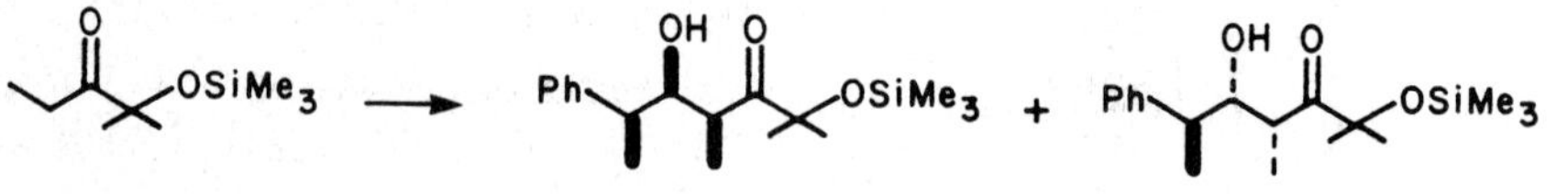

$$\text{23} \longrightarrow \text{29} \quad + \quad \text{30} \qquad (65)$$

that predicted by Cram's rule for asymmetric induction[46] and is often referred to as the *erythro*, Cram's rule product.[17] Similar behavior is shown by other chiral aldehydes. For example, O-benzyllactaldehyde reacts with **23** to give two *erythro* aldols (**31** and **32**) in a ratio of 2:1 (eqn. 66),[47] while the acetonide of glyceraldehyde reacts with **23** to give **33** and **34** in a ratio of 4.3:1 (eqn. 67).[47]

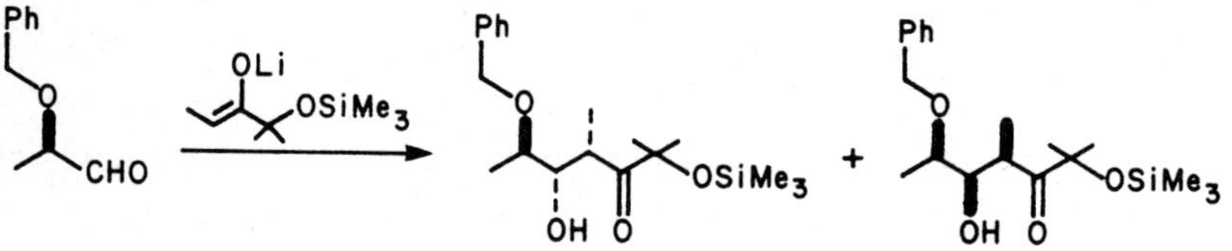

$$(66)$$

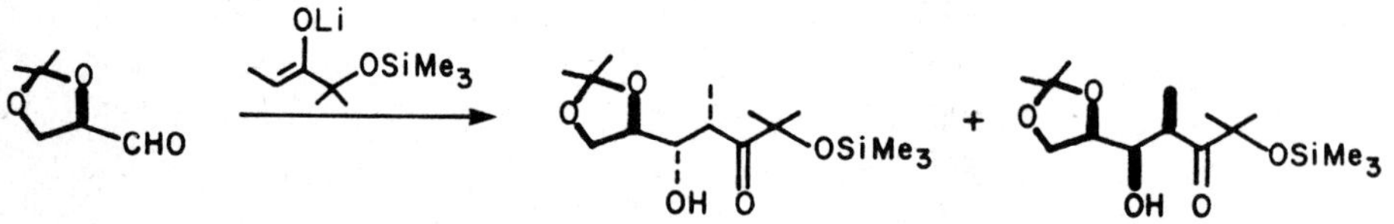

$$(67)$$

The major product in each of these condensations is that predicted by the Felkin model for asymmetric induction[46c] assuming that oxygen is the large group (Figure XII).[48]

One way in which higher diastereoface selectivity has been realized is by use of double stereodifferentiation.[49] To illustrate how this strategy is applied

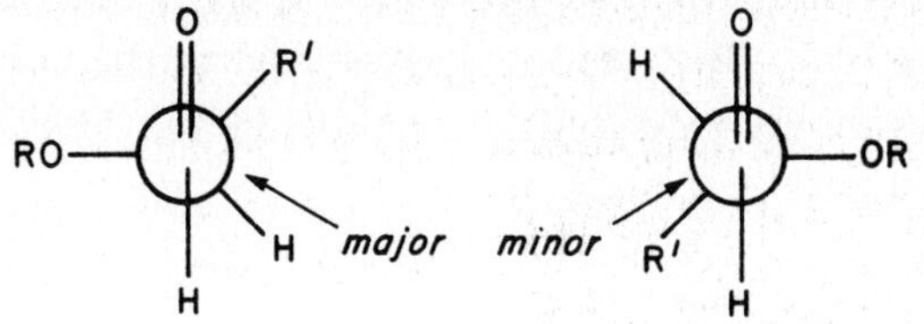

Figure XII. Inherent Diastereoface Selectivity of α-Alkoxy
Aldehydes in Reactions With Enolates

in an aldol condensation let us return to the two kinds of diastereoface selection
as illustrated in equations 63 and 64. Suppose that the (S)-enantiomer of a
given aldehyde gives mainly the (S,R,S)-aldol in its reaction with achiral enolates (eqn. 63). Suppose also that the (R)-enantiomer of a given ketone gives
mainly the (R,S,R)-aldol in its reactions with achiral aldehydes (eqn. 64). If
(S)-aldehyde is allowed to react with (R)-ketone, there are two *erythro* aldols
which may be formed, (S,R,S,R) and (S,S,R,R), as shown in eqn. 68. The

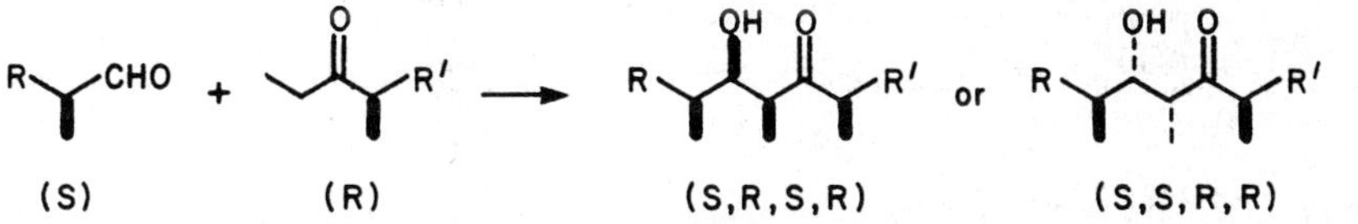

former diastereomer should greatly predominate, since the (R,S) configuration
at the two new centers is promoted by both aldehyde and enolate. To a first
approximation, the (S,R,S,R):(S,S,R,R) ratio in eqn. 68 should be the product
of the individual diastereoface selectivities of the aldehyde and the enolate in
their reactions with comparable achiral reaction partners. Thus if (S)-aldehyde
shows diastereoface selectivity of 5:1 favoring the (S,R,S)-aldol in reactions
with achiral enolates and if (R)-ketone shows diastereoface selectivity of 5:1
favoring the (R,S,R)-aldol in its reactions with achiral aldehydes, then the
(S,R,S,R):(S,S,R,R) ratio should be of the order of 25:1. Of course, reaction
of (S)-aldehyde with (S)-ketone would be expected to lead to *poorer* diastereoface selectivity from the standpoint of either reactant. For example, in the
hypothetical case being discussed, the (S,R,S,S):(S,S,R,S) ratio would be about
1:1. This form of aldol stereoselection has been termed *double stereodifferentiation*,[49b] since the new chiral centers are introduced under the
influence of two different chiral elements (in this example, the two reaction
partners). Of course, double stereodifferentiation may also be achieved by the
use of a chiral auxillary, such as the solvent.

Both types of double stereodifferentiation have been demonstrated.[49] In the most dramatic case, ethyl ketone **35** was used as the enolate precursor and the two enantiomeric glyceraldehyde acetonides **36** and **37** as the aldehyde com-

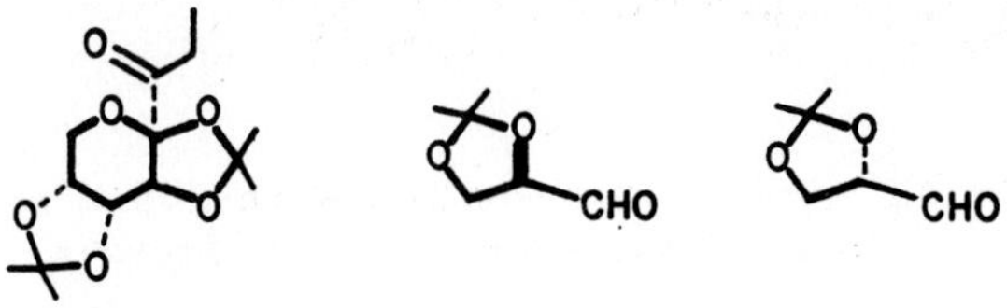

ponents. Ketone **35** reacts with benzaldehyde to give two *erythro* aldols in a ratio of 4:1 (eqn. 69). This ratio corresponds to the approximate diastereoface

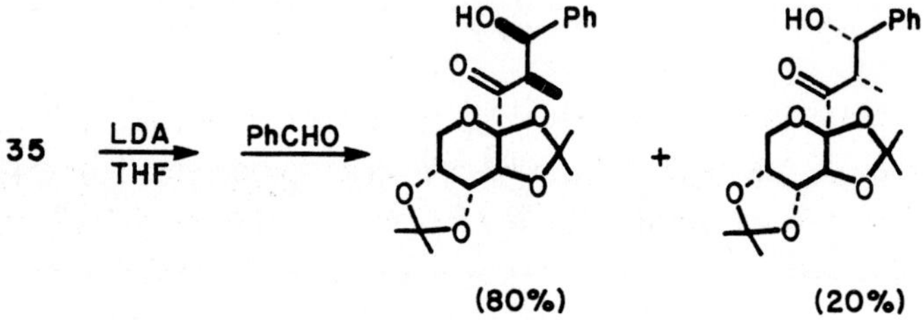

selectivity of **35**. Aldehydes **36** and **37** are known to exhibit inherent diastereoface selectivity of about 4.3:1 (eqn. 67).[47] Reaction of ketone **35** with aldehyde **36** gives two *erythro* aldols and a *threo* aldol in a ratio of 5.5:2.5:1 (eqn. 70). The *erythro* aldols have been shown to have structures **38** and **39**. On the other hand, ketone **35** reacts with the enantiomeric aldehyde **37** to give only a single *erythro* aldol, which has been shown to have structure **40** (eqn. 71)!

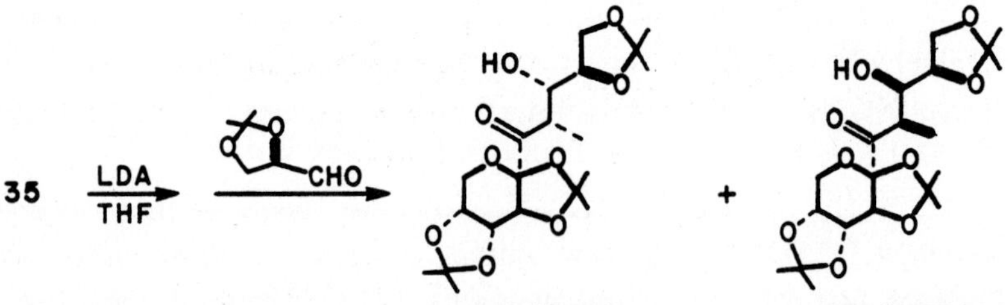

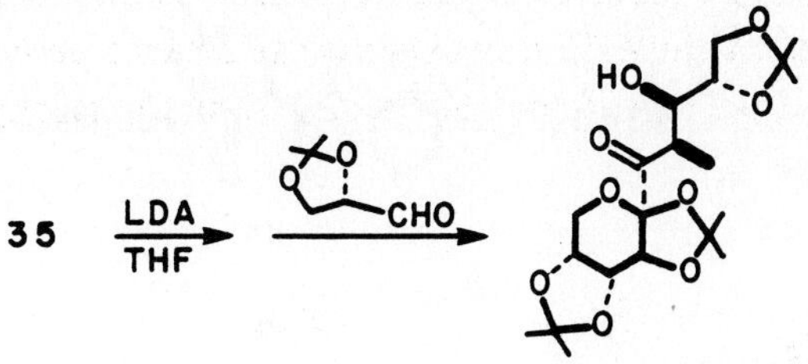

$$35 \quad \xrightarrow[\text{THF}]{\text{LDA}} \quad \text{(aldehyde) CHO} \quad \longrightarrow \quad \textbf{40} \ (\geq 97\%)$$

(71)

Similar experiments have been reported by Masamune and co-workers.[50a] These investigators employed the enantiomeric ketones **41** and **42** and studied

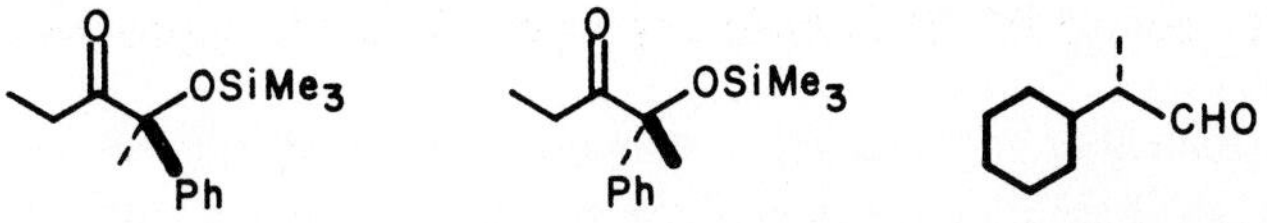

their reactions with aldehyde **43**. Reaction of **41** with **43** gives *erythro* aldols **44** and **45** in a ratio of 1:8 (eqn. 72). On the other hand, enantiomer **42** reacts with **43** to give *erythro* aldols **46** and **47** in a ratio of 1.5:1 (eqn. 73).*

---

* The normal diastereoface selectivity of aldehyde **43** is as shown in eqn. 74.[51] Qualitatively, this is the selectivity predicted by all the models for asymmetric induction.[46,48,52] Masamune and co-

$$43 \ + \ \text{MeMgI} \ \longrightarrow \quad (66\%) \quad + \quad (34\%)$$

(74)

workers report that, for some inexplicable reason, **43** reacts with enolates to give primarily the "anti-Cram" product (e.g., eqn. 75). The reason for this aberrant behavior remains to be clarified.

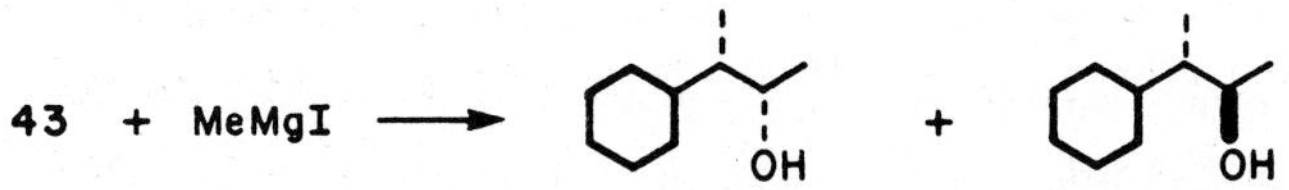

$$43 \ + \quad \longrightarrow \quad (73\%) \quad +$$

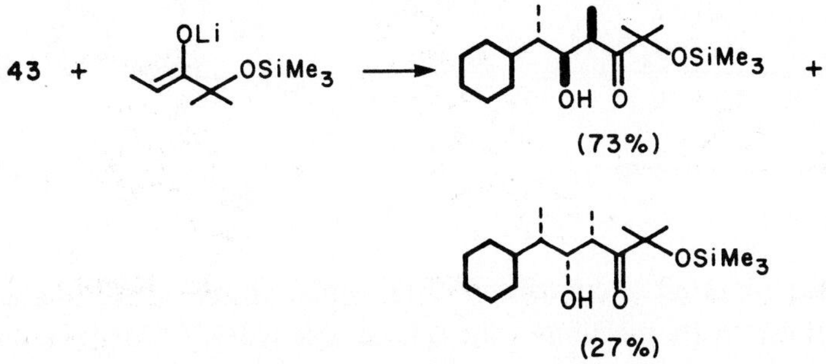

(75)

$$(27\%)$$

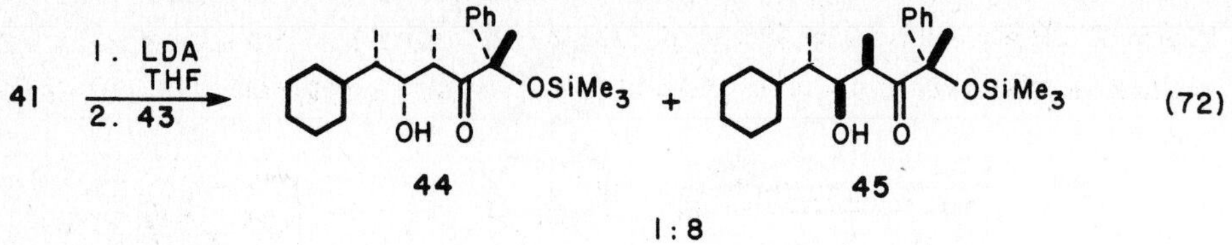

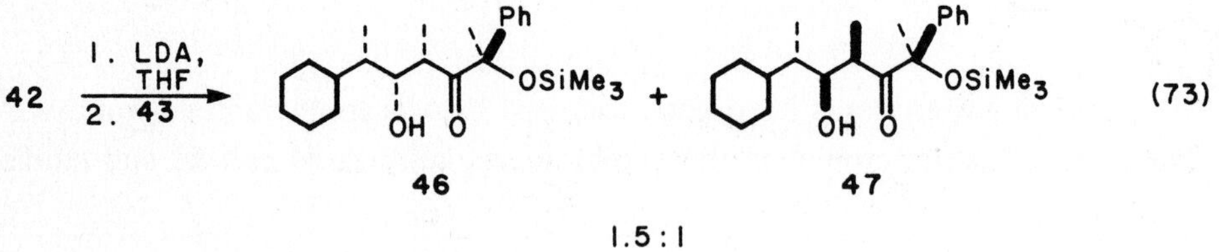

$$41 \xrightarrow[\text{2. 43}]{\substack{\text{1. LDA} \\ \text{THF}}} \quad 44 \quad + \quad 45 \qquad (72)$$

1 : 8

$$42 \xrightarrow[\text{2. 43}]{\substack{\text{1. LDA,} \\ \text{THF}}} \quad 46 \quad + \quad 47 \qquad (73)$$

1.5 : 1

Masamune has also introduced a related pair of chiral ketones (**48** and **49**) which show even higher diastereoface selectivity than do **41** and **42**.[50b] Compounds **48** and **49** are conveniently prepared in three-step sequences beginning with (R)- and (S)-mandelic acid, respectively. The boron enolates of **48** and **49**

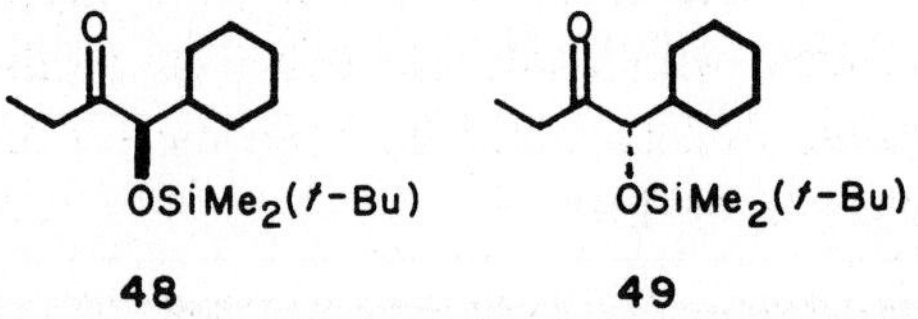

48   49

show exceptionally high inherent diastereoface selectivity in their reactions with several aldehydes. For example, the dicyclopentylboron enolate of **49** reacts with propionaldehyde to give, after desilylation and periodate oxidation, the (2R,3S)-hydroxy acid **50**. The enantiomeric and diastereomeric purity of **50** has been estimated to be greater than 99% (eqn. 76).[50b]

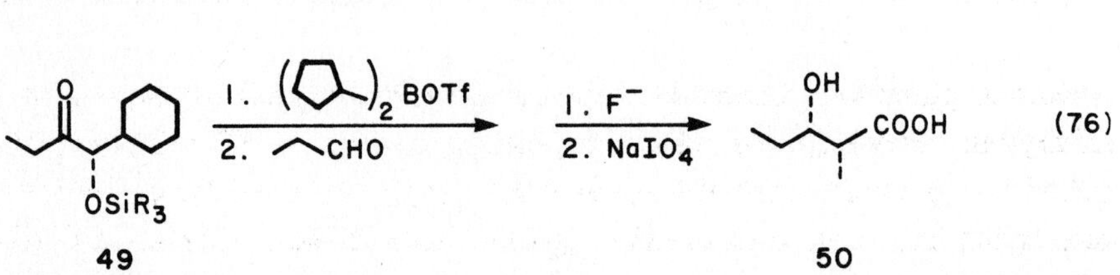

$$49 \xrightarrow[\text{2. } \diagup\text{CHO}]{\substack{\text{1. } (\text{⬠})_2 \text{ BOTf}}} \xrightarrow[\text{2. NaIO}_4]{\text{1. F}^-} 50 \qquad (76)$$

Evans has also reported a compound (**51**) which shows good inherent diastereoface selectivity in its reactions with achiral aldehydes.[53] Condensation of **51** with isobutyraldehyde *via* the lithium enolate gives two *erythro* aldols in a ratio of 70:30. However, if the dibutylboron enolate is employed, only one adduct is obtained (eqn. 77). The methyl ketone corresponding to **50** shows

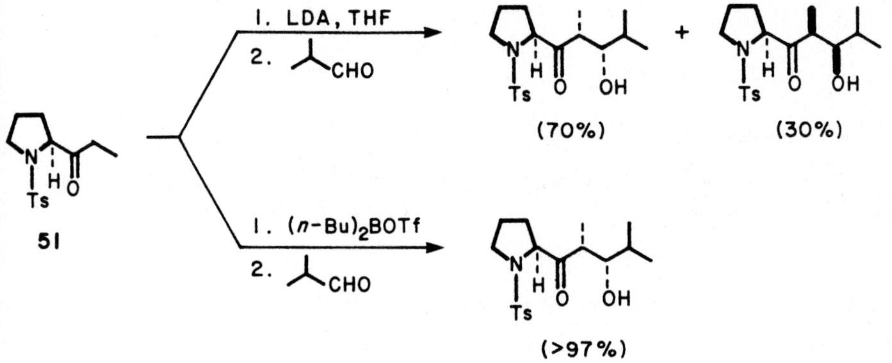

much less inherent diastereoface selectivity than does **50** (54:46 with lithium, 74:26 with boron).[*] However, the trend that the boron enolate shows higher inherent diasteroface selectivity than the lithium enolate is still observed. The importance of this observation is the finding that the inherent diastereoface selectivity is greater with the boron than with the lithium enolate. Although **51** has not been condensed with chiral aldehydes in double stereodifferentiation experiments, its boron enolate might be even more effective than the lithium enolates of **35**, **41/42**, or **48/49**.

Evans and his co-workers have also studied the aldol condensations of a series of propionimides such as **52** and **53**.[54] The enolates of these compounds

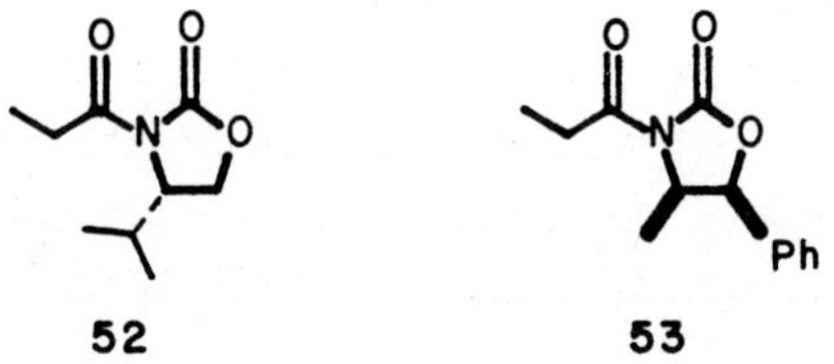

show exceptional inherent diastereoface selectivity in their reactions with achiral aldehydes. For example, the dibutylboron enolate of **52** reacts with benzaldehyde to give only a single *erythro* aldol (eqn. 78); the inherent diastereoface selectivity has been estimated at greater than 2500:1. Compound **53** shows similar behavior but with complementary stereochemistry (eqn. 79). This is an

----

[*] Investigations at Berkeley have also shown that enolates derived from chiral methyl ketones show less inherent diastereoface selectivity than do enolates derived from ethyl ketones.[47]

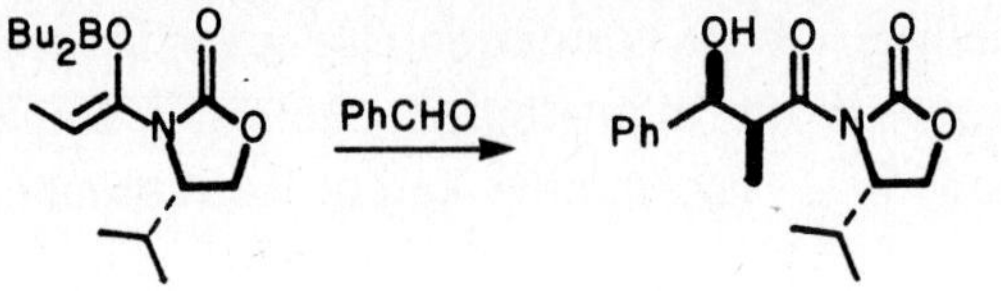

(78)

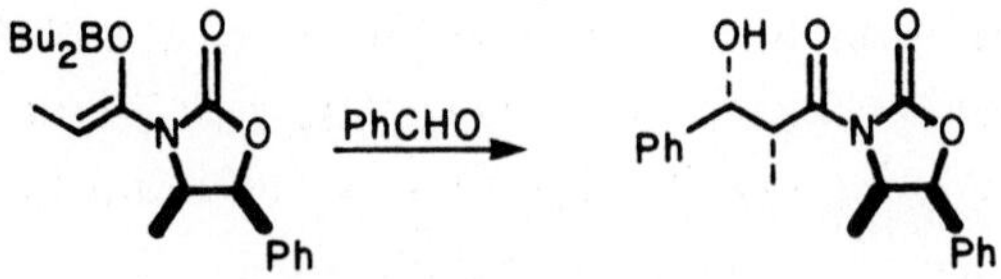

(79)

exciting development, as **52** and **53** are readily-available (from valine and (+)-norephedrine, respectively) and since the aldols may be manipulated so as to produce a variety of β-hydroxy acids and β-hydroxy aldehydes in good optical purity.[54]

Compounds such as **52** and **53** show such high inherent diastereoface selectivity that they essentially override the modest inherent diastereoface selectivity of most chiral aldehydes. Thus, the zirconium enolate of the related amide **54** reacts with both enantiomers of 3-benzyloxy-2-methylpropanal to give products having the same absolute configuration of the two newly created asymmetric carbons (eqn. 80).[55] The synthetic utility of these reagents is obvious.

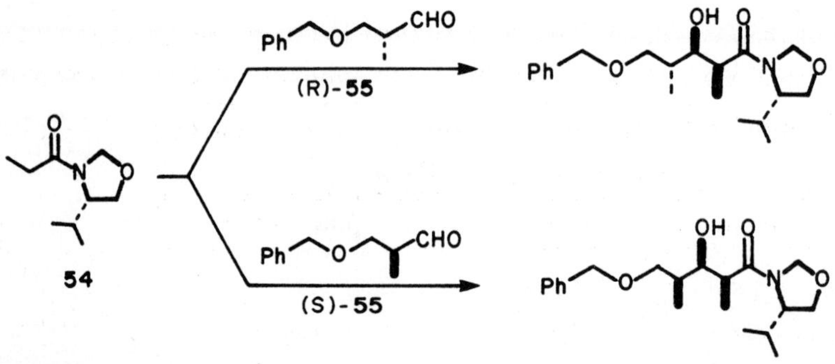

(80)

Because of their high selectivity, it is possible to control the *absolute* configuration of the newly created asymmetric carbons regardless of the chirality of the substrate aldehyde.

Double stereodifferentiation has also been demonstrated when one of the chiral elements is the solvent.[49] The reaction of ketone **23** with benzaldehye has been carried out in the enantiomeric 1,2,3,4-tetramethoxybutanes **56** [(+)-TMB] and **57** [(−)-TMB].[56] After oxidation of the aldols, the product hydroxy

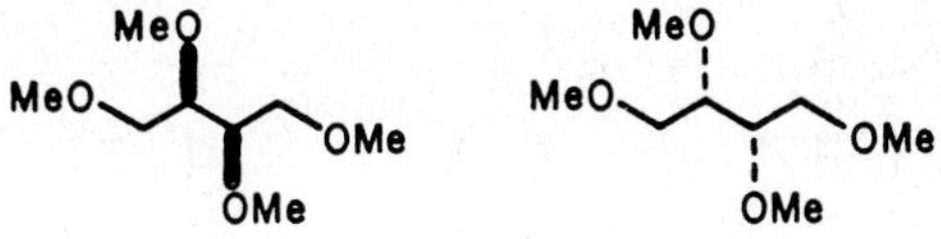

56   **57**

acids are found to have been formed with 7% enantiomeric excess (ee). The stereochemistry of the reaction is such that the (SS) enantiomer is favored when (−)-TMB is used as solvent (eqn. 81). The reaction of ketone **23** with (R)-glyceraldehyde acetonide (**36**) gives a mixture of **33** and **34** (eqn. 67). The

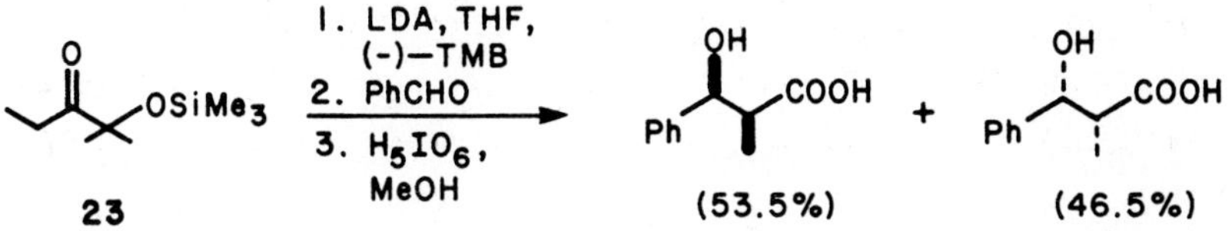

33:34 ratio is found to depend upon the chirality of the solvent. Data are shown in Table XIII. A small but distinct double stereodifferentiation is seen. The effect is small because the intrinsic stereodifferentiating ability of the tetra-methoxybutanes is small, at least in reactions of enolates having a neighboring alkoxy group.

Table XIII. Reaction of Ketone 23 with Aldehyde 36 in Various Solvents

Solvent	33:34	*Threo* Aldol
THF	4.3:1.0	4%
(-)-TMB	5.0:1.0	14%
(+)-TMB	3.6:1.0	17%

Recent experiments at Berkeley have shown that double stereo-differentiation may even be observed when both the enolate and the aldehyde are chiral and racemic.[57] Ketone **58** shows good to excellent inherent diastereo-face selectivity in its reactions with achiral aldehydes (eqn. 82, Table XIV).

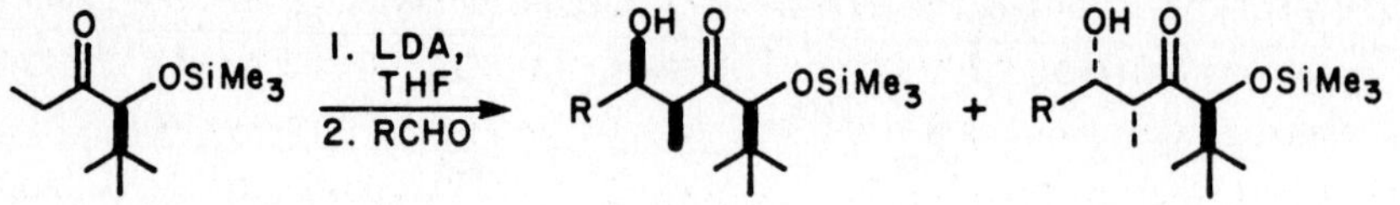

$$(82)$$

Table XIV

Condensation of Ketone 58 with Various Aldehydes

R	Diastereomer Ratio 59:60	Combined Yield, %
Ph_2CH	$\geq 9:1$	69
$PhCH_2$	87:13	75
Me_2CH	3:1	93
Me_3C	$\geq 95:5$	47
Ph	3:1	75

Reaction of racemic compound **58** with racemic α-phenylpropionaldehyde affords a single racemic aldol (eqn. 83)! Compound **58** shows similar behavior in its reaction with other racemic chiral aldehydes.

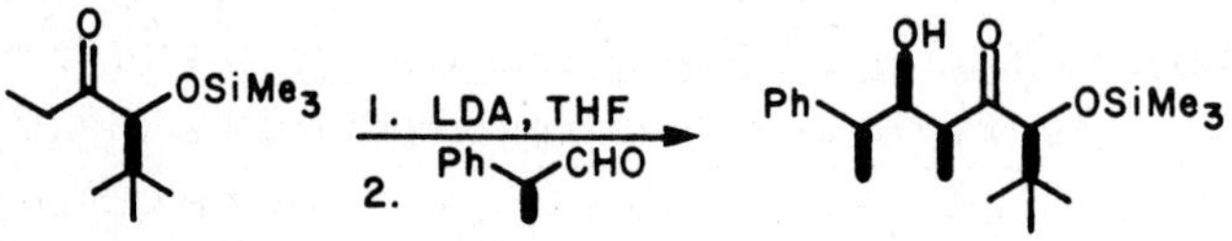

$$(83)$$

The high degree of mutual kinetic resolution which is displayed in the reactions of racemic **58** with various racemic aldehydes is, at first sight, startling. However, it can be shown that at least some of this mutual kinetic resolution may be anticipated on purely logical grounds as a consequence of the individual diastereoface selectivities of the two reaction partners. To illustrate this point, consider the reaction of racemic ketone **58** with racemic aldehyde **61**. Since the product aldols have four asymmetric carbons, a total of eight racemates are possible. Four of these are *erythro* and four are *threo* with respect to the two new centers which are created in the condensation. We may start with the assumption that *threo* diastereomers can be ignored, since ketones such as **58** have been shown to have *erythro:threo* kinetic stereoselectivity of about 80:1.[17] The four remaining possibilities are the *erythro* aldols **62**, **63**, **64**, and **65** (Figure XIII, note that only one enantiomer is depicted for each racemate).

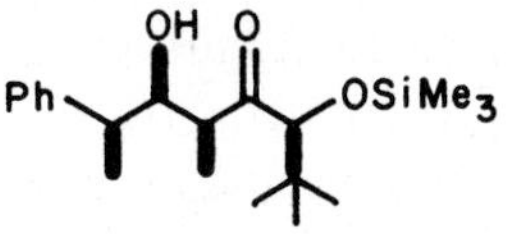

62

(25 × 10 = 250), 87.4%

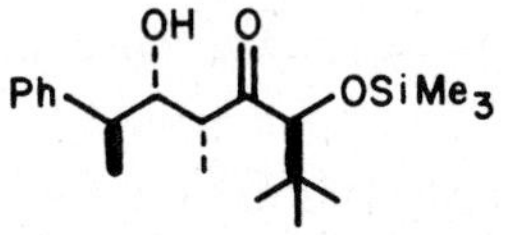

63

(1 × 1 = 1), 0.4%

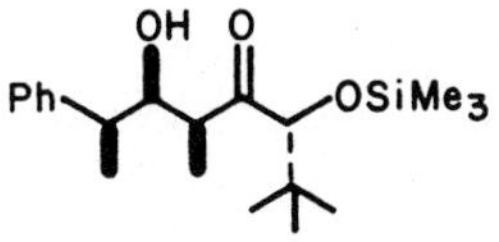

64

(10 × 1 = 10), 3.5%

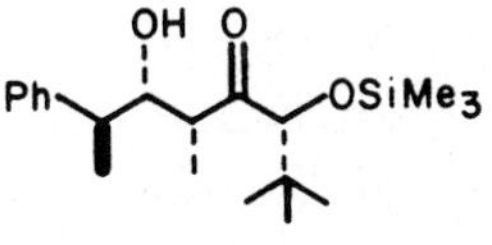

65

(1 × 25 = 25), 8.7%

Figure XIII. Prediction of Diastereomer Ratios for the Reaction of Aldehyde 61 With Ketone 58.

We may make the further assumption that the relative amounts of these four racemates may be *approximated* by using the inherent diastereoface selectivities of the reaction partners. Ketone **58** seems to show very high diastereoface selectivity in its reactions with aldehydes having a bulky group attached to the carbonyl (cf. Table XIV). Therefore, we shall assume an inherent diastereoface selectivity for **58** of 25:1 for reactions with such aldehydes. Aldehyde **61** shows diastereoface selectivity in the range 3:1 to 6:1.[52] However, there is a suggestion that this compound also may show higher diastereoface selectivity with more sterically demanding nucleophiles. For example, the selectivity in its reactions with various Grignard reagents is 2:1 for CH_3MgBr, 3:1 for C_2H_5MgBr and >4:1 for C_6H_5MgBr.[47] An example of the same trend with lithium enolates is seen in the reaction of α-phenylpropionaldehyde with pinacolone and ethyl *t*-butyl ketone, where the diastereoface selectivity is 3:1[58] and 6:1,[17] respectively. Thus, we shall assume an inherent diastereoface selectivity for aldehyde **61** of 10:1 for reactions with the bulky enolates derived from ketones such as **58**. Using these inherent diastereoface selectivities, we may estimate the **62:63:64:65** ratios to be $(25 \times 10):(1 \times 1):(1 \times 10):(25 \times 1) =$ 87.4:0.4:3.5:8.7. Thus, the ratio **(62+63):(64+65)** is predicted to be about 7:1. Note that this is the predicted kinetic resolution in the reaction as both **62** and **63** result from reaction of (S)-aldehyde with (S)-ketone, while **64** and **65** result from reaction of (S)-aldehdye with (R)-ketone.

Of course, aldols **62** and **64** both produce the same acid upon periodic acid oxidation, while aldols **63** and **65** would both produce the *erythro*, anti-Cram's rule diastereomer. The approximation being used here predicts that the eventual ratio of two acids would be $(87.4 + 3.5):(0.4 + 8.7)=10:1$, since the multiplicative approximation does not change the net diastereoface selectivity shown by either reaction partner. Thus, it is clear that, although the multiplicative approximation accounts for some of the mutual kinetic resolution observed in these double racemic aldol condensations, it does not account for the very high net diastereoface selectivity shown by the aldehydes. However, if there is some other factor which promotes mutual kinetic resolution, then these surprisingly high net diastereoface selectivities can be readily understood. For example, suppose the kinetic resolution in the **58 + 61** reaction is 35:1 instead of 7:1. That is, suppose there is an independent stereoselectivity favoring the reaction (S)-**58** + (S)-**61** over (S)-**58** + (R)-**61** by an additional factor of 5. Then the ratio of products **(62 + 63):(64 + 65)** would be 35:1. Note that, in view of the principle of double stereodifferentiation, the ratio **62:63** is expected to be very high (219:1), while the ratio of **64:65** is expected to be low (1:2.5). Thus, if mutual kinetic resolution is 35:1, the ratio **62:63:64:65** is expected to be 96.8:0.4:0.8:2.0 and the eventual ratio of diastereomeric acids is predicted to be 97.6:2.4.

IX. Application of Stereoselective Aldol Condensations in Natural Product Synthesis

An interest in synthesizing complex polypropionates such as the macrolide and polyether ionophore antibiotics has provided much of the stimulus for the extensive recent research on aldol stereoselection. Actual applications of stereoselective aldol condensations to these problems have just begun. In this section, we shall discuss selected examples in which the aldol condensation has been used to establish the desired stereochemistry in the synthesis of natural products. Examples have been chosen to illustrate various methods which have been discussed earlier in this review.

The simple alkaloid (±)-ephedrine (**66**) has been prepared by a route involving the *erythro*-selective condensation of reagent **23** with benzaldehyde.[17] This example illustrates an important point: the aldol condensation may be used to establish relative stereochemistry even in compounds which are not β-hydroxy carbonyl compounds.

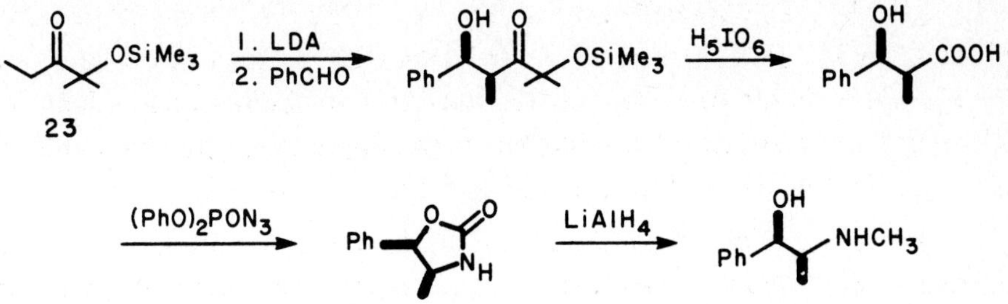

Compound **23** has also been employed as an *erythro*-selective reagent in Still's synthesis of monensin.[59] Condensation of the magnesium enolate of **23**

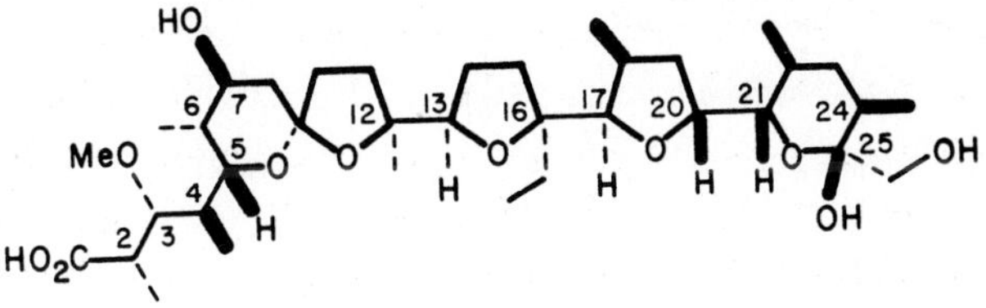

monensin

with aldehyde **53** gives a product (**67**) which is elaborated by a further sequence of steps into aldehyde **68**, which corresponds to C_1-C_7 of monensin.

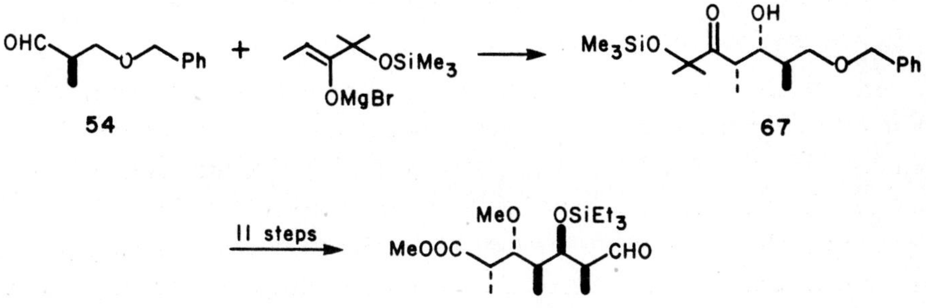

Masamune and co-workers have used chiral *erythro*-selective reagents **48** and **49** in a synthesis of 6-deoxyerythronolide B (**69**).[60] Compound **48** reacts

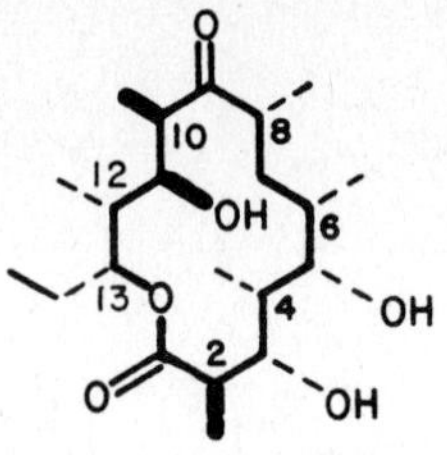

69

with propionaldehyde to give the *erythro* aldol, which is desilylated and oxidized to β-hydroxy acid **70**. This compound is further transformed into the protected β-hydroxy aldehyde **71**, which constitutes C_{11}-C_{15} of the aglycone. Compound

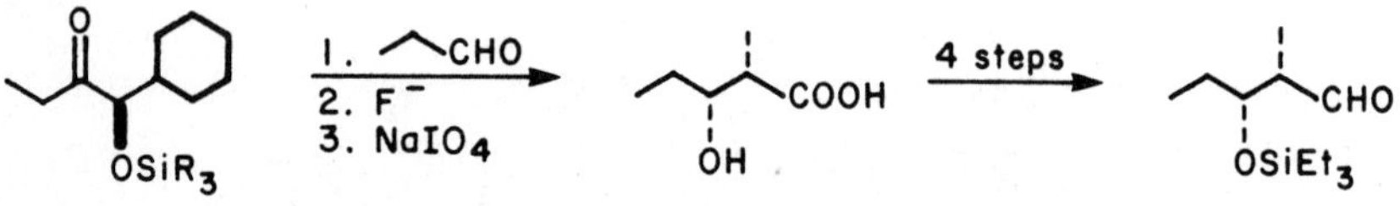

70   **71**

49 was used in a synthesis of the C_1-C_{10} segment. Condensation of **49** with the chiral ester aldehyde **72** gives aldol **73** of approximately 98% stereochemical purity.

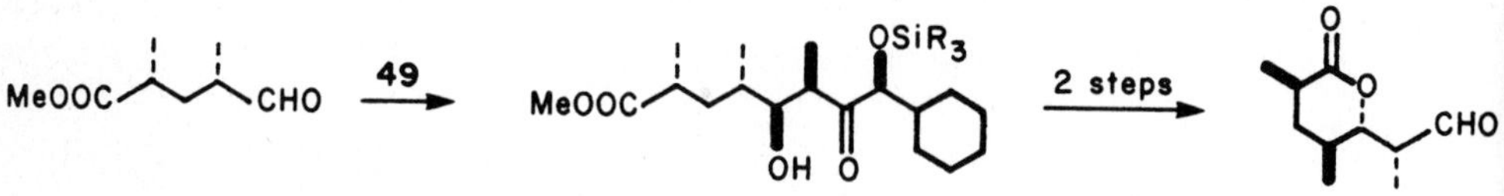

72   **73**   **74**

The next two carbons of the chain are again introduced by use of compound **49**; aldols **75** and **76** are obtained in an approximate 14:1 ratio. After

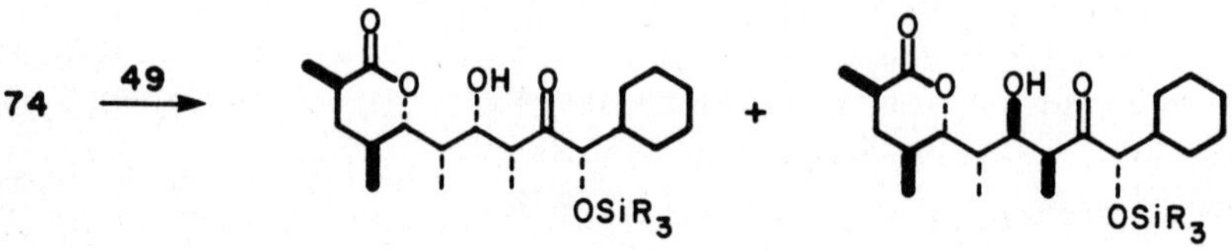

75   **76**

conversion of **75** to ethyl ketone **77**, the two segments are coupled by way of the *cis* lithium enolate. Two *erythro* aldols, **78** and **79**, are obtained in a ratio of 17:1. The major isomer is converted into aglycone **69** by a seven-step sequence. This is an important demonstration of the power of stereoselective aldol condensations in synthesis. In all, four condensations are involved and the overall stereochemical yield appears to be on the order of 84%.

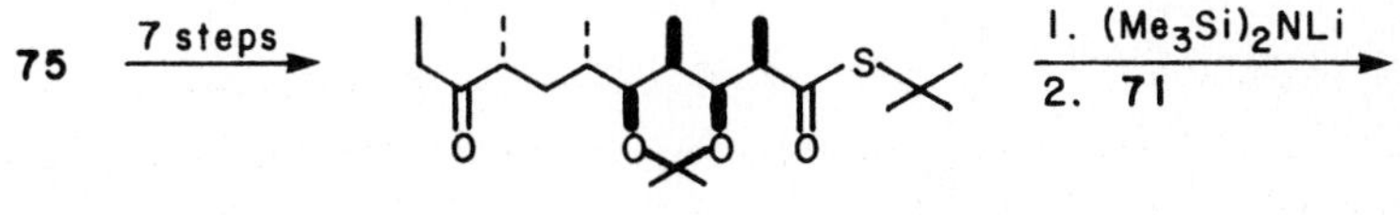

77

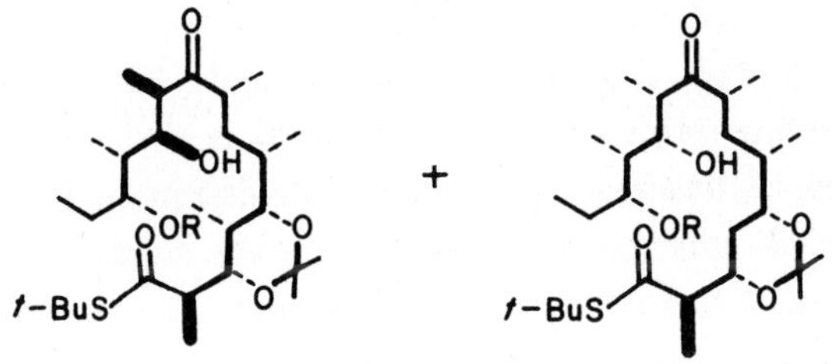

78   +   **79**

An approach to the total synthesis of erythronolide A (**80**) utilizing stereoselective aldol condensations is underway at Berkeley.[61] The Berkeley

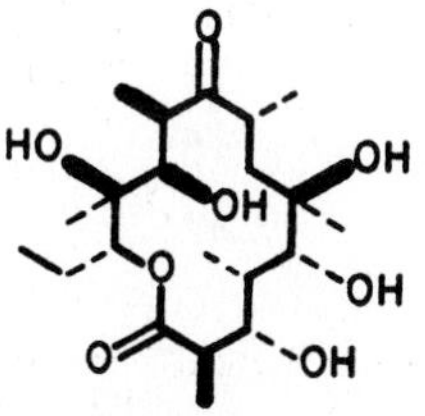

80

approach begins with aldehyde **81**, which is condensed with reagent **23** to give *erythro* aldol **82** of 94% stereochemical purity. Elaboration of **82** gives an aldehyde (**83**) which condenses with reagent **28** (see eqn. 61) to give aldol **84**, of 85% stereochemical homogeniety. Straightforward transformations convert **84** into aldehyde **85**, which comprises about one-half of the erythronolide A chain, and contains five asymmetric centers.

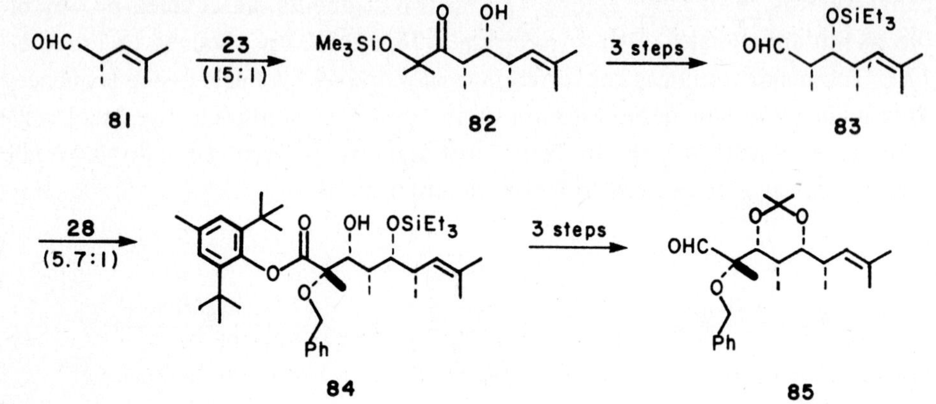

For synthesis of the other half of the chain, enantiomerically-homogeneous β,γ-unsaturated aldehyde **86** is condensed with reagent **28** to give aldol **87**. As with β,γ-unsaturated aldehyde **81**, compound **86** shows excellent diastereoface preference, and aldol **87** is the only stereoisomer that has been detected in this reaction. After conversion of **87** into protected β-hydroxy aldehyde **88**, addition of ethyllithium gives a diastereomeric mixture of alcohols, which is oxidized and the resulting ketone reduced to obtain diastereomerically-homogeneous **89**. Protection of the secondary hydroxy group and ozonolysis of the double bond gives ketone **90**, which comprises the

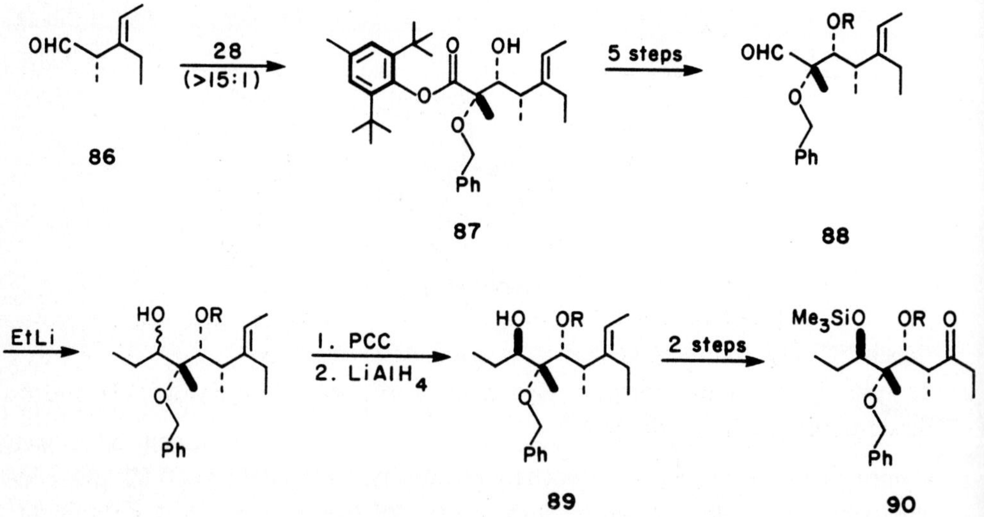

C-8 to C-15 segment of erythronolide A, and contains four of the five remaining stereocenters. In recent work, aldehyde **85** has been joined with ketone **90**, and the resulting mixture of aldols dehydrated, to obtain enone **91**, in which the entire backbone of erythronolide has been assembled.

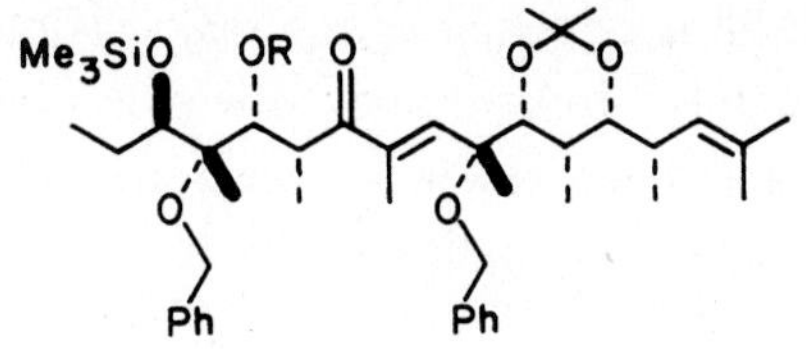

91

Double stereodifferentiation has been used advantageously in a synthesis of (±)-blastmycinone, a degradation product of the antibiotic antimycin A_3.[57b] As was discussed earlier, O-benzyllactaldehyde shows poor diastereoface selectivity in its condensation with reagent **23** (see eqn. 66). However, with compound **92**, it gives the two *erythro* aldols in a ratio of 10:1. The major isomer has been converted into (±)-blastmycinone (**93**) by a straightforward three-step sequence.

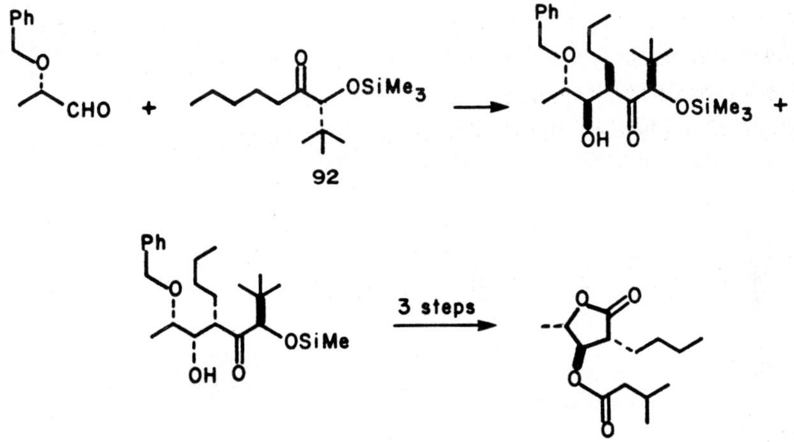

92

93

The naturally occurring hydroxy acid corynomycolic acid has been synthesized using the *threo*-selective reagent **94**.[44] Reaction of **96** with *n*-hexadecanal, followed by methanolysis, gives methyl corynomycolate (**95**) and its diastereomer **26** in a ratio of 2:1.

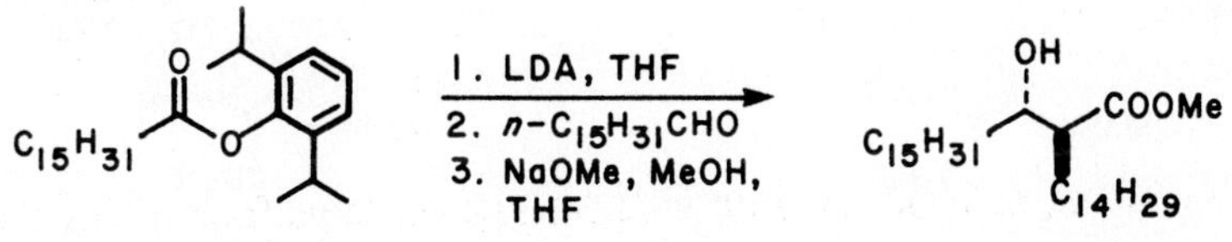

94

1. LDA, THF
2. n-$C_{15}H_{31}CHO$
3. NaOMe, MeOH, THF

95

Several syntheses have employed thermodynamically controlled aldol condensations to establish *threo* relative configuration. In Kishi's synthesis of lasolocid-A (**98**)[62], ethyl ketone **96** was coupled with aldehyde **97** by the House method.[14] *Threo* aldol **98** is obtained as the major isomer of a 40:10:7:3 mixture of the four possible aldols. Ireland and co-workers have also prepared ketone **96** and aldehyde **97** and have raised a question regarding the stereochemical outcome of the reported aldol coupling in the Kishi synthesis.[63]

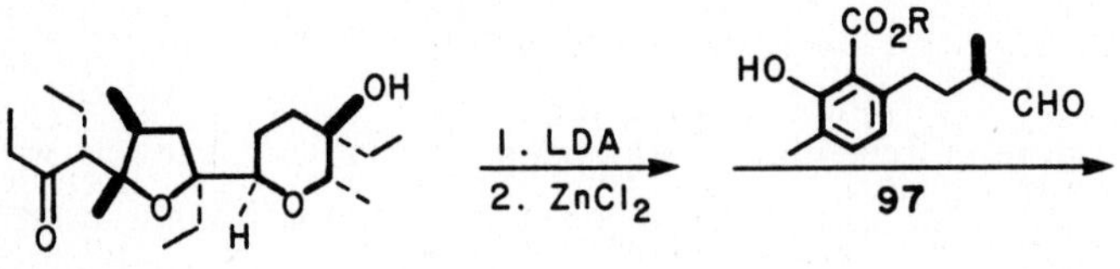

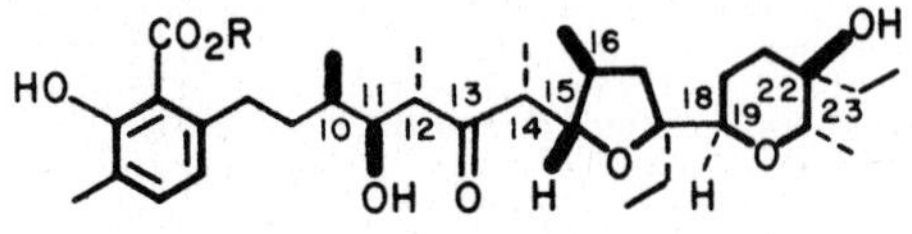

Evans' synthesis of ionophore A-23187 (**102**) serves as a final example of the use of stereoselective aldol technology in complex synthesis [64]. The crucial condensation joins ethyl ketone **99** with aldehyde **100**. The coupling gives a mixture of *threo* and *erythro* aldols (*threo:erythro*=70:30) from which ionophore A-23187 (**102**) is obtained in approximately 20% yield.

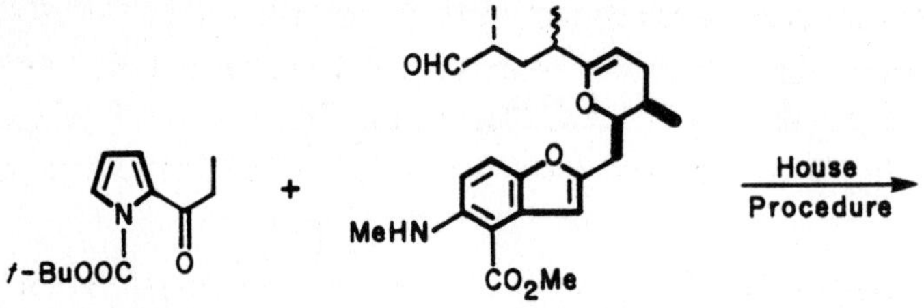

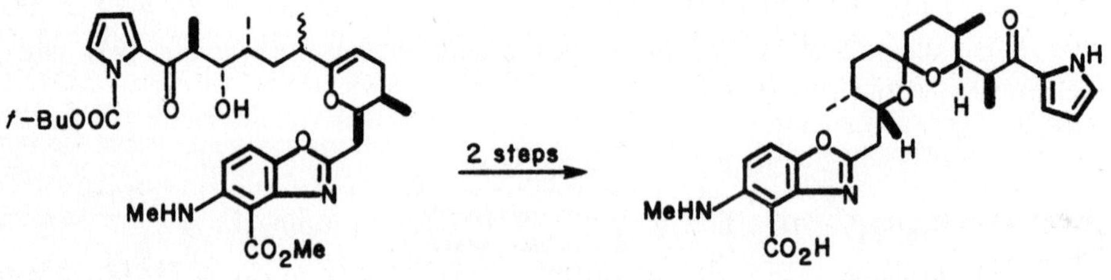

X. Conclusion, Further Problems

As has been seen in this review, progress in controlling the stereochemistry of the aldol condensation has been good. Several reagents are now available which show good simple diastereoselection. However, better reagents are still needed, since some of the currently available ones require several additional steps to convert the initial aldol into the more desirable β-hydroxy acid or β-hydroxy aldehyde. Several such "second-generation" reagents are currently under development at Berkeley and elsewhere.

There is also room for improvement in controlling the "Cram's rule" problem. Double stereodifferentiation appears to be a good solution when it is the *erythro* aldol which is desired. However, no suitable *threo*-selective reagents have yet been developed which are amenable to this strategy. Evans' highly selective propionimides and the optically active α-trialkylsilyloxy ethyl ketones used by Masamune provide an excellent way to enhance or even reverse the inherent diastereoface selectivity of a chiral aldehyde, but these reagents also give only *erythro* products. There is still no suitable method for preparing the *threo*, anti-Cram diastereomer.

Nevertheless, the venerable aldol condensation has undergone considerable maturation during the past decade. It is certain that these problems will be solved, and it is equally certain that the aldol condensation will continue to be one of the most useful methods for establishing relative stereochemistry in the synthesis of acyclic compounds.

Acknowledgements: Much of the work discussed in this article was carried out in Berkeley by a talented cadre of graduate students and postdoctoral associates, whose names are mentioned in the various references. I also thank Professors David Evans and Phillip Stotter for informing me of their results prior to publication.

Note added in proof. The foregoing article was written by invitation of the Editors during December, 1980 and January, 1981. Due to certain problems not under the control of the author, publication was delayed for more than two years. Thus, it is unfortunate that the article is somewhat dated, although it was possible to make minor corrections in March, 1983. For more up-to-date treatment of some aspects of aldol stereoselection see references 65-67.

References

(1) R. Kane, *Ann. Physik Chem.*, [2], **44** (1838) 475; *J. Prakt. Chem.*, **15** (1838) 129.

(2) A.T. Nielsen and W.J. Houlihan, *Organic Reactions, Volume 16*, John Wiley & Sons, Inc., New York 1968.

(3) J.W. Cornforth in "Perspectives in Organic Chemistry," Edited by A. Todd, Interscience, New York 1956.

(4) F.C. Frostick, Jr., and C.R. Hauser, *J. Am. Chem. Soc.*, **71** (1949) 1350.

(5) J.M. Hamell and R. Levine, *J. Org. Chem.*, **15** (1950) 162.

(6) a. G. Wittig, H.D. Frommeld and P. Suchanek, *Angew. Chem. Internat. Edn. Engl.*, **2** (1963) 683. b. G. Wittig and H. Reiff, *ibid.*, **7** (1968) 7. c. G. Wittig, *Rec. Chem. Prog.*, **28** (1967) 45.

(7) C. Kowalski, X. Creary, A.J. Rollin, and M.C. Burke, *J. Org. Chem.*, **43** (1978) 2601.

(8) W.R. Dunnavant and C.R. Hauser, *J. Org. Chem.*, **25** (1960) 503.

(9) C.R. Kruger and E.G. Rochow, *J. Organometal. Chem.*, **1** (1964) 476.

(10) M.W. Rathke, *J. Am. Chem. Soc.*, **92** (1970) 3222.

(11) a. J.E. Dubois and M. Dubois, *Tetrahedron Lett.*, (1967) 4215. b. J.E. Dubois and M. Dubois, *Chem. Commun.*, (1968) 1567. c. J.E. Dubois and M. Dubois, *Bull. Soc. Chem. Fr.*, (1969) 3120. c. J.E. Dubois and M. Dubois, *Bull. Soc. Chim. Fr.*, (1969) 3553.

(12) J.E. Dubois and P. Fellman, *C.R. Acad. Sci. Ser. C.*, **274** (1972) 1307.

(13) J.E. Dubois and P. Fellman, *Tetrahedron Lett.*, (1975) 1225.

(14) H.O. House, D.S. Crumrine, A.Y. Teranishi and H.D. Olmstead, *J. Am. Chem. Soc.*, **95** (1973) 3310.

(15) W. Fenzl and R. Koster, *Liebigs Ann. Chem.*, (1975) 1322.

(16) W.A. Kleschick, C.T. Buse, and C.H. Heathcock, *J. Am. Chem. Soc.*, **99** (1977) 247.

(17) C.H. Heathcock, C.T. Buse, W.A. Kleschick, M.C. Pirrung, J.E. Sohn, and J. Lampe, *J. Org. Chem.*, **45** (1980) 1066.

(18) H. Zimmerman and M. Traxler, *J. Am. Chem. Soc.*, **79** (1957) 1920.

(19) S. Masamune, S. Mori, D. Van Horn, and D.W. Brooks, *Tetrahedron Lett.*, (1978) 1665.

(20) a. D.A. Evans, E. Vogel, and J.V. Nelson, *J. Am. Chem. Soc.*, **101** (1979) 6120. b. D.A. Evans, J.V. Nelson, E. Vogel and T.R. Taber, *ibid.*, **103** (1981) 3099.

(21) E.A. Jeffrey, A. Meisters and T. Mole, *J. Organometallic Chem.*, **74** (1974) 373.

(22) K.K. Heng and R.A.J. Smith, *Tetrahedron*, **35** (1979) 425.

(23) K.K. Heng, J. Simpson, R.A.J. Smith and W.T. Robinson, *J. Org. Chem.*, **46** (1981) 2932.

(24) A.I. Meyers and P. Reider, *J. Am. Chem. Soc.*, **101** (1979) 2501.

(25) a. J. Mulzer, G. Brüntrup, Jürgen Finke and M. Zippel, *J. Am. Chem. Soc.*, **101** (1979) 7723; b. J. Mulzer, M. Zippel, G. Brüntrup, J. Segner, and J. Finke, *Liebigs Ann. Chim.*, (1980) 1108.

(26) R. Noyori, I. Nishida, J. Sakata, and M. Nishizawa, *J. Am. Chem. Soc.*, **102** (1980) 1223.

(27) R. Noyori, I. Nishida, and J. Sakata, *J. Am. Chem. Soc.*, **103** (1981) 2106.

(28) C.H. Heathcock and John Lampe, *J. Org. Chem.*, submitted for publication.

(29) John E. Sohn, unpublished results.

(30) R. Noyori, K. Yokayama, J. Sakata, I. Kuwajima, E. Nakamura, and M. Shimizu, *J. Am. Chem. Soc.*, **99** (1977) 1265.

(31) J. Mulzer, J. Segner, and G. Brüntrup, *Tetrahedron Lett.*, (1977) 4561.

(32) a. Hirama and S. Masamune, *Tetrahedron Lett.*, (1979) 2225. b. D.E. Van Horn and S. Masamune, *Tetrahedron Lett.*, (1979) 2229. c. M. Hirama, D.S. Garvey, L.D.-L. Liu, and S. Masamune, *Tetrahedron Lett.*, (1980) 3937.

(33) a. J. Hooz and L. Linke, *J. Am. Chem. Soc.*, **90** (1971) 5936, 6891. b. J. Hooz and D.M. Gunn, *Chem. Commun.*, (1969) 139. c. J. Hooz and D.M. Gunn, *J. Am. Chem. Soc.*, **91** (1969) 6195. d. J. Hooz and D.M. Gunn, *Tetrahedron Lett.*, (1969) 3455. e. J. Hooz and G.F. Morrison, *Can. J. Chem.*, **48** (1970) 868.

(34) T. Mukaiyama and T. Inoue, *Chem. Lett.*, (1976) 559.

(35) D. Maruoka, S. Hashimoto, Y. Kitagawa, H. Yamamoto, and H. Nozaki, *J. Am. Chem. Soc.*, **99** (1977) 7705.

(36) a. D.A. Evans and L.R. McGee, *Tetrahedron Lett.*, (1980) 3975. b. D.A. Evans, private communication.

(37) Y. Yamamoto and K. Maruyama, *Tetrahedron Lett.*, (1980) 4607.

(38) K. Saigo, M. Osaki, and T. Mukaiyama, *Chem. Lett.*, (1975) 989.

(39) T.H. Chan, T. Aida, P.W.K. Lau, V. Gorys, and D.N. Harpp, *Tetrahedron Lett.*, (1979) 4029.

236

(40) J.P. Hagen, unpublished results.

(41) C.T. White and C.H. Heathcock, *J. Org. Chem.*, **46** (1981) 191.

(42) M.C. Pirrung, unpublished results.

(43) M.C. Pirrung and C.H. Heathcock, *J. Org. Chem.*, **45** (1980) 1727.

(44) C.H. Heathcock, M.C. Pirrung, S.H. Montgomery, and J. Lampe, *Tetrahedron*, **37** (1981) 2290.

(45) C.H. Heathcock, J.P. Hagen, E.T. Jarvi, M.C. Pirrung, and S.D. Young, *J. Am. Chem. Soc.*, **103** (1981) 4972.

(46) a. D.J. Cram and F.A. Abd Elhafez, *J. Am. Chem. Soc.*, **74** (1952) 5828; b. G.J. Karabatsos, *J. Am. Chem. Soc.*, **89** (1967) 1367; c. M. Cherest, H. Felkin, and N. Prudent, *Tetrahedron Lett.*, (1968) 2201.

(47) C.H. Heathcock, S.D. Young, J.P. Hagen, M.C. Pirrung, C.T. White, and D. VanDerver, *J. Org. Chem.*, **45** (1980) 3846.

(48) N.T. Anh and O. Eisenstein, *Nouv. J. Chem.*, **1** (1977) 61.

(49) a. C.H. Heathcock and C.T. White, *J. Am. Chem. Soc.*, **101** (1979) 7076; b. C.H. Heathcock, C.T. White, J.J. Morrison, and D. VanDerveer, *J. Org. Chem.*, **46** (1981) 1296.

(50) a. S. Masamune, S.A. Ali, D.L. Snitman, and D.S. Garvey, *Angew Chem. Internal. Edn. Engl.*, **19** (1980) 557. b. S. Masamune, W. Choy, F.A.J. Kerdesky, and B. Imperali, *J. Am. Chem. Soc.*, **103** (1981) 1556.

(51) D.J. Cram and F.D. Greene, *J. Am. Chem. Soc.*, **75** (1953) 6005.

(52) J.D. Morrison and H.S. Mosher, "Asymmetric Organic Reactions", Prentice-Hall, Inc., Englewood Cliffs, New Jersey 1971, pp. 81-132.

(53) D.A. Evans and T.R. Taber, *Tetrahedron Lett.*, (1980) 4675.

(54) D.A. Evans, J. Bartoli, and T.L. Shih, *J. Am. Chem. Soc.*, **103** (1981) 2127.

(55) D.L. Evans and L.R. McGee, *J. Am. Chem. Soc.*, **103** (1981) 2876.

(56) D. Seebach, H.-O. Kalinowsky, B. Batsani, G. Cross, H. Daum, N. DuPreez, V. Ehrig, W. Langer, C. Nüssler, H.-A. Oei, and M. Schmidt, *Helv. Chim. Acta.*, **60** (1977) 301.

(57) a. C.H. Heathcock, M.C. Pirrung, C.T. Buse, J.P. Hagen, S.D. Young, and J.E. Sohn, *J. Am. Chem. Soc.*, **101** (1979) 7077; b. C.H. Heathcock, M.C. Pirrung, J. Lampe, C.T. Buse, and S.D. Young, *J. Org. Chem.*, **46** (1981) 2290.

(58) C.T. Buse, unpublished observations.

(59) D.B. Collum, J.H. MacDonald III, and W.C. Still, *J. Am. Chem. Soc.*, **102** (1980) 2117, 2118, 2120.

(60) S. Masamune, M. Hirama, S. Mori, S.A. Ali, and D.S. Garvery, *J. Am. Chem. Soc.*, **103** (1981) 1568.

(61) J.P. Hagen and S.D. Young, unpublished results.

(62) T. Nakata, G. Schmid, B. Vranesic, M. Okigawa, T. Smith-Palmer, and Y. Kishi, *J. Am. Chem. Soc.*, **100 (1978) 2933.**

(63) R.E. Ireland, G.J. McGarvey, R.C. Anderson, R. Badouch, B. Fitzsimmons, and S. Thaisrivongs, *J. Am. Chem. Soc.*, **102** (1980) 6180.

(64) D.A. Evans, C.E. Sacks, W.A. Kleschick, and T.A. Taber, *J. Am. Chem. Soc.*, **101** (1979) 6789.

(65) D.A. Evans, J.V. Nelson and T.R. Taber, in "Topics in Stereochemistry, Volume 13," N.L. Allinger, E.L. Eliel and S.H. Wilen, Eds., Wiley-Interscience, New York, 1982.

(66) C.H. Heathcock, in "Current Trends in Organic Synthesis," H. Nozaki, Ed., Pergamon Press, Oxford and New York, 1983.

(67) C.H. Heathcock, in "Asymmetric Synthesis, Volume 3," J.D. Morrison, Ed., Academic Press, Inc., New York, 1983.

CHAPTER 5

TECHNIQUES IN CARBANION CHEMISTRY

TONY DURST

OTTAWA-CARLETON INSTITUTE FOR RESEARCH AND GRADUATE STUDIES IN
CHEMISTRY

DEPARTMENT OF CHEMISTRY, UNIVERSITY OF OTTAWA, CANADA   K1N 9B4

CONTENTS:

I. INTRODUCTION

The reactions of carbanions represent the most important method for constructiong carbon-carbon bonds. Historically this type of reaction includes the many condensations involving enolates of carbonyl compounds which were prepared using bases such as hydroxides, alkoxides, sodium or potassium amide and sodium hydride. The development of alkyllithiums as reagents for the generation of highly basic carbanions began with the work of Gilman in the 1940's. This has greatly accelerated in the past twenty years due to the availabilty of relatively inexpensive, commercially available reagents and the introduction of a variety of the non-nucleophilic, but highly basic lithium dialkylamides such as lithium diisopropylamide (LDA) and lithium tetramethylpiperadide (LTMP).

Today, by judicious choice of the metallating agent(s), solvent system and temperature, virtually any type of carbanion centre can be created. In the authors opinion, one of the more remarkable examples which illustrates the above point comes from the work of Beak and co-workers(1) who have been able to generate an anion centre α to the sulfur and oxygen atoms in the otho-substituted esters 1 and 2 using sec-BuLi tetramethylenediamine (TMEDA) in THF at $-78°$.

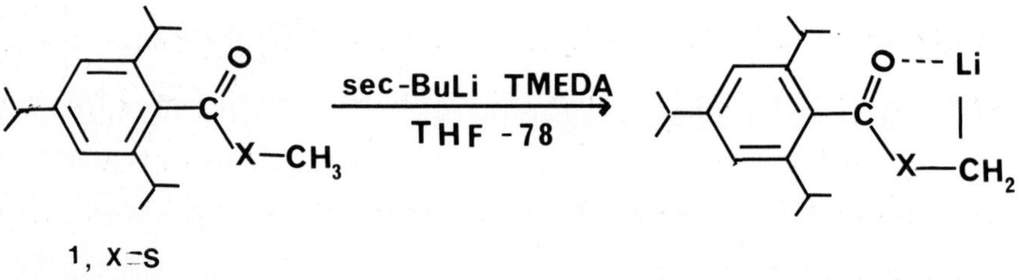

1, X＝S
2, X＝O

The purpose ot this article is the summarize the most commonly used combinations of bases, solvents and temperature for the generation of widely used types of carbanions. Representative examples of experimental proceedures are included at the end of the article. The article also summarizes several methods for titrating commercial solutions of alkyllithiums, drying and storing of commonly used solvents and reagents. It is recognized that the latter items are not of particular interest or value to experienced workers in the field but may be of use to those who occasionly use these techniques. Methods of preparation of ultrapure samples of alkyllithiums for careful structural studies or physicochemical measurements are considered outside the scope of this article.

II. HANDLING OF REAGENTS AND SOLVENTS.

Reactions involving highly basic carbanions must be carried out under inert atmosphere conditions using dry equipment since most of these reagents are destroyed by both water and oxygen. The most commonly used inert atmosphere system is high purity, dry nitrogen. Argon, despite its higher unit cost, is gaining increasing acceptance over nitrogen because its greater than air density offers extra protection. When judiciously used, the cost of argon per experiment is quite small.

All glassware and other equipment, including syringes, should dried for several hours in an oven at 125 , cooled in a vacuum dessicator, assembled and flushed with the inert gas prior to the introduction of solvents and reagents. The extra precaution of flaming the equipment after assembly under a stream of inert gas can also be taken.

A variety of equipment setups have been described for carrying out reactions under a dry inert atmosphere. This can be as simple as a roundbottom flask equipped with a rubber septum, filled with a nitrogen atmosphere and a magnetic stirrer for mixing of the reagents. Solvents and reagent solutions are introduced via syringes. A second needle connected to a bubbler filled with mineral oil serves as a vent. Such a system serves quite well for many small scale reactions carried out in 10-50 mL flasks and requiring only a few mL of solvent.

The equipment can however be consideraby more sophisticated and involve motor driven syringes, double tip needles for transfer from one reaction flask to another, pressure valves, etc. Amongst others, the Aldrich Chemical Co., Milwaukee, Wisc., distributes free of charge a small, but excellent information booklet describing methods and equipment for handling air-sensitive compounds.(2) More sophisticated details are also availble in Shriver's monograph, "The manipulation of air-sensitive Compounds (3) and in H.C.Brown's "Organic Synthesis via Boranes".(4)

III. PREPARATION AND STORAGE OF COMMONLY USED SOLVENTS AND REAGENTS.

Excellent general guides to solvent purification and drying are available. These include a chapter in the Weissberger series "Physical Methods in Chemistry, Vol. II, (5), and more recently, an evaluation of a number of common dessicants used by organic chemists for the drying of seveal types of solvents such as hydrcarbons, ethers and acetonitrile (6), based on a new and very sensitive tritiated water tracer method for determining water content.(7)

Tetrahydrofuran (THF) is by far the most commonly used solvent for the preparation of highly basic carbanions. Other solvents which are occasionally used to good advantage are dimethoxy ethane (DME), diethyl ether and, less frequently, benzene or hexane. Carbanion reactions carried out in DMSO using the methylsulfinyl carbanion as base(8) will not be dealt with in this chapter.

Tetrahydrofuran is most readily obtained in dry form suitable for alkyllithium and carbanion reactions upon distillation from sodium or potassium using the intense blue colour of the benzophenone ketyl (7) as an indicator of dryness. Distillation from $LiAlH_4$ does not seem to be as efficient a method for drying ethers (6), and is also more expensive than the $Na.Ph_2CO$ method. Typically, in either of these methods the THF employed should be reagent grade quality and thus contain rather small amounts of moisture or peroxides. Poorer quality THF should first be stored over KOH pellets prior to drying as described above. Obviously, care should be taken not to allow the distillation pot to approach "dryness", and in the recharching of the pot with fresh solvent. The use of sodium is preferred over potassium due to lesser cost and easier and safer handling.

Generally, dry THF is not stored, but used directly as generated. A continuous distillation apparatus, either of commercial origin , or readily constructed in-house can provide a continuous supply of high quality dry THF. A version of such an apparatus is shown in Fig. 1. If good reagent grade THF is used to recharge the resevoir, a few grams of sodium or potassium will provide many liters of dry solvent without the necessity of cleaning and restarting the distillation system.

Dry diethyl ether, dimethoxyethane or dioxane can also be prepared by distillation from Na/Ph_2CO, (9) or from $LiAlH_4$. (10) These solvents can be stored for reasonable periods under argon over sodium wire or calcium hydride. If used frequently a

FIGURE 1.

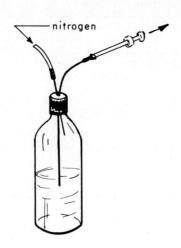

FIGURE 2.

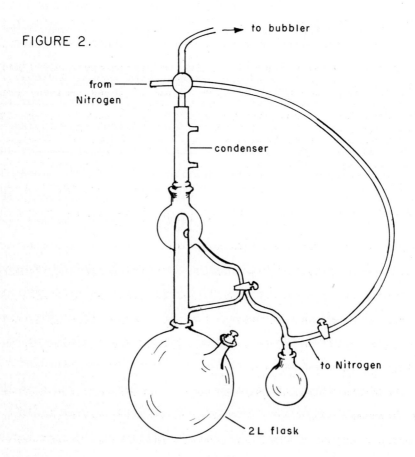

continuous still arrangement should be considered.

Hexamethylphosphoramide (HMPA) is a highly polar solvent and is often used as a promoter of carbanion reactions, especially alkylations. Distillation from CaH_2 or BaO under reduced pressure followed by storage over molecular sieves 3A°(11), or 4A° (10), under argon assures a good supply of this solvent over a period of weeks.

Benzene and hydrocarbons such as hexanes are easily dried by stirring over Al_2O_3 .(6) Distillation to remove higher boiling impurities or chemical treatment to remove alkenes from hexanes (5) may also be necessary depending on the quality of the solvents to be dried.

Diisopropylamine, and 2,2,6,6-tetramethylpiperidine are the most commonly used amines for the preparation of the very useful non-nucleophilic amide bases i.e. lithium diisopropylamide (LDA) and lithium tetramethylpiperidide (LTMP). These amines are usually distilled from calcium hydride and stored over 3A molecular sieves under an argon atmosphere. Tetramethylethylene diamine, an amine which when added in small amounts greatly increases the reactivity of alkyllithiums and thus their metallating capabilities, can be similarly distilled and stored.

IV. HANDLING OF ALKYLLITHIUM REAGENTS.

A variety of alkyllithiums including CH_3Li, n-BuLi, sec-BuLi, and t-BuLi are commercially available as 1.5-2.5M solutions in hydrocarbons such as pentane or hexane. Preparations of methyllithium are sold as ether solutions due to the insolubility of methyllithium in hydrocarbons.

All commercial samples are sold with a variety of self-sealing rubber septums. For typical laboratory scale reactions the required amount of reagent is removed via syringe. An effective method of filling a syringe using a positive nitrogen pressure is shown in Fig.2. Such a proceedure is particularly useful when relatively large amounts of solution are withdrawn. Alternatively, the syringe is first filled with nitrogen which is injected into the vessel containing the alkyllithium prior to the removal of the reagent. A slight positive pressure over the sealed stock solution will help prevent the entry of moisture or air during the repeated puncturing of the septums. Covering the septum with "PARAFILM" after each use is also helpful.

It is generally recommended that a slight excess of reagent and a small amount of the nitrogen atmosphere be removed with the syringe from the stock bottle. As soon as the syringe needle leaves the septum, the syringe should be turned upside down (needle pointing up in the hood) and the plunger pushed back until the entire syringe chamber is filled wwith the alkyllithium solution. The excess can then be expelled into a beaker containing a small amount of ethyl acetate which provides for a mild method of destroying these reagents.

V. ASSAYING OF ORGANOLITHIUM REAGENTS.

The titre reported on the label of commercially available alklyllithium solutions is often quite accurate but should not be taken at face value since the quality may be affected by shipping and handling proceedures. Alkyllithium solutions should be titrated regularly commencing with the first use of a new bottle and after about every ten uses or a one month period, whichever comes first.

A number of simple methods of determining the concentration of active alkyllithiums have been described beginning with the double titration method of Gilman and Cartledge.(12) This approach requires titration of one aliquot against standard acid to yield a total base concentration (RLi and ROLi), and treatment of a second aliquot with either benzyl chloride or 1,2-dibromoethane to yield the ROLi concentration. The alkyllithium titre is then obtained by difference. The Gilman method, although more cumbersome than those described below nevertheless continues to represent the standard by which the others are evaluated.

Each of the five methods which are described below require only a single titration. In each case a simple, easily dectected endpoint is observed thus giving these methods good reproduceability.

Eppeley and Dixon (13) utilized the red colour of triphenyl-methyllithium in DMSO as the indicator. Their method involves titration of a known amount of a standard organic acid (benzoic acid) in a DMSO-monoglyme-hydrocarbon mixture. The equations involved are:

$$PhCO_2H + RLi \longrightarrow PhCO_2Li \text{ (colourless)}$$
$$RLi + Ph_3CH \longrightarrow RH + Ph_3CLi \text{ (red)}$$

The endpoint occurs after all of the benzoic acid has been converted to lithium benzoate and a trace of triphenylmethyl-lithium has been formed. DMSO is a necessary component. In its absence a white precipitate is formed rather than a red colour change. Alkoxides which may be present in the alkyllithium solution do not interfere.

The method is reproduced verbatim.(13,14) "The indicator solution is prepared by dissolving 0.50g of triphenylmethane in 400mL of DMSO and 100mL of monoglyme and dried to the first permanent red colour by addition of an alkyllithium solution and stored in a serum capped bottle. A 50 mL Erlenmeyer flask fitted with a septum and containing a magnetic stirring bar was evacuated, flamed and flushed with purified nitrogen three times. Ten mL of the indicator solution was added with a syringe and dried to the first permanent endpoint. The flask was immediately charged with weighed benzoic acid, or an accurately measured aliquot of standard acid solution in monoglyme. The amount of acid was chosen so that the titration would consume about 2.00 to 2.50 mL of base solution. Needle inlet and exhaust ports were inserted and purified nitrogen was admitted in a constant stream. The solution was immediately titrated to the red endpoint with the unknown organolithium solution added from a 2.50 mL Hamilton gas tight syringe."

Watson and Eastham eveloped a method based on strongly coloured complexes fromed between alkyllithiums and 1,10-phenanthroline, 3, or 2,2'biquinoline, 4.(15). The method involves addition of a few mg of 3 or 4 to a solution of n-BuLi in hexane to 20 mL of benzene which produces a yellow-green or chartreuse coloured solution for 4 and a rust-red complex for 3. After titration with 1.0M sec-butyl alcohol solution in xylene the solution is clear and colourless; the colour fades sharply after the addition of one equivalent of titrant. This method is not applicable to the assaying of methyllithium in ether since ether solvents interfere with the formation of the coloured complexes.

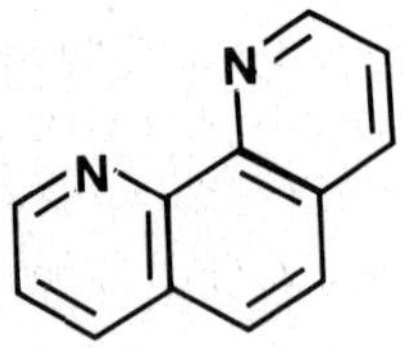

3 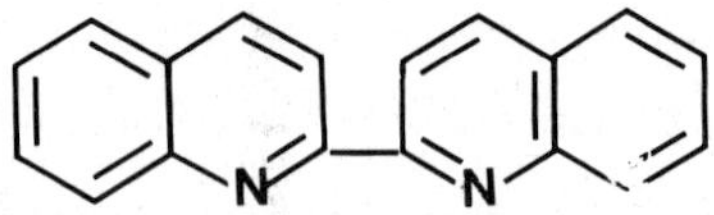   4

The use of diphenylacetic acid as an indicator has been developed by Kofron and Baclawski. (16) The method is based on the observation that the dianion 5 is yellow. " A weighed sample of diphenylacetic acid (typically 0.50g) is placed in an Erlenmeyer flask and dissolved in THF (10 mL) and the alkyllithium is run in from a syringe until the yellow endpoint is reached. The yellow colour indicates formation of lithium α-lithiodiphenylacetate after all the carboxyl proton is consumed."

$$Ph_2CHCO_2H \quad + \quad RLi \quad \longrightarrow \quad Ph_2CHCO_2Li \quad + \quad RLi \quad \longrightarrow \quad Ph_2\underset{Li}{C}CO_2Li$$

(colourless) (yellow)

2,5-Dimethylbenzyl alcohol can also be used as an indicator for the determination of alkylithium concentrations.(17) Reaction of a solution of the alcohol dissolved in THF, ether or benzene with various alkyllithiums gives a colourles slution of the ithium alkoxide. After one equivalent has been added, the addition of a small amount (less than 0.01 equiv.) causes the development of an intense red colour. In a typical titration a small amount of the alcohol (100-200 mg) was weighed into a small flask fitted with a magnetic stirrer and saled with a septum. The flask is flushed with nitrogen and the alcohol is dissolved in 1-4 mL of the appropriate dry solvent. The organolithium reagent was added dropwise a syringe until the solution was just charged with a pink colour. The results obtained by this method compared closely with those using other indicators such as the 1,8-phenanthroline.

Very recently Lipton and coworkers reported that dibenzyl tosylhydrazone serves as an excellant indicator in the tirations of alkylithiums.(17) This compound, which is readily available in pure form, and like diphenylacetic acid is not hygroscopic, reacts with one equivalent of an alkyllithium to give a colourless solution of the N-lithio derivative 6. Upon addition of more than one equivalent red dianion 7 is rapidly formed. The results using this method are claimed to be highly reproduceable and applcable to n-BuLi, MeLi, PhLi and t-BuLi. The proceedure is as follows: "An oven dried 50 mL Erlenmeyer flask containing a magnetic stirring bar is charged with an accurately weighed portion of tosyl hydrazone. The flask is fitted with a rubber serum cap, an inlet for scrubbed nitrogen and a vent and is flushed with the inert gas. Anhydrous THF (10 mL) is then added and stirring is continued until the tosyl hydrazone disslves. The flask is cooled to $0°C$ by means of an ice bath. (This minimizes errors associated with the reaction of the alkyllithium with the solvent) and the alkyllithium is added dropwise (rapid addition causes premature colouration and loss of the endpoint sharpness) via a 1.00 mL syringe until the orange colour of dianion formation persists."

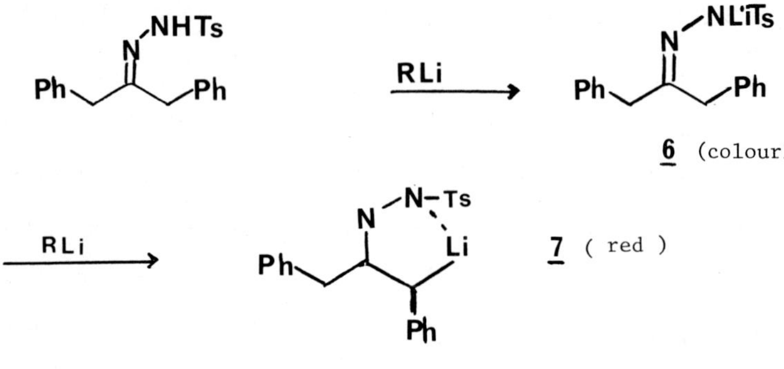

6 (colourless)

7 (red)

Comparison of the results indicates that all of the above methods give consistent and reproducible results. The choice thus depends on the indicator available. Since relatively small amounts of indicator are actually employed the cost is of minor importance. The diphenylacetic method seems to be most commonly.

VI. CHOICE OF REAGENTS AND REACTION CONDITIONS.

A. GENERAL.

This aspect is often quite crucial but often still a matter of trial and error. A number of well-known researchers were asked how they decided on the base, solvent, and other reaction conditions to be used in a particular situation. Probably the best advice was: "We look for a literature precedent and if that is not successful we try a variety of conditions until we find one that works". A few of the general guidelines that can be gleaned from the literature are described below. An excellent up-to-date set of examples concerning the use of a particular base can be found in the various volumes of "Reagents in Organic Chemistry" by Fieser and Fieser.

Anions α to ketones, esters, nitriles acids, etc., are best generated with non-nucleophilic bases. For these purposes the lithium dialkylamides are generally the reagents of choice. Typically, these secondary amides should be hindered to minimize any nucleophilic tendencies. The lithium salts of diisopropyl-amine, isopropylcyclohexylamine, dicyclohexylamine, 2,2,6,6-tetramethylpiperdine, and hexamethyldisilazane seem to best serve for this purpose. Lithium diethylamide is generally no longer used in the above context. Prior to the development of the dialkylamides, sodium triphenylmethide was often employed as a non-nucleophilic base; its use for this purpose is now relatively rare.

Solutions of the dialkylamides are readily prepared in situ simply by addition of an alkyllithium, such as n-BuLi, to the amine dissolved in THF, or other suitable solvent at temperatures varying from -78° to 0°, and above. The formation of the lithium amides, especially in the ether solvents, is very rapid and essentially irreversible. The lithium bases are suffiently strong bases to deprotonate and cleave solvents such as THF, diethyl ether, and especially DME and HMPA at temperatures above 0 ; solutions of these bases are usually not stored for extended periods. Dilute solutions of LDA in hexane are stable.(19) A method for preparing molar quantities of LDA in ether using styrene as hydrogen acceptor has been described.(20)

$$2 \ (CH_3)_2CH_2NH \ + \ 2 \ Li \ + \ PhCH=CH_2 \longrightarrow 2 \ LDA \ + \ PhCH=CH_2$$

In the formation of the ketone or ester enolates, the substrate is added slowly to the LDA solution. This minimizes the self-condensation reaction. A number of examples of the generation of enolate, α-nitrile, α-carboxylate and α-imine anions chosen from the recent literature are given in Table 1. These examples are not meant to be complete but only representative of the types of carbanions which may be prepared with LDA. Detailed descriptions of the generation of various ketone and imine enolates and the stereochemistry of the reactions of these species with carbonyl compounds and alkylating agents are described in two other chapters of this Volume.

B. LITHIUM DIISOPROPYLAMIDE. (LDA)

Amongst the lithium dialkylamides LDA is certainly used the most widely, mainly because of its ready availabilty, low cost, the ease with which the resultant amine, b.p. 84°, is removed from the reaction mixture by simple evaporation, and the general effectiveness of this reagent under a variety of conditions.

The potassium salt of diisoppropylamine (KDA), 8, has been prepared by the reaction of n-BuLi and KOtBu with the amine in

TABLE 1

Generation of Anions with LDA and Reactions with Electrophiles

Substrate	Electrophile	Product (%)	Ref.
$CH_3CH_2\overset{\overset{O}{\|\|}}{C}\underline{t}Bu$	PhCHO	$\overset{OH}{Ph}\overset{\blacktriangledown}{CH}CH\overset{\overset{O}{\|\|}}{C}\underline{t}Bu$ over $\overset{\blacktriangle}{CH_3}$ erthro 78(%)	21
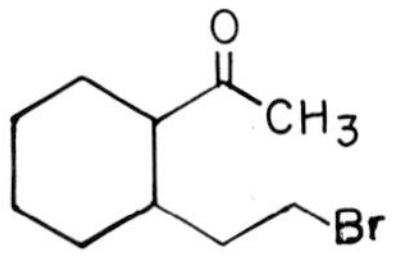	—— 1	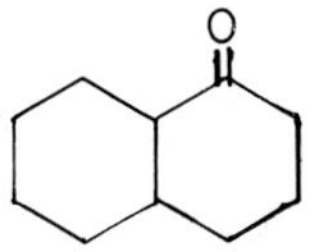 (80)	19
$CH_3CH_2\overset{\overset{O}{\|\|}}{C}-N$	RI [1]	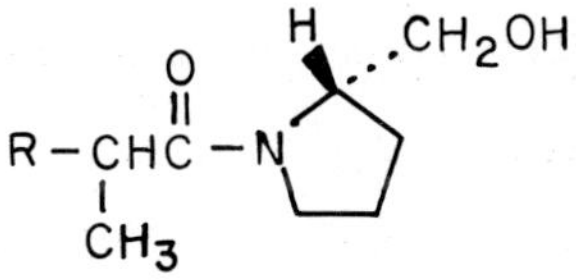 $R-\overset{\overset{O}{\|\|}}{C}H\overset{\overset{O}{\|\|}}{C}-N$ over $\overset{\blacktriangle}{CH_3}$ (75-99)(~95% ee)	22
$\overset{OH}{CH_3CHCH_2CO_2Et}$	$CH_2=CHCH_2Br$	$\overset{OH}{CH_3CHCHCO_2Et}$ over $\overset{\blacktriangle}{CH_2CH=CH_2}$ (80) (96% threo)	23
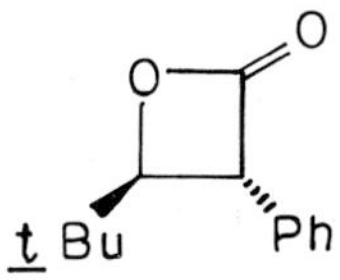	RX	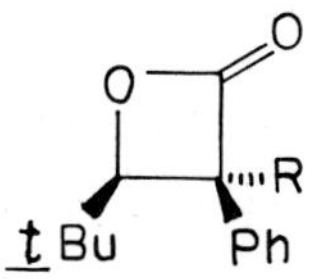	24
$CH_3CO_2\underline{t}Bu$	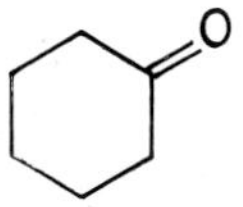	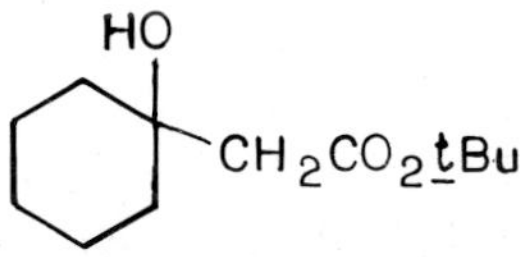 (100)	25

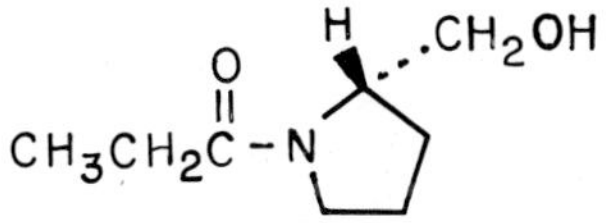

Substrate	Electrophile	Product (%)	Ref.
$CH_3CH=CHCO_2Et$	RX [1]	$CH_2=CH-\underset{\underset{R}{\mid}}{C}HCO_2Et$ (>90)	26
	RX [1]	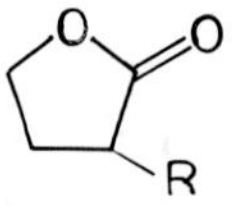 (>90)	27
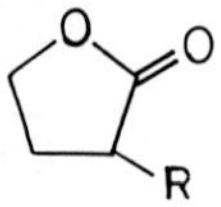	$R'X$ [1]	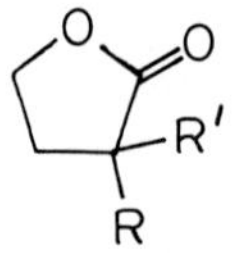	27
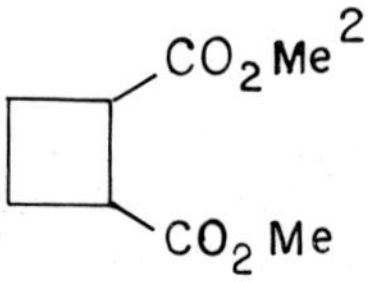 CO_2Me [2]	D_2O	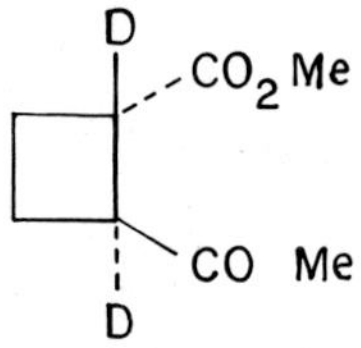 (90)	28
$EtO_2C-\underset{\underset{HO}{\mid}}{\overset{\overset{H}{\mid}}{C}}-CH_2CO_2Et$ (R)	CH_3CH_2I	$EtO_2C-\underset{\underset{HO}{\mid}}{\overset{\overset{H}{\mid}}{C}}-\underset{\underset{CH_2CH_3}{\mid}}{\overset{\overset{H}{\mid}}{C}}-CO_2Et$	29
$CH_3\overset{\overset{S}{\parallel}}{C}-S\ MgC\ell$	$SO_2(OCH_3)_2$	$CH_2=C\overset{\diagup SCH_3}{\diagdown SCH_3}$ 60 %	30
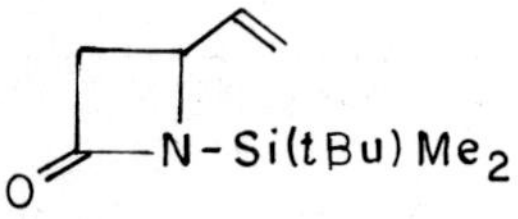	$CH_3\overset{\overset{O}{\parallel}}{C}H$	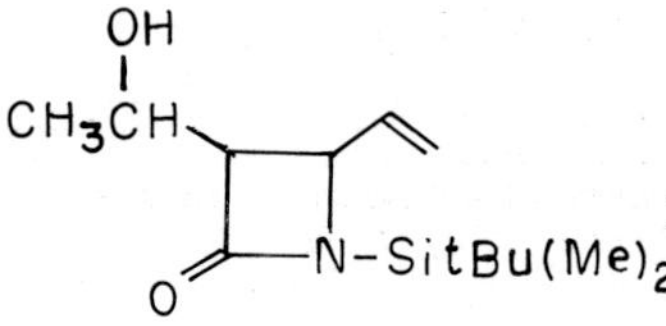	31

Substrate	Electrophile	Product (%)	Ref.
$CH_3CH=\overset{\overset{\displaystyle CH_3}{|}}{C}-CO_2H$	$CH_2=CH-CH_2Br$	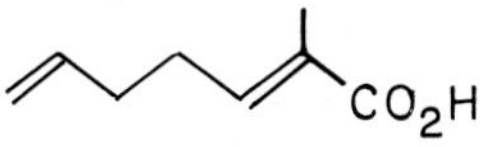 CO_2H	32
$PhOCH_2CO_2H$	RX	$PhO\overset{\underset{\displaystyle R}{|}}{C}HCO_2H$ (50–70)	33
$R-\overset{\underset{\displaystyle CN}{|}}{C}H\overset{\underset{\displaystyle CH_3}{|}}{C}HOEt$	$R'X$	$R\overset{\overset{\displaystyle R'}{|}}{C}O\overset{\underset{\displaystyle CN}{|}}{C}H\overset{\underset{\displaystyle CH_3}{|}}{C}HOEt$	34
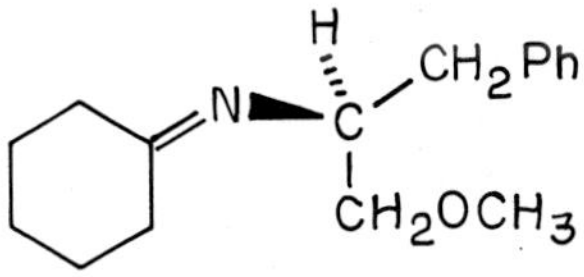	RX	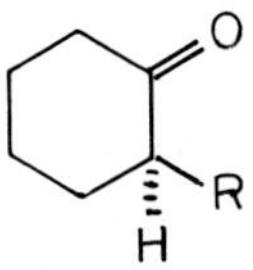 $(60-85^2)\,(80-99\%\ ee)$	35
$(CH_3)_2CHCH_2\overset{\overset{\displaystyle O}{\|}}{S}CH_3$	$MeS-SMe$	$(CH_3)_2CHCH_2\overset{\overset{\displaystyle O}{\|}}{S}CH_2SCH_3$ (65)	36
$(CH_3)_2\,CH\,\overset{\overset{\displaystyle O}{\|}}{S}e\,Ph$	$Ph_2C=O$	$(CH_3)_2\overset{\underset{\displaystyle Ph_2C-OH}{|}}{C}-\overset{\overset{\displaystyle O}{\|}}{S}e\,Ph$	36 a
$PhSCH=CH_2$	RX	$PhS\overset{\underset{\displaystyle R}{|}}{C}H=CH_2$	37
$H\overset{\overset{\displaystyle O}{\|}}{C}N\overset{\underset{\displaystyle R}{|}}{C}H_2OCH_3$	$R_1\overset{\overset{\displaystyle O}{\|}}{C}R_2$	$R_1-\overset{\underset{\displaystyle R_2}{|}}{C}-\overset{\overset{\displaystyle O}{\|}}{C}-\overset{\underset{\displaystyle R}{|}}{N}CH_2OCH_3$ with OH on first C	38

1. HMPA was added with the alkylating agent to promote the alkylation step.
2. After hydrolysis.

THF.(39) It is reported to be more effective than LDA and allowed the preparation of the α-selenocarbanion 9 by deprotonation of the phenyl vinyl selenide 10. 2-Methyl-1-phenyl-selenoalkenes, 11, in contrast, give allylic deprotonation to afford 12. The reagent 8 is stable in THF only at temperatures considerably below 0^0. KDA has also been found to be effective for the deprotonation of nitrosamines.(40)

$$KOtBu \; + \; \left[(CH_3)_2CH\right]_2NH \; + \; n\text{-BuLi} \longrightarrow KDA \; (\underline{8})$$

$$PhSeCH=CH_2 \; + \; \underline{8} \longrightarrow PhSeC=\underline{C}H_2$$

$$\underline{10} \qquad\qquad\qquad \underline{9}$$

$$PhSeCH=CH-CH_3 \; + \; 8 \longrightarrow PhSeCH-CH-CH$$

$$\underline{11} \qquad\qquad\qquad\qquad \underline{12}$$

C. LITHIUM 2,2,6,6-TETRAMETHYLPIPERIDIDE. (LTMP)

Lithium 2,2,6,6-tetramethylpiperidide, 13, was introduced by Olofson and Dougherty in 1973 (41) as a highly basic non-nucleophilic reagent which has since proven to be highly effective for the generation of carbenoids via α-eliminaton reactions. A comparison of LTMP vs several other lithium amides including LDA, lithium piperidide, and lithium isopropyl-cyclohexylamide showed it to be superior for the preparation of 7-phenylnorcarane from benzyl chloride and cyclohexene.

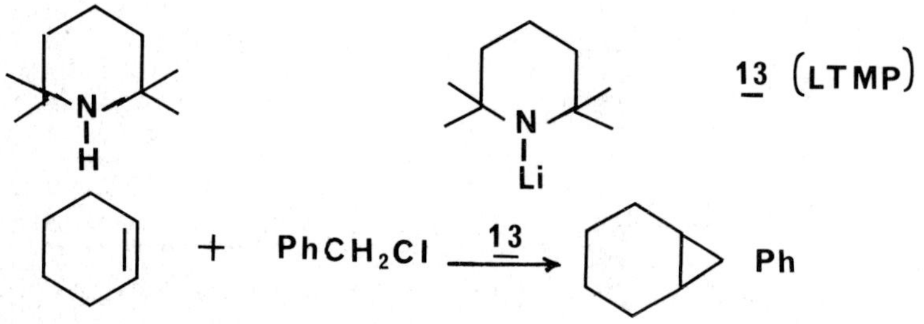

LTMP has since been employed by Olofson's group for the

prepartion of variety of substituted cyclopropanes including the alkoxy derivative 13, (42) acyl cyclopropanols 14, (43) and the trimethylsilyl and trimethylstannyl compounds 15.(44)

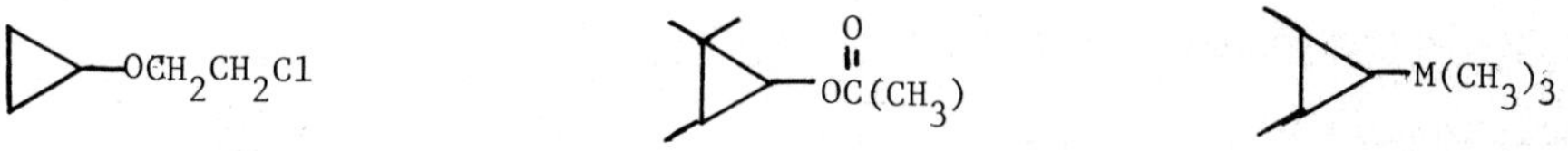

OCH_2CH_2Cl
$OC(CH_3)$
$M(CH_3)_3$

14 (55-80%) 14a (35%) 15 (M = Si,Sn) (21-23%)

Incidentally, reaction of chlorotrimethylsilane, 16, the precursor of the carbenoid leading to 15 (M= $Si(CH_3)_3$, with sec -BuLi/TMEDA led cleanly to the anion 16 which was trapped efficiently with the usual electrophiles.(45)

$$(CH_3)_3SiCH_2Cl \quad + \quad \text{sec-BuLi} \xrightarrow{\text{TMEDA}} (CH_3)_3Si\underset{\displaystyle Li}{\underset{|}{C}}HCl \qquad (16)$$

Danheiser et al (46) have trapped the carbenoid from chloromethyl β -chloroethyl ether with 1,3-dienes to obtain 2-vinylcyclopropyl ethers, 18. Subsequent halogen-metal exchange using n-BuLi afforded the lithio derivative 19 which readily rearranged to cyclopent-3-en-1-ols, 20.

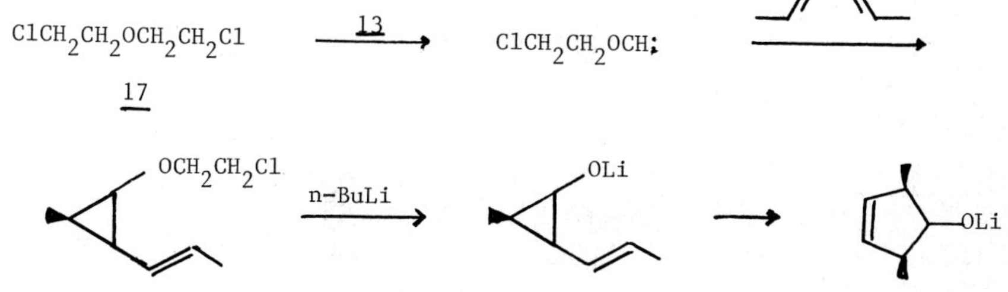

$ClCH_2CH_2OCH_2CH_2Cl$
13
$ClCH_2CH_2OCH$:
OCH_2CH_2Cl
n-BuLi
OLi
OLi

17 18 19 20

LTMP is also very useful for the conversion of halobenzenes to benzynes. For example, reaction of phenylacetylene and bromobenzene in THF with an excess of LTMP afforded tolan, 21, in excellent yield.(41) Dehydrochlorination of chlorobenzene with LTMP in the presence of N-trimethylslylpyrrole gave 22 in moderate yield. (47)

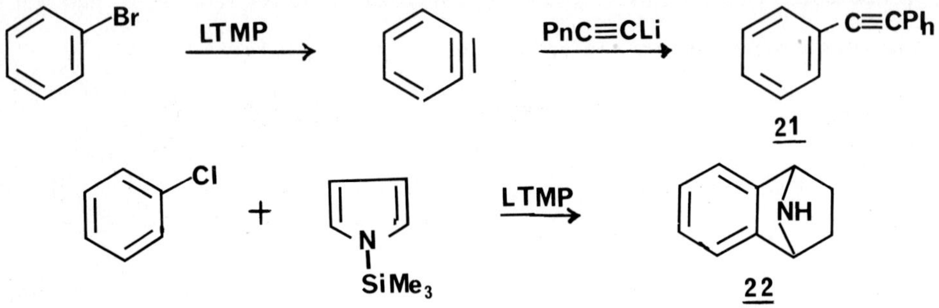

Treatment of bromobenzene with excess LTMP in refluxing THF afforded anthracene in 65% yield.(48) This amazing reaction involves the LTMP induced fragmentation of tetrahydrofuran to the enolate of acetaldehyde, 23, coversion of bromobenzene to benzyne, reaction of 23 with benzyne to yield the benzocyclo-butanol derivative 24, possible opening of 24 to 25 and trapping of the latter with another molecule of benzyne and, finally, on workup, dehydration. No anthracene was obtained when the reaction was carried out at 0^{0} since no 23 is generated from THF and LTMP at that temperature; instead, a 75% yield of N-phenyl-2,2,6,6- tetramethylpiperidide was isolated.

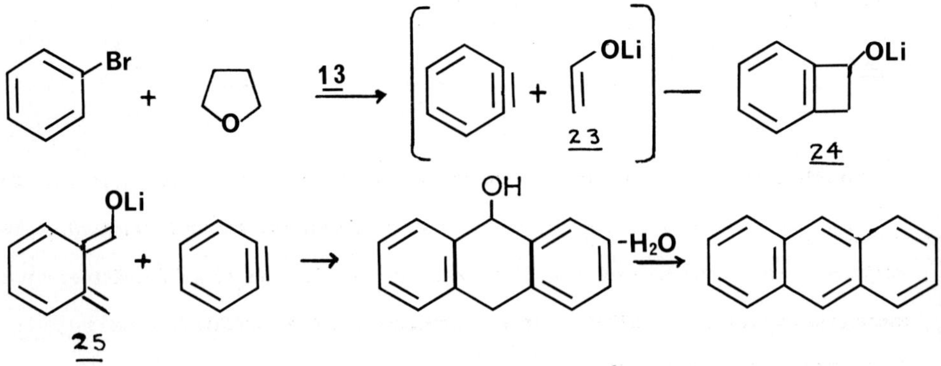

LTMP is sufficiently basic to deprotonate the ortho position of ethyl benzoate, 26. The anion thus generated reacts with the unionized starting material, 26, to afford o-carbethoxybenzophenone.(50)

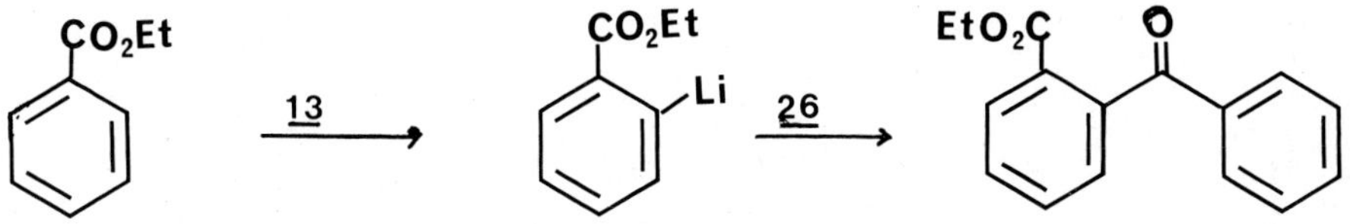

26

Rickborn and coworkers (50) have used LTMP to induce the 1,4-elimination of methanol from methyl o-methylbenzyl ether, 27, thereby generating o-quinodimethane, 28. When this elimination was carried out in hexane in the presence of unactivated alkenes, the expected Diels-Alder adducts could be obtained in low to moderate yields depending on the structure of the alkene. Intramolecular trapping of the o-quinodimethane was also realized. LTMP was found to be approximately twice as reactive as LDA in causing the 1,4-elimination reaction from 27.

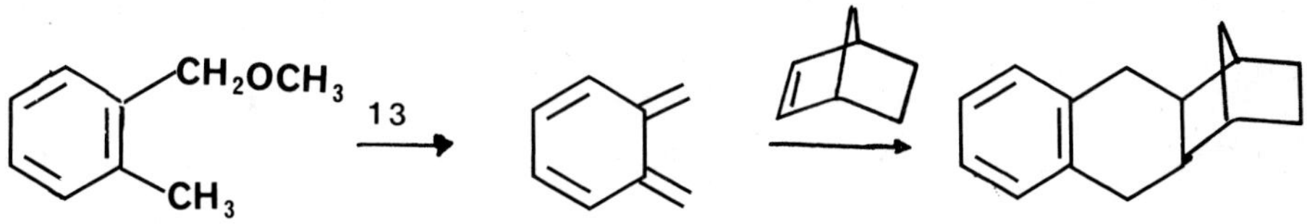

27   28

Heathcock and coworkers have shown that the deprotonation of the ethyl ketones 29 with lithium dialkylamides generates a variety of cis-trans ratios of the enolates 30.(51) Typically amongst LDA, LTMP, and lithium bistrimethyl (LBTMSA), the latter reagent can be used to generate the highest amount of the cis enolate, LTMP gives the most trans and LDA an intermediate amount of the two enolates. (See also Chapter 4 , this Volume).

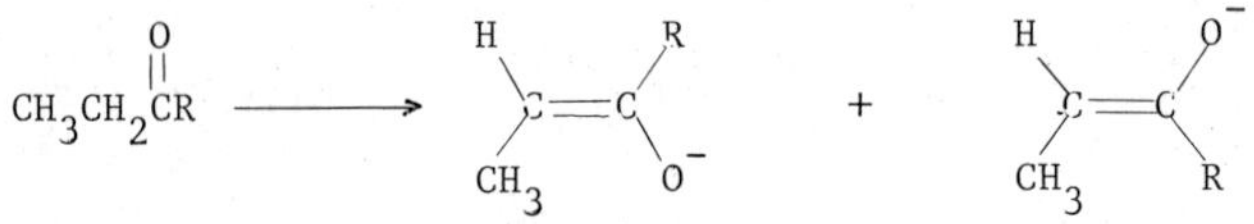

$$CH_3CH_2\overset{\overset{\text{O}}{\|}}{C}R \longrightarrow \underset{\underset{\underline{30a}}{}}{\overset{H}{\underset{CH_3}{}}C=C\overset{R}{\underset{O^-}{}}} \quad + \quad \underset{\underline{30b}}{\overset{H}{\underset{CH_3}{}}C=C\overset{O^-}{\underset{R}{}}}$$

$$\underline{29}$$

D. LITHIUM N-ISOPROPYLCYCLOHEXYLAMIDE (LICA, $\underline{31}$).

Rathke and Lindert reported that lithium N-isopropylcyclo-hexylamide (LICA) was an exceptionally good reagent for the generation of ester enolates, being superior to LDA, lithium di-n-butylamide, and several monosubstituted lithium amides. (52) The ester enolate was generated by the usual inverse addition proceedure. Alkylations of these enolates was most successful when the enolate solution was added to the alkylating agent dissolved in DMSO. Quenching of the enolate solution with D_2O was shown not to be a reliable measure of the extent of metallation, presumably due to competition from internal return.

$$\underline{31} \quad + \quad H-\overset{|}{\underset{|}{C}}-CO_2Et \longrightarrow -\overset{|}{\underset{|}{C}}-CO_2Et \quad \xrightarrow{R-X} \quad R-\overset{|}{\underset{|}{C}}-CO_2Et$$
$$(50-80\%)$$

This reagent is also useful for the generation of the enolate of ethyl crotonate and methyl-2-butynoate in 4:1 THF-HMPA mixtures. When these species are reacted with electrophiles , -unsaturated esters and allenes, respectively are produced.(53)

$$\underline{31} \quad + \quad CH_3CH=CH-CO_2Et \longrightarrow CH_2=CH-CH-CO_2Et \quad \xrightarrow{CH_3I} \quad CH_2=CH-\underset{\underset{CH_3}{|}}{CH}-CO_2Et$$
$$(87\%)$$

$$\underline{31} \quad + \quad CH_3C\equiv C-CO_2Et \longrightarrow CH_2=C=\underset{=}{C}-CO_2Et \quad \xrightarrow{H_2O} \quad CH_2=C=CH-CO_2Et$$
$$(60\%)$$

LICHA has also been used effectively to generate enolates from γ-lactones (54) and to prepare the enolates 32 which undergo the ester enolate Claisen rearrangement.(55) Interestingly, the solvent plays a major role in determining the stereochemistry of the rearrangement products since it affects the stereochemistry of the enolization reaction. Assuming a chair-like transition state for the rearrangement, the results indicate that in THF a Z-enolate but in the THF-HMPA mixture the E-enolate is formed preferentially.

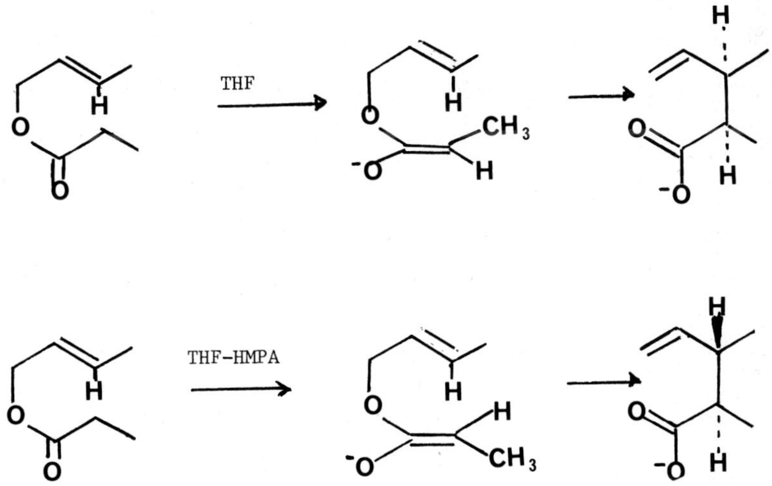

In a number of the cases reported, for example, ref. 54 and 55, LICA and LDA were used interchangeably. The two reagents renerate essentially the same cis to trans enolate ratio in the deprotonation of a series of ethyl ketones.(51) Generally there appears to be relatively little advantage in the use of LCHA vs LDA and recent references to the former reagent are becoming relatively rare.

E. BIS-TRIMETHYLSILYLAMIDES, (HEXAMETHYLDISILAZIDES).

The alkali salts of these amides 33 are easily prepared by reaction of hexamethyldisilazane with KNH (56) to form K^+33, or NaNH (Na^+ 33) (57), or alkyllithiums (Li^+ 33) (58), in a suitable solvent. The sodium and lithium salts are commercially available as a solid sample and as standard solutions in several solvents. Typically these bases have shown high regio- and stereoselectivity in proton abstractions. For example treatment of the ketone 34^+ with Li^+ 33 at -78^0 gave preferential deprotonation of the methyl rather than the methtylene group.(59) Furthermore, no cleavage of the trimethylsilyl ether function occurred. Oxygen-silcon bond cleavage in trimethylsilyl ethers often competes with proton removal with most alkyllithiums and dialkylamides. Generally the protection of alcohols and phenols as t-butyldimethylsilyl ethers avoids such complications. The use of Li^+ 33 for the formation of enolates from the ethyl ketones 29 gave mainly the cis enolates 30a.(51)

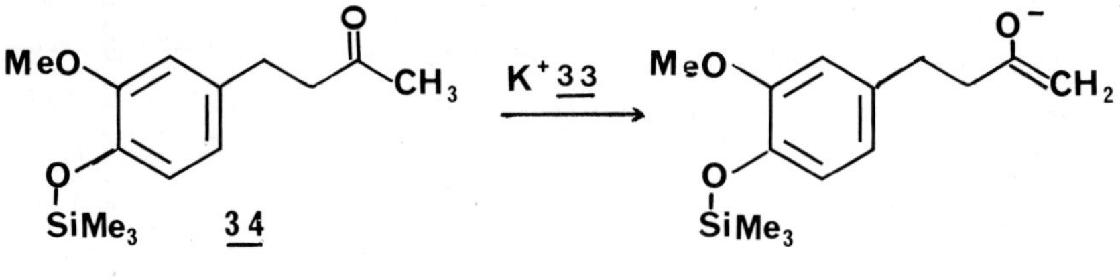

A principle advantage of these series of salts is that counterion variations are readily made without changing any of the other reaction conditions. Such variations can have considerable effect on the rate, regio- and stereochemistry of a particular reacation.

For example, Stork and coworkers found that the salts of 33 were particularly effective in a number of cyclizations of α-nitrile anions.(56, 60-62) Thus the reaction of 35 with the potassium salt of 33 afforded the cyclohexanone 36 in 80% yield.(56) Interestingly, the biscyclization of 37 with K⁺33 afforded mainly the trans decalin 38b, while cyclization with Li⁺ 33 gave mainly the cis isomer 38a.(60) The explanation for the observed difference was made in terms of a tighter ion paired intermediate in the case of the lithium salt as compared to the potassium salt. A similar change in the product ratio occurred when the sol vent was changed from benzene to THF.

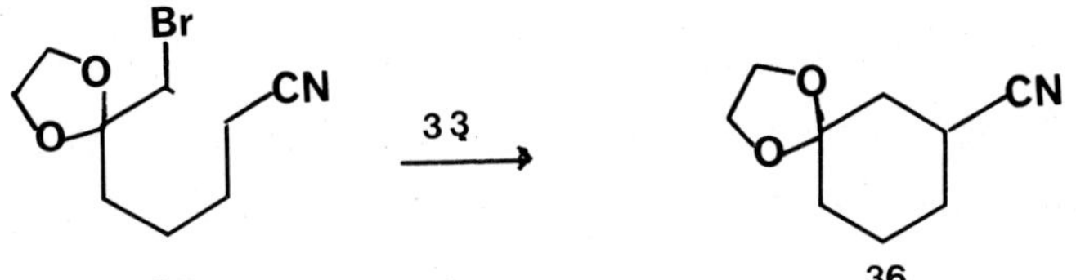

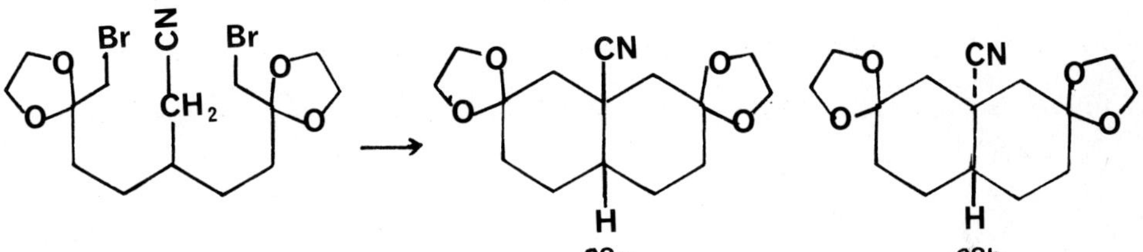

Epoxynitrile cyclizations have also been carried using he sodium salt of 33 or LDA in THF. The grandisol precursor 39 was obtained with a 95/5 stereoselectivity when the cyclization was performed using Li⁺ 33 in benzene.(61) Cyclization of the cyanohydrins of :ω-haloaldehydes 40 with Na⁺ 33 in THF gave good yields of the cyanohydrins 41 of both cyclopropanone and cyclobutanone.(62)

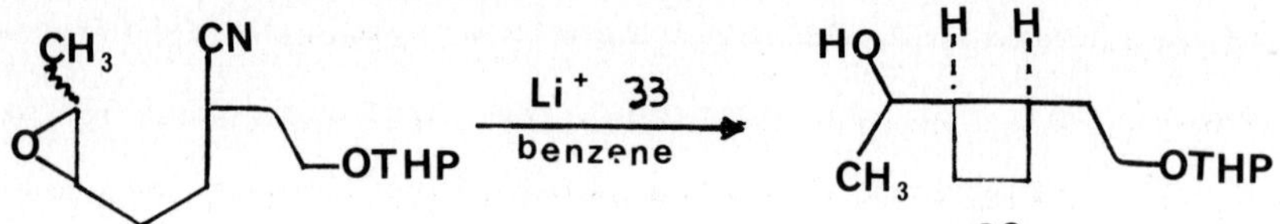

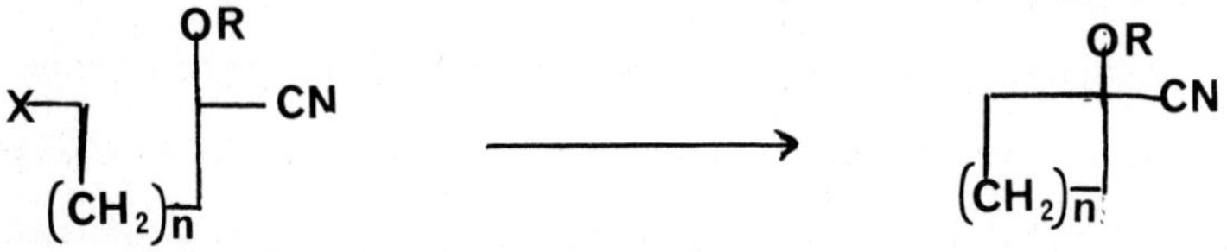

Magnus and Gallagher reported that the alkylation of the ketosulfide anion 42 generated with K$^+$33 in THF at 0 reacted with allyl bromide with complete retention of configuration to afford 43.(63) This transformation was part of a sequence describing the synthesis of members of the kopsane alkaloid family.

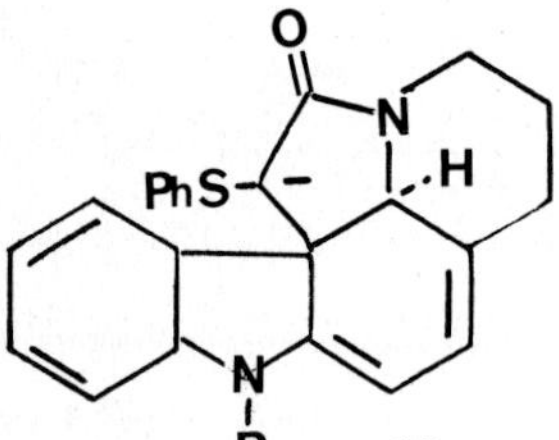

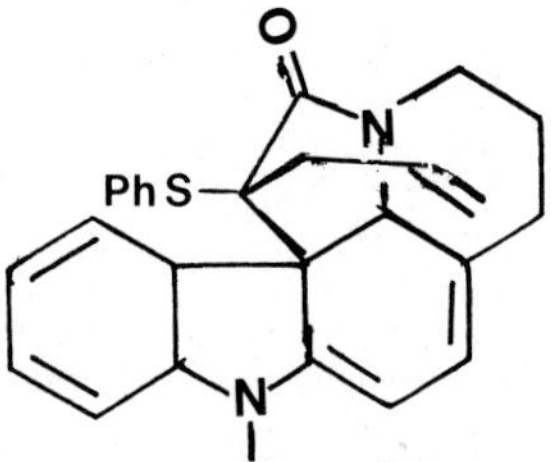

Other examples where salts of 33 were effective include the generation of monobromo and monochloro carbenes from dibromo- and dichloromethane (64), chloromethylcarbene from 1,1-dichloroethane(65). Piers et al (66) reported that 33 was very effective in the formation of the bicyclic diketone 44.

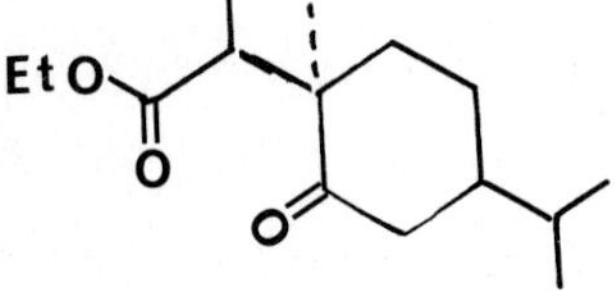

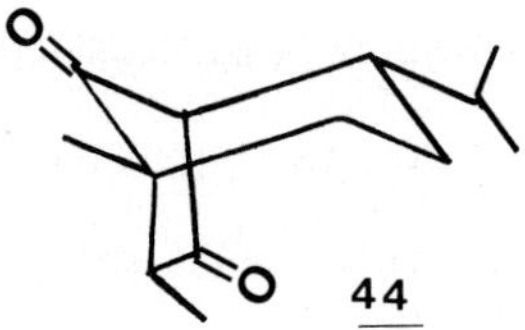

F. ALKYLLITHIUMS.

The most commonly used alkylithium reagent is n-BuLi in hexane. It is often the reagent of choice for most situations including proton abstractions to a variety of functional groups which are not susceptible to addition reactions, halogen metal exchange, and exchange reactions with other groups such as PhSe in 1,1-diselenides and R Sn in organotin derivatives. Representative examples of all these types of reactions are given in Table II. the potency of n-BuLi and sec-BuLi is increased by the presence of TMEDA. DABCO (1,4-diazabicyclo-octane), an early favourite (67), is now rarely used for this purpose. n-BuLi potentiated by TMEDA has been used to prepare mono- and polyanions from alkenes and dienes. (68)

sec-BuLi, alone, or in the presence of TMEDA Is gaining increasing favour as a somewhat stronger and more hindered reagent. It appears to be the reagent of choice in many ortho-metallation reactions especially in the metallation of N,N-diethylbenzamides (69) and in the preparation of a variety of allylic carbanions bearing sulfur, (70) silicon, (71) oxygen (72,74) and halogen (73) substituents. In the case of the allylic ethers, alkylation gave mainly the -product, whereas reaction with ketones led to -substitution. When the lithio counterion was changed to zinc, the -hydroxyalkylation product dominated again. (72) Examples of the various types of carbanions which have been generated with sec-BuLi are given in Table 3.

α-Lithioalkoxides such as 45 are also quite readily prepared by reaction of the corresponding tri-n-butylstannanes with n-BuLi in THF at -78^{0} .(75) Interestingly, this exchange reaction was shown to be stereospecific, presumeably occurring with retention of configuration, since 45 gave only one product upon lithiation and reaction with acetone.

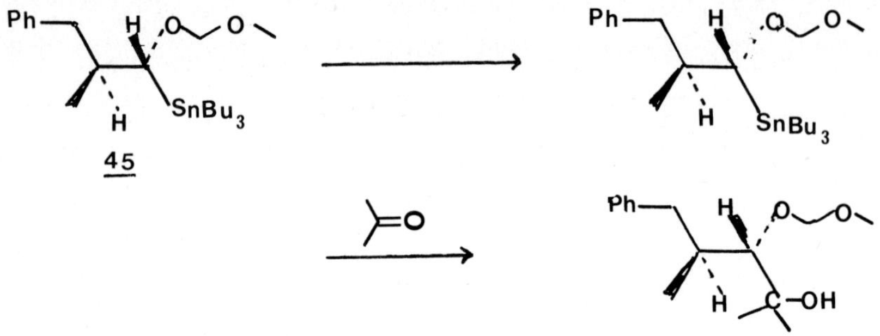

45

Very recently Cohen and Matz (76) have prepared a variety of α-lithio ethers such as **46** by reductive lithiation of α-phenylthioethers with lithium 1-dimethylaminonaphthalide. in the case of 46, reaction with propionaldehyde followed by treatment with dilute acid afforded brevicomin in 40% yield.

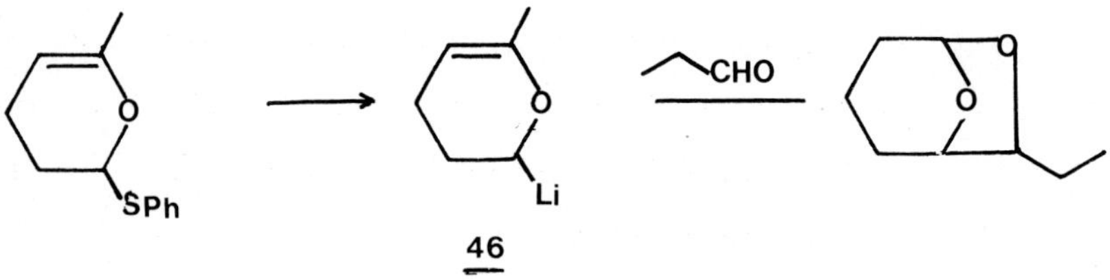

46

sec-BuLi seems particularly valuable in the deprotonation of a variety of silicon reagents. Magnus and coworkers (77) have shown that 47 reacts in a different manner with each of n-BuLi, sec-BuLi, and t-BuLi with the first causing displacement at silicon to give n-butyltrimethylsilane and dimethylsulfonium methylid and the last, surprisingly, causing demethylation to 48. In contrast, sec-BuLi cleanly gave the desired α-silylated ylid. Methoxymethyl trimethylsilane 49 could be converted to the α-methoxy- α-trimethylsilyl anion 50 only with sec-BuLi. (76) Again, the use of n-BuLi caused Si-C bond cleavage while t-BuLi gave deprotonation at one of the methyl groups, probably the result of kinetic control in the reaction.

$$Me_3SiCH_2 \overset{+}{S}Me_2 \ I \quad \underset{47}{}$$

$$\begin{array}{l} \xrightarrow{n-BuLi} Me_3Sin-Bu \\ \xrightarrow{t-BuLi} Me_3SiCH_2CH_2SMe \quad (\underline{48}) \\ \xrightarrow{sec-BuLi} Me_3Si\overset{-}{C}H\overset{+}{S}Me_2 \end{array}$$

$$Me_3SiCH_2OCH_3 \quad \underset{49}{}$$

$$\begin{array}{l} \xrightarrow{n-BuLi} Me_3Sin-Bu \\ \xrightarrow{t-BuLi} \overset{-}{C}H_2Si(Me_3)CH_2OCH_3 \\ \xrightarrow{sec-BuLi} Me_3Si\overset{-}{C}HOCH_3 \quad (\underline{50}) \end{array}$$

A major use of t-BuLi involves halogen-metal exchange of vinylic and aromatic halides. Most examples are restricted to the exchange of iodides and bromides. A number of examples are shown in Table IV. As mentioned above, n-BuLi is also an effective reagent for many similar halogen-metal exchanges. Table II. Two equivalents of t-BuLi are required per mole of vinylic or aromatic halide since the initially formed product of exchange, t-BuBr, rapidly consumes a second mole of t-BuLi to give isobutylene and 2-methylpropane.

$$RCH{=}CHBr \ + \ 2 \ t{-}BuLi \longrightarrow RCH{=}CHLi \ + \ (CH_3)_2C{=}CH_2 \ + \ (CH_3)_3CHCH_3$$

t-BuLi is effective for the preparation of α-lithiovinyl ethers from methyl vinyl ethers (78) and dihydropyran (79). the latter compound could not be peprotonated with either n-BuLi or sec-BuLi and gave trans-1- hydroxy - 4-nonene, the result of an addition elimination sequence.(130) the allylic anion 50 bearing the opptically active pyrrolidino group hgas been metallated with t-BuLi in petroleum ether at -78°.(80) It reacted with primary alkyl halides to give the enamines 51 in 55-75% yield with 50-65% e.e. When the reaction was carried out in THF the e.e. was reduced to less than 40%. A tighter ion pair in petroleum ether vs THF was suggested to cause this difference. Similar results were obtained with respsect to the optical purity of the product when t-BuLi/t-BuOK (80) was used as base.

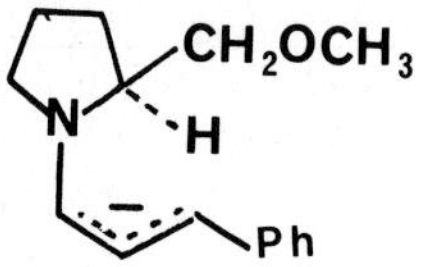

50

A number of authors have carried out comparison studies of the regiochemistry of metallation of aromatics bearing several "ortho-metallating groups". Meyers and Avila (81) reported that the position of metallation in 52 was dependent not only on the nature of the alkyllithium but also the solvent. For example, sec-BuLi/TMEDA in DME at -45⁰ gave a 90:10 ratio of the intermediates 53 and 54 as judged by trapping with methyl iodide. In THF the same reagent gave a 2:1 miixture while in ether the ratio climbed again to 93:7. The rate of metallation of 52 in ether was however quite low and thus the yields of the final methylated products were poor. n-BuLi in DME and THF gave 85:15 and 22:78 ratios, respectively, while t-BuLi afforded a 64:36 ratio in DME and a 22:78 mixture in THF. Clearly, "predicting sites of metallation in multiply-substituted aromatics is difficult and shoul be approached with caution". (81)

Ronald and Winkler have made similar observations with respect to the position of metallation of substituted methoxymethoxy arenes.(82) the results which show considerable variations in the ratio of the metallation at the two different positions are shown below.

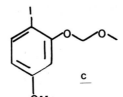

Solvent	RLi	Ratio 52:53
Et$_2$O	sec-BuLi	93:0
DME	sec-BuLi	90:0
THF	sec-BuLi	67:33
THF	n-BuLi	22:78
THF	t-BuLi	22:78
DME	t-BuLi	64:36

Metallating Conditions	Ratio a:b	Overall Yield (%)
t-BuLi-Et$_2$O, 0°	99:1	78
n-BuLi-Et$_2$O, 0°	82:18	44
n-BuLi-hexane, RT	2:98	44
t-BuLi-hexane, RT	57:43	--

Metallating Conditions	Ratio c:d	Overall Yield (%)
t-BuLi-hexane, 0°	3:97	78
t-BuLi-ether, 0°	41:59	95
t-BuLi-TMEDA, ether, −78°	5:95	93

TABLE 2

Carbanion Generation with n-BuLi and trapping will Electrophiles.

a) Halogen-metal Exchange Reactions

Substrate	Reaction Conditions	Electrophile	Product (%)	Ref.
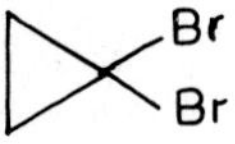 Br, Br (dibromocyclopropane)	THF / 90°	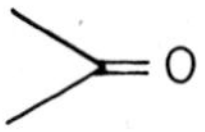 $\rangle$=O	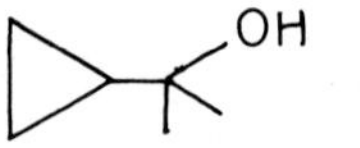 OH	83
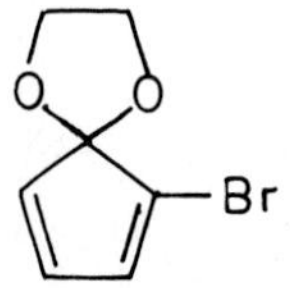 —Br	a	CH_2O	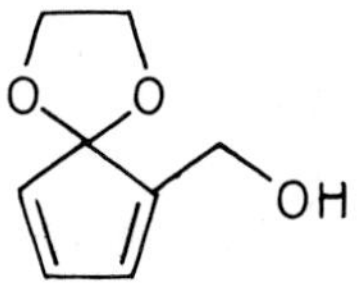 OH (84)	84
$Ph_3SiC=CH_2$ | Br	a	$R_1\overset{O}{\overset{\|}{C}}R_2$	$R_1-\overset{OH}{\underset{R_2}{\overset{\|}{C}}}-\overset{SiPh_3}{\underset{CH_2}{C}}$	85
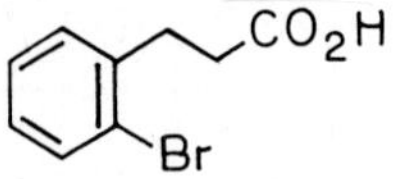 CO₂H, Br	a	—	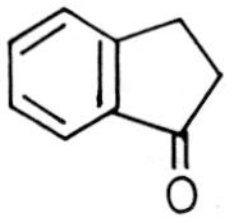 O (65)	86
CO₂H, Br	THF / -100°	$Ar\overset{O}{\overset{\|}{C}}C\ell$	CO₂H, Ar, O	87
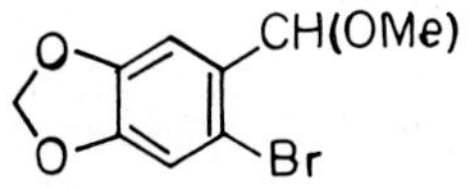 CH(OMe), Br	a	ArCHO	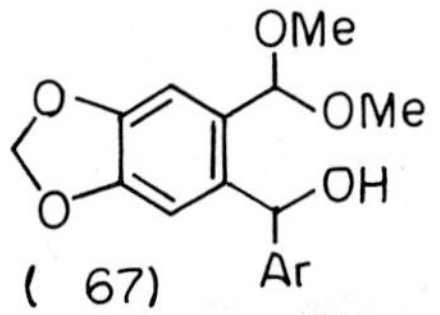 OMe, OMe, OH, Ar (67)	88 a
(epoxide), Br	a, MgBr₂	—	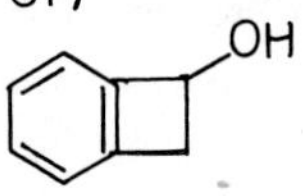 OH (83)	88 b

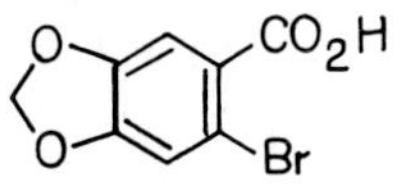

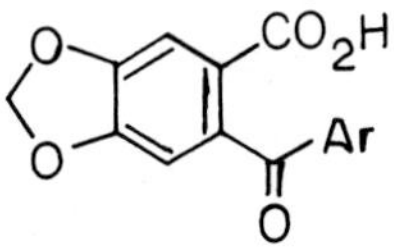

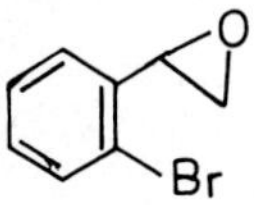

(b) Lithium–Non metal Exchanges

Substrate	Reaction Conditions	Electrophile	Product (%)	Ref.
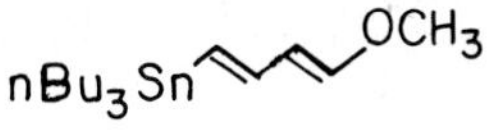 nBu_3Sn ~~~ OCH_3	a	PhCHO	Ph ~~~ CHO[b] (90)	89 a
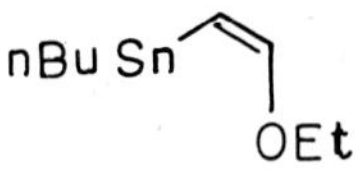 $nBu\,Sn$ ~ / OEt	a	RCHO	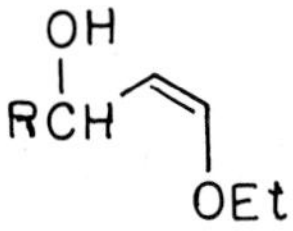 OH / RCH ~ OEt	89 b,c
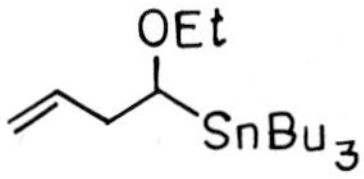 OEt / $SnBu_3$	a	PhCHO	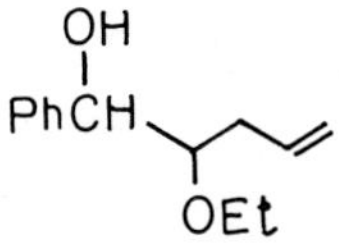 OH / PhCH ~ OEt (81)	90
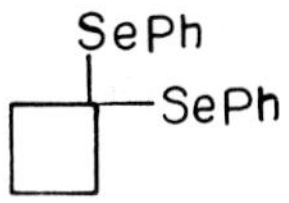 SePh –SePh	a	MeCHO	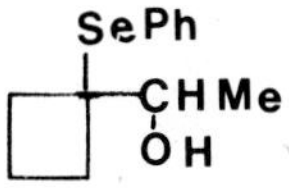 SePh –CHMe OH	91
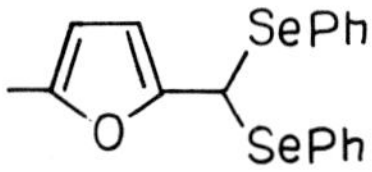 SePh / SePh (furan)	a	Me_3SiCl	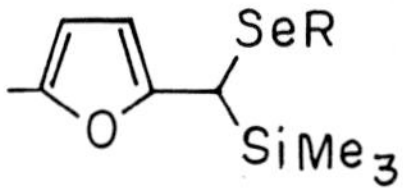 SeR / SiMe_3 (furan)	92

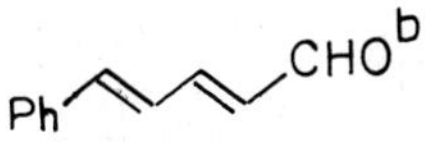

(c) Deprotonations

 ~~~ (butadiene)	THF/−60°	Me_3SiCl	~~~ $SiMe_3$ (85)	93
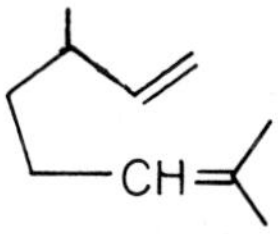 (diene chain) –CH=	hexane, TMEDA	H_2O	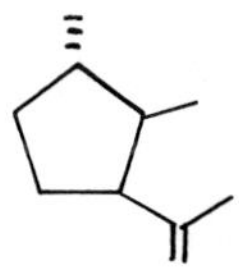 (cyclopentane ring)	94
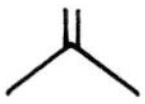 (isobutylene)	2 nBuLi	C_2H_5Br	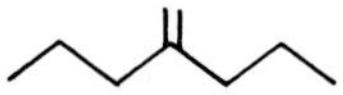 ~~~ (alkene) (100)	68

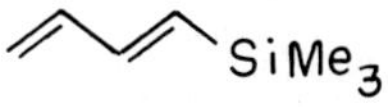

Substrate	Reaction Conditions	Electrophile	Product (%)	Ref.

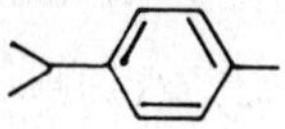

KOtBu — FB(OMe)$_3$ — CH$_2$OH — (44)^c — 95

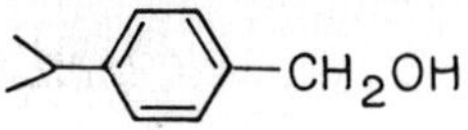

H$_2$C=C(H)(OMe) — a — cyclohexenone (O) — OH, OMe (86) — 96

CH$_2$=C=C(H)(O–O–) — THF/-85° — Me$_3$SiCl — CH$_2$=C=C(SiMe)(OEE) (86) — 97

— a — MeI — (S=O ring) — 98

PhSO$_2$ (epoxide) — a — — — PhSO$_2$ (cyclopropyl)OH (>95) — 99

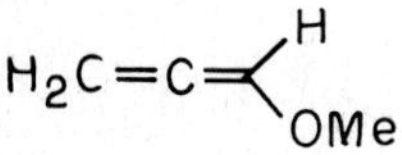

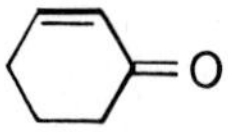

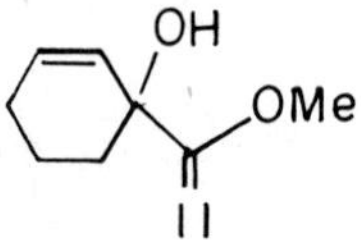

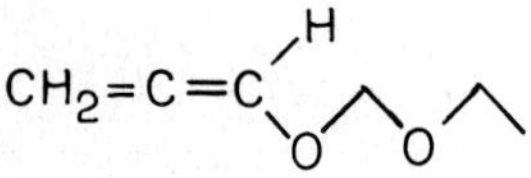

cyclopropyl-SPh — THF/0° — R$_1$CR$_2$ (C=O) — R$_1$-C(OH)(R$_2$)-cyclopropyl, SPh (80-95) — 100

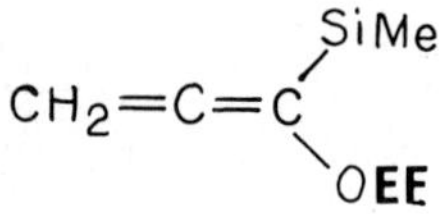

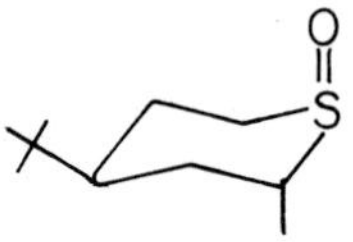

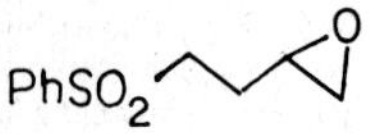

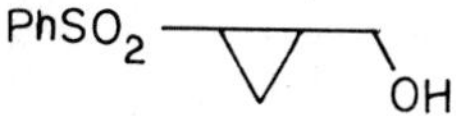

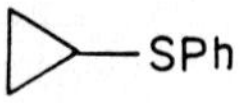

PhS — THF/HMPA — cyclopentenone — (with SPh allyl) (85) — 102

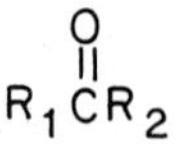

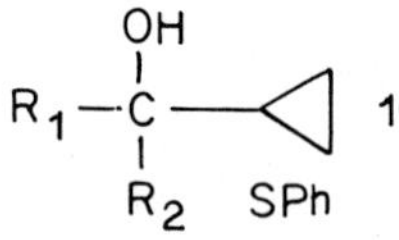

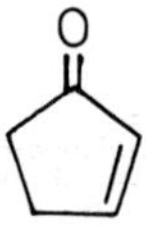

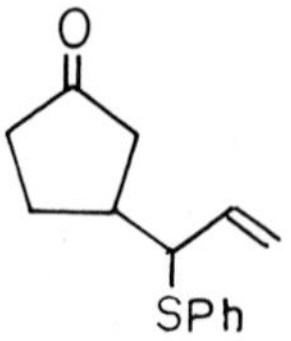

PhSO$_2$ — THF/HMPA — cyclohexenone — SO$_2$Ph (73) — 103

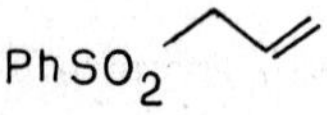

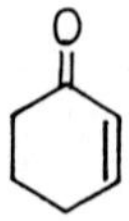

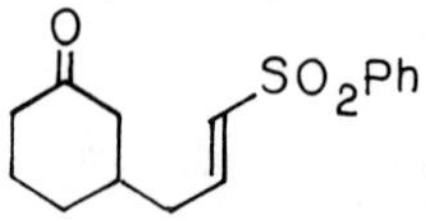

Substrate	Reaction Conditions	Electrophile	Product (%)	Ref.
Me₃Si⌇⌇	TMEDA	PhCHO	PhCH(OH)⌇⌇SiMe₃	104
	TMEDA/MgBr₂	PhCHO	PhCH(OH)⌇⌇ with SiMe₃	104
(benzyl-SiMe₃, CH₂NMe₂)	a	R X	(R, SiMe₃, CH₂NMe₂)	105
epoxide-H, SiPh₃	a	MeI	epoxide-SiPh₃, Me (73)	106
Ph₂C(=O)CH₃	a	R₂CO₂Et	Ph₂P(=O)CH₂C(=O)R₂ (75-90)	107
Me₂N=C(Me)Me	a	Br⌇⌇OCH₂Ph	Me₂N=C(Me)⌇⌇OCH₂Ph (68)	108
Ph-tetrahydropyridine-N-Me	a	Br(CH₂)₃Cl	Ph-ring-Cl, N-Me	109
oxazoline-Cl	a	MeI	oxazoline-CH₃, Cl (>95)	110

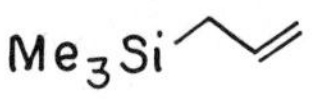

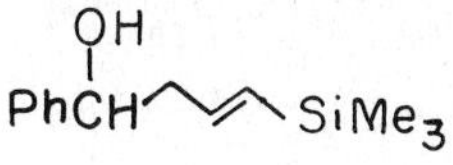

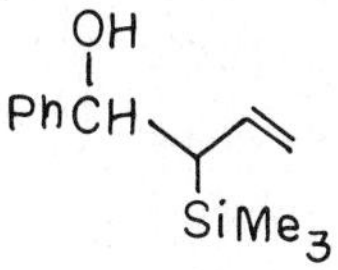

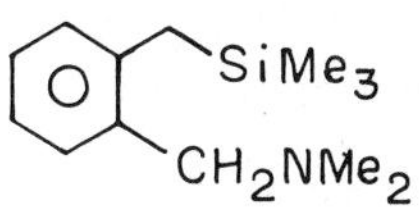

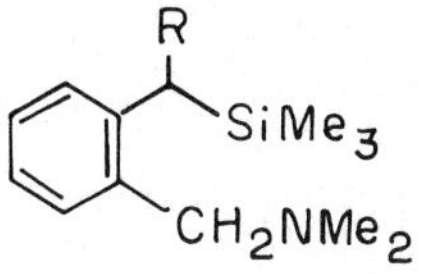

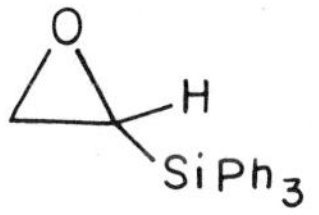

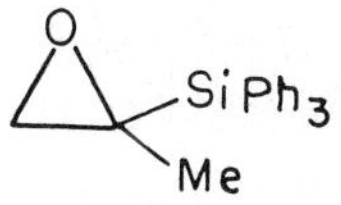

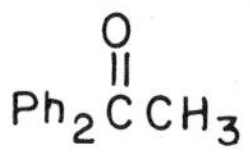

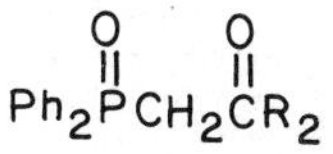

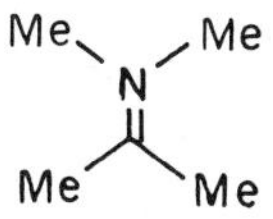

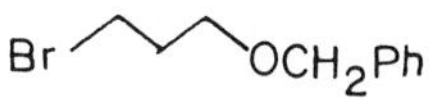

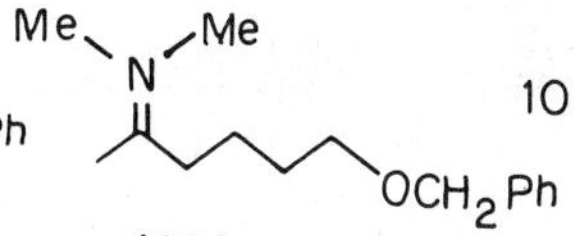

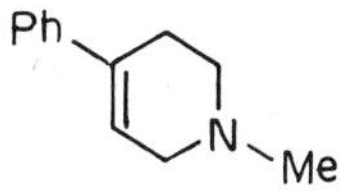

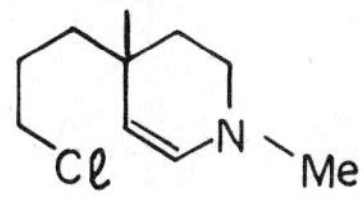

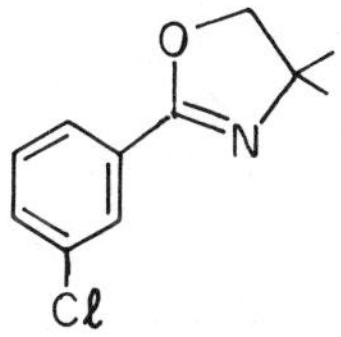

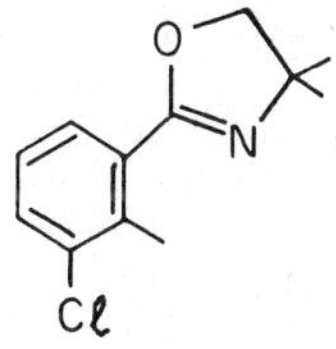

Substrate	Reaction Conditions	Electrophile	Product (%)	Ref.
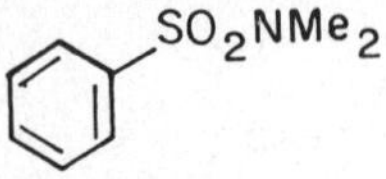 (benzene with SO₂NMe₂) SO_2NMe_2	−45°	MeI	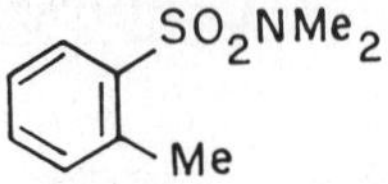 (benzene with SO_2NMe_2 and Me, ortho) (88)	111
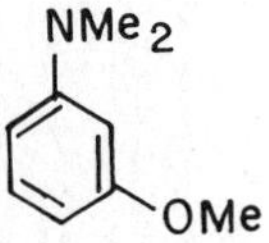 (NMe_2 / OMe benzene)	a	Ph_2CO	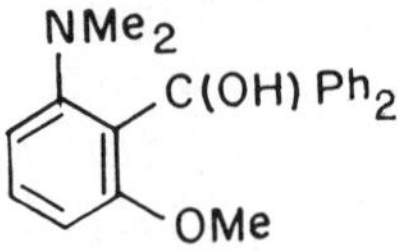 (NMe_2, $C(OH)Ph_2$, OMe benzene) (71)	112

(d) Dianion Formation

Substrate	Reaction Conditions	Electrophile	Product (%)	Ref.
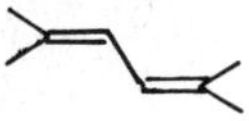 (diene structure)	hexane, RT KOtBu	MeI	$C_{10}H_{16}$ isomers	113
$RCHC\equiv CH$, OH	a	CO_2	$RCHC\equiv CCO_2H$, OH (73)	114
$PhSCH_2CO_2H$	2n BuLi	$CH_3CH_2CH_2ONO_2$	$PhCHCO_2H$, NO_2 (53)	115
$CH_3\overset{O}{\overset{\|}{C}}CH_2CO_2Et$	NaH/n BuLi	EtBr	$C_3H_7\overset{O}{\overset{\|}{C}}CH_2CO_2Et$ (84)	116
$CH_3\overset{O}{\overset{\|}{C}}CH_2CO_2Et$	NaH/n BuLi	(epoxide)	(tetrahydrofuran with CO_2Et vinylidene) (55)	117
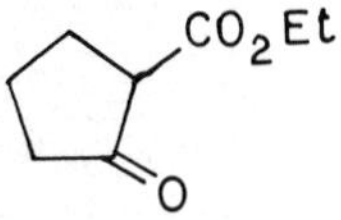 (cyclopentanone with CO_2Et)	NaH/nBuLi	PhSSPh	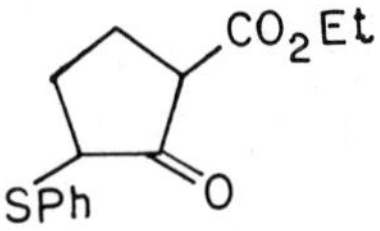 (cyclopentanone with CO_2Et and SPh)	118
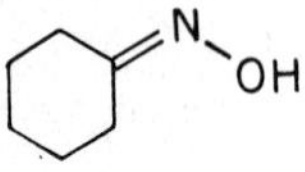 (cyclohexanone oxime, N–OH)	2n BuLi	$HCONMe_2$	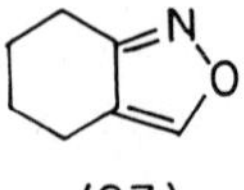 (bicyclic isoxazole structure) (87)	119

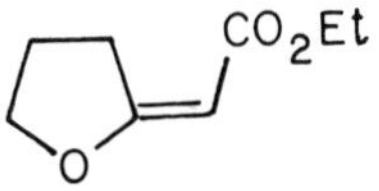

Substrate	Reaction Conditions	Electrophile	Product (%)	Ref.
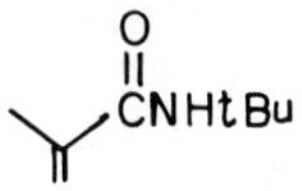 $\underset{\text{CNHtBu}}{\overset{O}{\parallel}}$	2nBuLi	nBuI	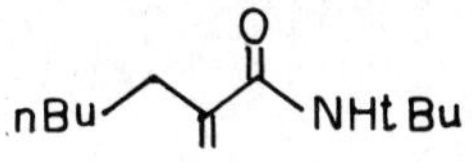 $nBu\text{-}\underset{\overset{O}{\parallel}}{}\text{-NHtBu}$ (60-80%)	120
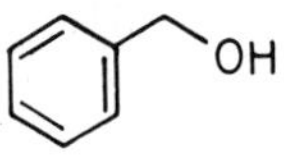 (benzyl) OH	TMEDA	R_1X	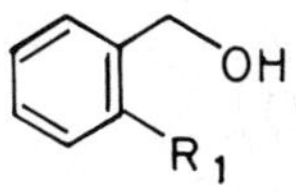 OH, R_1 (20-60%)	121
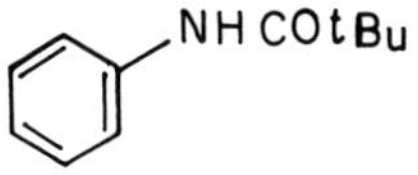 NHCOtBu	0°	MeSSMe	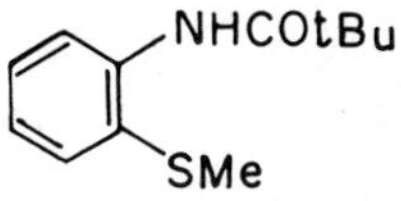 NHCOtBu, SMe (79)	127

a) Represents the most commonly used conditions: nBuLi in THF at - 78°C. In all other cases the indicated solvent or base was used in addition to the nBuLi and THF.

TABLE 3

Generation of Carbonions with sec-BuLi

Substrate	Additive[1]	Electrophile	Product (%)	Ref.

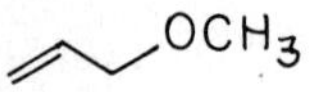

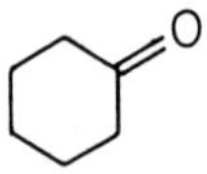

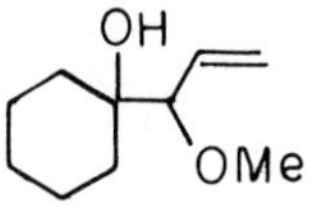

Substrate: allyl–OCH₃ (CH₂=CH–CH₂–OCH₃); Additive: —; Electrophile: cyclohexanone; Product (92); Ref. 72

Substrate: allyl–SPh; Additive: —; Electrophile: ethylene oxide; Product with OH, SPh (78); Ref. 122

Substrate: allyl–SiMe₃; Additive: HMPA; Electrophile: ketone (bicyclic); Product (782); Ref. 71a

Substrate: (vinyl)CH(SiMe₃)(SR); Additive: HMPA; Electrophile: cyclohexenone; Product SiMe₃ (78); Ref. 71b

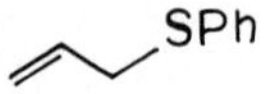

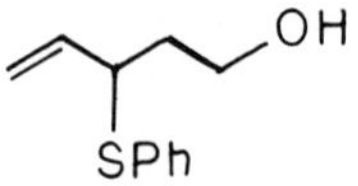

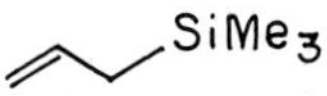

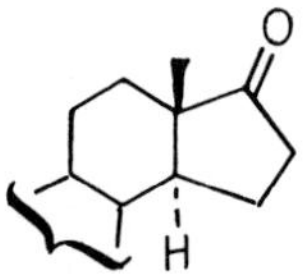

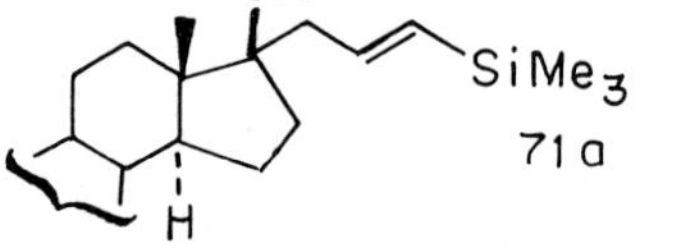

Substrate: Me₃SiCH₂OMe; Additive: —; Electrophile: cyclohexanone; Product OH, SiMe₃, OMe; Ref. 74b

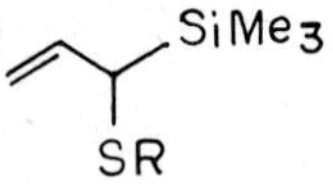

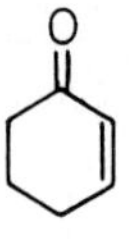

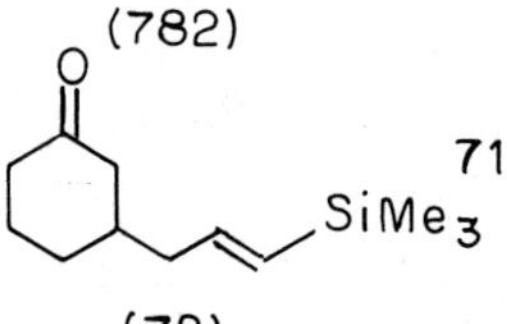

Substrate: Me₃SiCHMe–Cl; Additive: —; Electrophile: cyclohexanone; Product epoxide SiMe₃, Me (80); Ref. 73

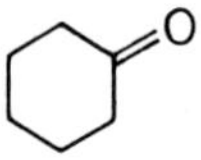

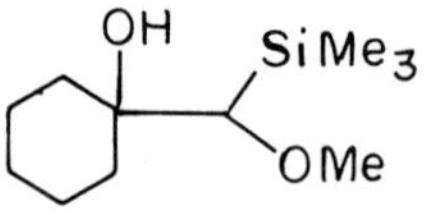

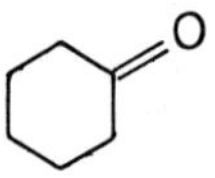

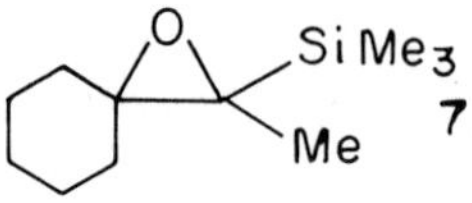

Substrate: Me₃SiCH₂S⁺Me₂ I⁻; Additive: —; Electrophile: cyclohexenone; Product SiMe₃ (55); Ref. 77

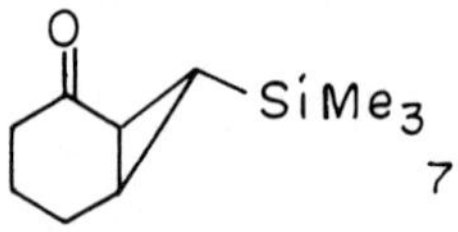

Substrate	Additive	Electrophile	Product (%)	Ref.
Ph₂P CH₂OMe	—			74a
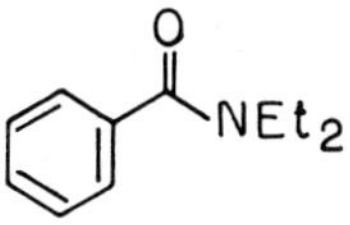	TMEDA	D₂O	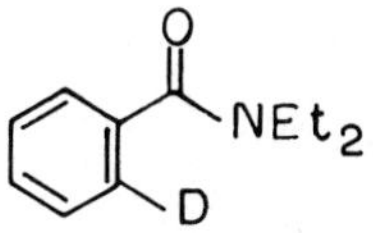 (88)	69a
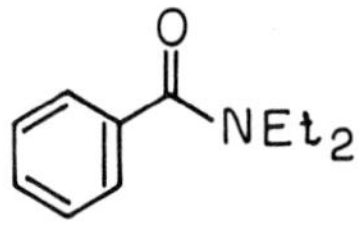	TMEDA	PhCHO	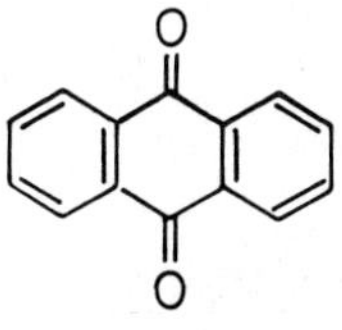 (43)	69c
	TMEDA − 98°		(37)	1d
	—	C₇H₁₅I		1d

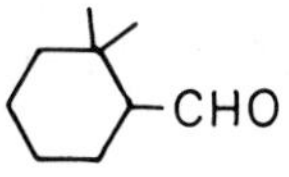

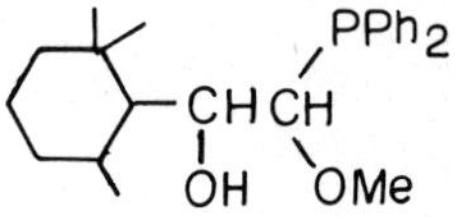

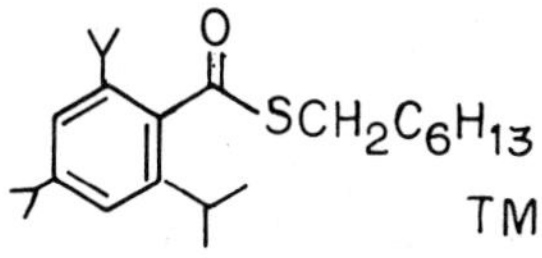

Substrate	Additive	Electrophile	Product (%)	Ref.
$Ph_2P\,CH_2OMe$	—	(1,1-dimethylcyclohexyl)–CHO	(1,1-dimethylcyclohexyl)–CH(OH)CH(OMe)PPh$_2$	74a
o-benzoyl NEt_2 (PhC(=O)NEt$_2$)	TMEDA	D_2O	2-deutero-benzamide, $C(=O)NEt_2$, D (88)	69a
PhC(=O)NEt$_2$	TMEDA	PhCHO	anthraquinone (43)	69c
(2,4,6-trisubstituted aryl)C(=O)$SCH_2C_6H_{13}$	TMEDA − 98°	$CH_3\overset{O}{\overset{\|}{C}}CH_3$	$Ar\,\overset{O}{\overset{\|}{C}}sCH-\overset{OH}{\underset{C_6H_{13}}{C}}(CH_3)_2$ (37)	1d
$Me_2N-\overset{O}{\overset{\|}{C}}-SCH_2CH_3$	—	$C_7H_{15}I$	$MeN\overset{O}{\overset{\|}{C}}-\underset{Me}{S}CHC_7H_{15}$	1d

b) Halogen – Metal – Exchanges

Substrate	Electrophile, or (Metallating condition)	Product (%)	Ref.

$CH_2=CHBr$	PhCHO	$\overset{OH}{CH_2=CHCHPh}$ (76)	123
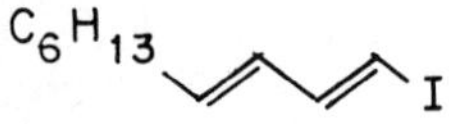 C_6H_{13} —CH=CH—CH=CH—I	$\left(\begin{array}{c}THF/pentane\\-120°\end{array}\right)$	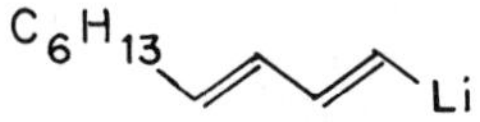 C_6H_{13} —CH=CH—CH=CH—Li	124
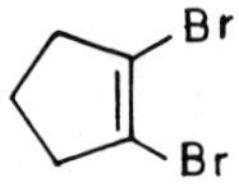 (cyclopentene with Br, Br)	Te (THF, −80°)	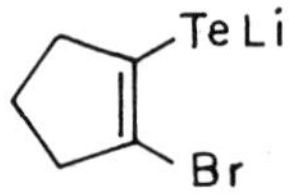 (cyclopentene with TeLi, Br)	125
(cyclohexene with O⁻, Br, Me)	ClSiMe₃	(cyclohexene with $OSiMe_3$, $SiMe_3$, Me) (71)	126
(methylenedioxy benzene with Br and CH₂CH₂Cl)		(benzocyclobutene methylenedioxy) (86)	128
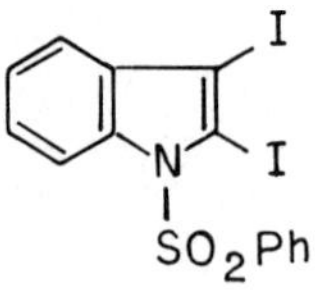 (indole with 3-I, 2-I, N-SO₂Ph)	(THF, −100°, 4 $\underline{t}$BuLi)	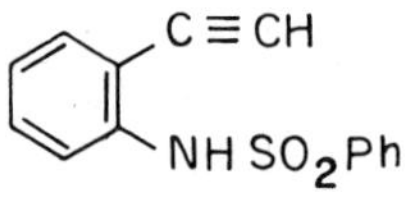 (benzene with C≡CH, NHSO₂Ph) (81)	129

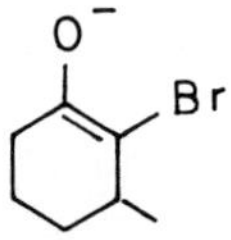

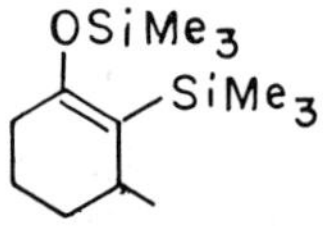

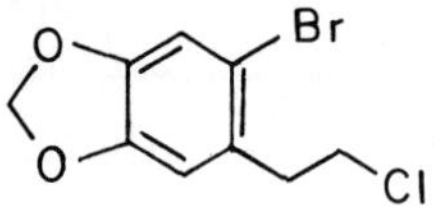

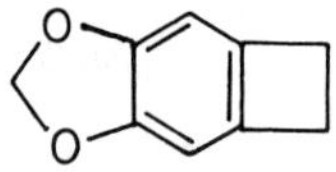

TABLE 4
Carbanion Generation with t BuLi and Reaction with Electrophiles

a) Deprotonation

Substrate	Electrophile, or (Metallating condition)	Product (%)	Ref.
$CH_2=CHOCH_3$	PhCHO	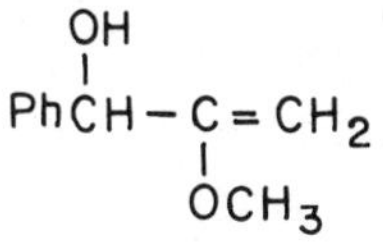 $PhCH-\underset{OCH_3}{\overset{OH}{\mid}}C=CH_2$ (78)	78 a
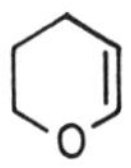	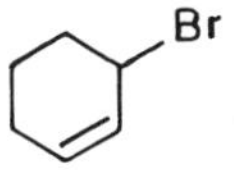	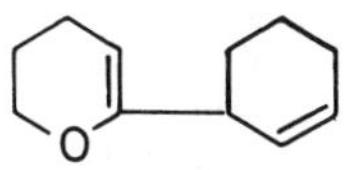 (61)	79
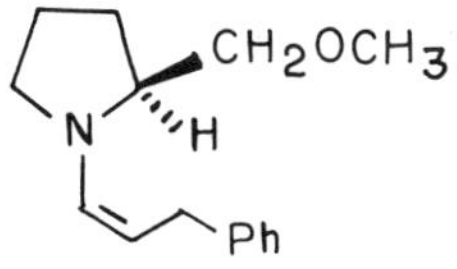	RI	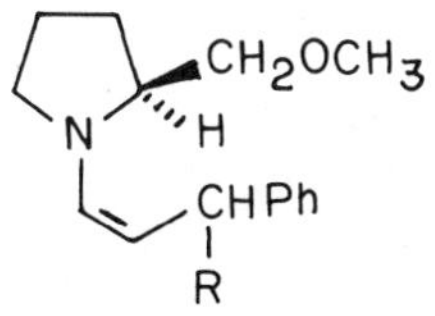 (55–75) (30–65 % ee)	80
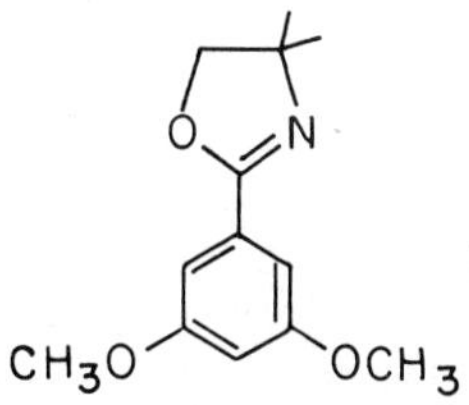	CH_3I	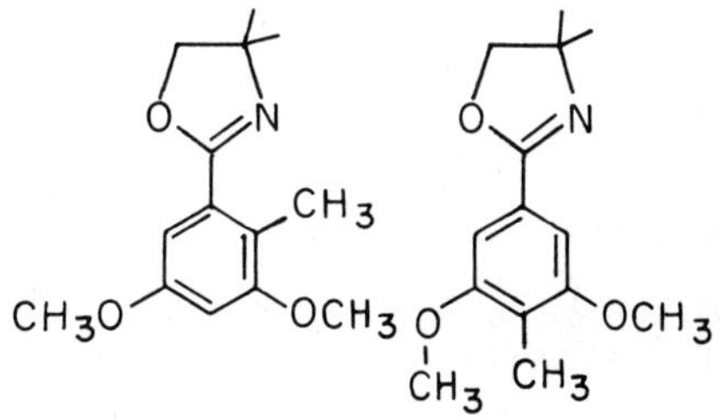 28 : 72 (26 % overall)	81

VII. EXPERIMENTAL.

A.CARBANION GENERATION WITH LDA. METALLATION AND ALKYLATION OF CYCLOHEXANONE IMINES. (14,35)

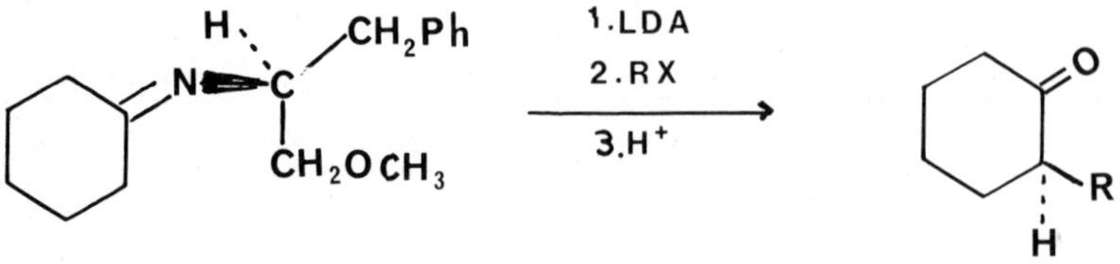

An oven-dried 50 mL flask equipped with a stir bar, a pressure equalized addition funnel, and a rubber septum was charged with 20 mL of anhydrous THF under a nitrogen atmosphere. Freshly distilled diisopropylamine (1.47 mL. 10.5 mmol) was added via syringe and the solution cooled to $0\,^{0}$ C. Butyllithium (4.4 mL) of a 2.4 M solution in hexane was added and the solution stirred at $0\,^{0}$C for 15 min and then cooled to $-30\,^{0}$C. The chiral imines of various ketones (10 mmol) in 10 mL of dry THF were added over 15 min and allowed to stir for 1-1.5 h. The solution was then cooled to $-78\,^{0}$C and the alkyl halide (10.5 mmol) was added in a solution of THF (3-4 volumes) over a period of 1 h, and the mixture allowed to stir at $-78\,^{0}$C for 1.5 h. The total volume of the THF was such as to make the final concentration of the imine about 0.25 M. The light yellow cold solution was poured into 100 mL of saturated salt solution and extracted 3 X with ether. The combined ether extracts were washed with brine, then dried (magnesium sulfate), and concentrated in vacuo, and the product, a light yellow oil was subjected to immediate hydrolysis to form the desired alkylated cyclohexanones.

B.CARBANION FORMATION WITH n-BUTYLLITHIUM. ORTHO LITHIATION OF ARYLOXAZOLINES. REACTION WITH DIPHENYL DISULFIDE. (14, 130)

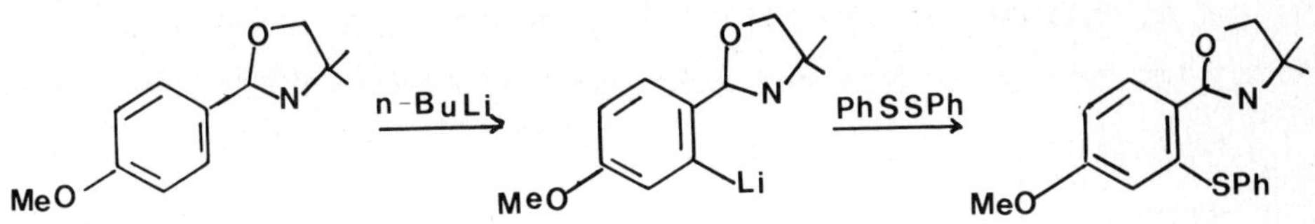

A solution of the above oxazoline (3.07 g, 15 mmol) in 65 mL of ether was cooled under a nitrogen atmosphere in an ice bath. Then 10.3 mL of a 1.6 M solution of n-BuLi (16.5 mmol) was added. After stirring at ice-bath temperatures for 4 h, a solution of 3.65 g (16.5 mmol) of diphenyl disulfide in 30 mL of ether was added at once and the mixture stirred at room temperature for 16 h. After work-up (water, brine sodium sulfate), the residue was recrystallized from ether-hexane to give 4.2 g (89%) of product.

C.CARBANION GENERATION WITH n-BUTYLLITHIUM. GENERATION OF THE DIANION OF METHYL ACETOACETATE AND SUBSEQUENT ALKYLATION. (14, 116)

$$CH_3\overset{O}{\overset{\|}{C}}\text{-}CH_2\text{-}\overset{O}{\overset{\|}{C}}OCH_3 \xrightarrow{NaH} CH_3\overset{O}{\overset{\|}{C}}\text{-}\overline{C}H\text{-}\overset{O}{\overset{\|}{C}}OCH_3 \xrightarrow{n\text{-}BuLi} \overline{C}H_2\overset{O}{\overset{\|}{C}}\text{-}\overline{C}H\text{-}\overset{O}{\overset{\|}{C}}OCH_3$$

Approximately 25 mL of tetrahydrofuran was distilled from lithium aluminum hydride into a 50-mL flask containing 0.54 g of NaH (50% mineral oil, 11 mmol). The flask was stoppered with a serum cap, flushed with nitrogen, and cooled in ice. Then 1.16 g of methyl acetoacetate (10 mmol) was added dropwise and the colourless solution was stirred at 0^0 C for 10 min. To this solution was added dropwise 4.8 mL of 2.2 M n-BuLi (10.5 mmol) in hexane and the yellow to orange solution was stirred at $0^{\circ}C$ for an additional 10 min prior to the addition of the alkyl halides.

To a solution of the dianion prepared as above was added 11 mmol of alkyl halide dissolved in 2 mL of THF. This reaction mixture was allowed to warm slowly to room temperature with stirring. The colour of the dianion faded immediately on addition of the alkylating agent. Apprioximately 15 min after the addition of the alkylating agent, the reation mixture was quenched with 2 mL of conc. HCl in 5 mL of water and 15 mL of ethyl ether. The aqueous layer was further extracted with 2 X 10 mL of ethyl ether. The extracts were combined, washed with water until neutral, dried over magnesium sulfate, and filtered. The solvents were removed under reduced pressure and the crude products purified in the appropriate manner. The yields of C-alklated products were inthe 65-85% range.

D.CARBANION GENERATION WITH sec-BUTLLITHIUM. PREPARATION OF LITHIOALLYL PHENYL SULFIDE AND REACTION WITH ETHYLENE OXIDE. (14, 122)

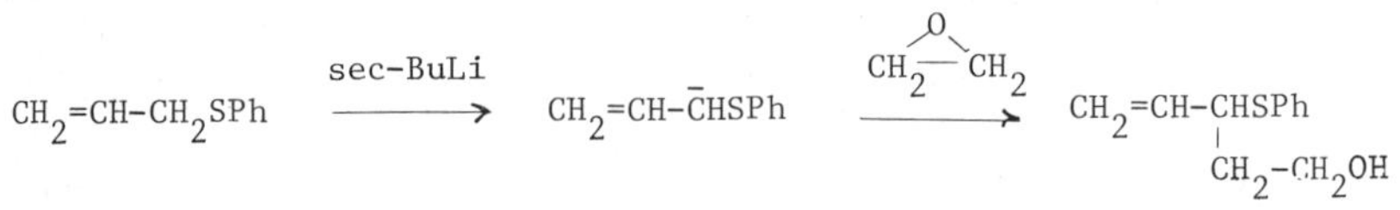

$$CH_2=CH-CH_2SPh \xrightarrow{\text{sec-BuLi}} CH_2=CH-\bar{C}HSPh \xrightarrow{\overset{\displaystyle O}{CH_2\diagdown CH_2}} \begin{array}{c} CH_2=CH-CHSPh \\ | \\ CH_2-CH_2OH \end{array}$$

To 2.47 g (16.5 mmol) of allyl phenyl thioether in 33 mL of anhydrous THF under a nitrogen atmosphere at -78^0C was added 13.3 mL of 1.23 M (18 mmol) of sec-BuLi in cyclohexane. After the mixture was stirred for 3 h at -78^0C a solution of 2.2 g (50 mmol) of ethylene oxide in 5 mL of anhydrous THF was added over a 2 min period and the solution was stirred for an additional 1 h at -78^0 C. The reaction mixture was quenched with 5 mL of water, diluted with 100 mL of ether, washed successively with 50 mL of 1 M aqueous HCl, 50 mL of water and 50 mL of brine and dried over anhydrous magnesium sulfate. Evaporation of the solvents afforded a pale yellow liquid which was fractionally distilled to yield 4.97 g og the desired alcohol.

E. CARBANION GENERATION WITH sec-BUTYLLITHIUM. PREPARATION OF ALLYL ETHER ANIONS AND REACTION WITH ALKYL HALIDES IN THE PRESENCE OF HMPA. (14, 72B)

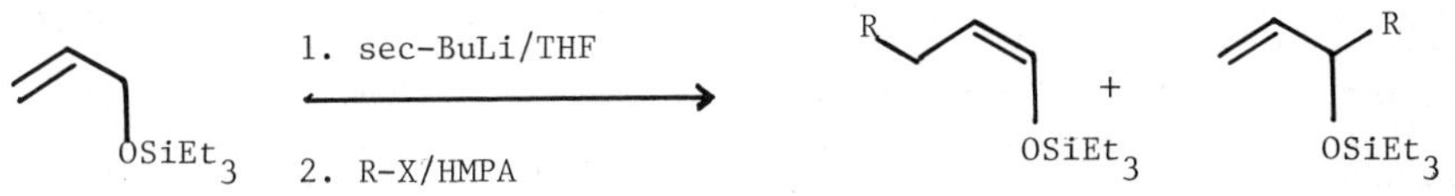

A solution of 2 mmol of the allyl trimethylsilyl ether in 1 mL of dry THF was added dropwise with stirring to a -78^0 C solution of 2.4 mmol of sec-BuLi in 10 mL of anhydrous, deoxygenated THF under a nitrogen atmosphere. When the addition was complete 0.5 mL of HMPA is added and the stirring is continued for an additional 15 min. The alkyl halide (2.4 mmol) is then added. Stirring is continued for 15 min more and the solution is allowed to warm to room temperature. The reaction mixture is then poured into pentane, washed with saturated ammonium chloride solution, and then water, and dried (magnesium sulfate). Solvent evaporation yields the crude product.

F. CARBANION GENERATION WITH t-BUTYLLITHIUM. PREPARATION OF METHOXYVINYLLITHIUM AND REACTION WITH CARBONYL COMPOUNDS. (14, 78a)

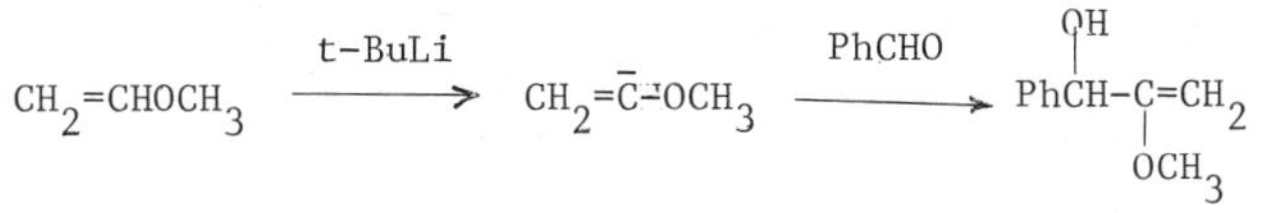

t-BuLi (100 mmol, 62.5 mL, 1.6 M in pentane) was added dropwise to a solution of methyl vinyl ether (9.2 g 160 mmol) in dry THF at -65^0 C. The reaction is exothermic and the addition should be slow enough to maintain the internal temperature between -65 and -55^0 C. After removal of the cooling bath the yellow precipitate (a 2:1 complex of t-BuLi and THF) dissolves and the solution becomes colourless between -5 and 0^0 C. The solution which contains quantitatively methoxyvinyllithium was cooled to -65^0 C. For most reactions the electrophile was added as a THF solution at that temperature. The reaction mixture was quenched at 0^0 C with aqueous 20% ammonium chloride solution if the vinyl ether product is required and worked up in the typical manner. If the corresponding carbonyl compound is required, the reaction mixture was stirred with aqueous methanolic (0.02 N) HCl for 30 min at 20^0 C.

G.   METAL-HALOGEN   EXCHANGE   WITH   t-BULI.   PREPARATION   OF VINYLLITHIUM FROM VINYL BROMIDE. (14,123)

$$CH_2=CHBr \ + \ 2 \ t\text{-BuLi} \ \longrightarrow \ CH_2=CHLi \ + \ (CH_3)_2C=CH_2 \ + \ (CH_3)_3CH$$

A magnetically stirred solution of 10 mmol of vinyl bromide in 42 mL of a 4:1:1 mixture of THF:ether:pentane was cooled under an argon atmosphere in a -120^0 C bath (ligroin, b.p. 30-50^0; isopropanol; acetone; 4:1:1/liquid nitrogen) and combined within 10 min with 20 mmol of t-BuLi in pentane. The temperature was kept between -120 and -110^0 C for 1 h, and then raised to -90^0 C. Benzaldehyde (9.9 mmol) was added and the stirring continued for 15 min at -78^0C and for 20 min at room temperature. the mixture was quenched by pouring into a separatory f;unnel containing 10 mmol of acetic acid, saturated NaCl, and methylene chloride. Evaoration of the organic layer, followed by purification gave 76% of 1-phenyl-2-propen-1-ol.

REFERENCES

1. (a) P. Beak and B.G. McKinnie, J. Amer. Chem. Soc., 99 (1977)
    5213; (b) D.B. Reitz, P. Beak, R.F. Farney, and L.S. Helmick,
    ibid., 100 (1978) 5428; (c) P. Beak, M. Baillargeon and
    L. Carter, J. Org. Chem., 43 (1978) 4255; (d) P. Beak and
    P.D. Becker, ibid., 47 (1982) 3885; and references therein.

2. C.F. Lane and G.W. Kramer. "Handling of Air-sensitive
    Compounds." Aldrich Chimica Acta, 10 (1977) 11.

3. D.F. Shriver, "The Manipulation of Air-sensitive Compounds."
    McGraw Hill, New York, N.Y., 1969.

4. D.F. Shriver, A.B. Levy and M.M. Midland in "Organic Synthesis
    via Boranes." H.C. Brown ed., John Wiley and Sons.

5. J.A. Riddick and W.B. Bunger in "Physical Methods in
    Chemistry", Vol. II, "Organic Solvents", A. Weissberger ed.,
    3rd edition, Wiley-Interscience, New York, 1970.

6. D.R. Burfield, K.-H Lee and R.H. Smithers. J. Org. Chem.,
    42 (1976) 3060.

7. D.R. Burfield, Anal. Chem., 48 (1976) 3060.

8. T. Durst in "Advances in Organic Chemistry" E.C. Taylor ed.,
    John Wiley and Sons, New York, N.Y. 1968.

9. D. Seyferth and R.J. Spohn, J. Amer. Chem. Soc., 91 (1969)
    3037.

10. G. Korbrich, W.E. Breckhoff, H. Heinemann and A. Ahktar, J.
    Organomet. Chem., 3 (1965) 492.

11. D. Seebach, Private Communication.

12. H. Gilman and F.K. Cartledge, J. Organomet. Chem., 2 (1964)
    447.

13. R.F. Eppely and J.A. Dixon, J. Organomet. Chem., 8 (1967)
    176.

14. This author would like to thank the authors and the Journals
    for permission to reproduce the various procedures.

15. S.C. Watson and J.F. Eastham, J. Organomet. Chem., 9 (1967)
    167.

16. W.G. Kofrom and L.M. Baclawski, J. Org. Chem., 41 (1976)
    1879.

17. M.R. Winkle, J.M. Lansinger and R.C. Ronald, J.C.S. Chem.
    Commun., (1980) 87.

18. M.L. Lipton, C.M. Sorensen, A.C. Sadler and R.M. Shapiro,
    J. Organomet. Chem., 186 (1980) 155.

19. H.O. House, W.V. Phillips, T.S.B. Sayer, and C.-C Yau,
    J. Org. Chem., 43 (1978) 700.

20. M.T. Reetz, Liebigs Ann., (1980) 1471.

21. C.H. Heathcock, C.T. Buse, W.A. Kleschick, M.C. Pirrung,
    J.E. Sohn and J. Lampe, J. Org. Chem., 45 (1980) 1066;
    M.C. Pirrung and C.H. Heathcock, ibid., 45 (1980) 1727.

22. D.A. Evans and J.M. Takacs, Tetrahedron Lett., 21 (1980).
    4233; P.E. Sonnet and R.R. Heath, J. Org. Chem., 45 (1980)
    3137.

23. G. Frater, Helv. Chim. Acta, 62 (1979) 2825, 2829.

24. J. Mulzer and T. Kerkmann, J. Amer. Chem. Soc., 102 (1980)
    3620.

25. M.W. Rathke and D.F. Sullivan, J. Amer. Chem. Soc., 95
    (1973) 3050.

25a. R.A. Gorski, G.J. Wolber and J. Wemple, Tetrahedron Lett.,
     (1976) 2577.

26. J.L. Hermann, G.R. Kieczykowski and R.H. Schlessinger,
    Tetrahedron Lett., (1973) 2433.

27. J.L. Hermann and R.H. Schlessinger, J. Chem. Soc. Chem.
    Commun., (1973) 711.

28. P.J. Garratt and R. Zahler, J. Amer. Chem. Soc., 100 (1978)
    7753; K.G. Bilyard, P.J. Garratt, R. Hunter and E. Leete,
    J. Org. Chem 47 (1982) 4731.

29. M.J. Miller, J.S. Bajwa, P.G. Mattingly and K. Peterson,
    J. Org. Chem., 47 (1982) 4731.

30. R. Kaya and N.R. Bellen, J. Org. Chem., 46 (1981) 196.

31. F.A. Bouffard and B.G. Christensen, J. Org. Chem., 46
    (1981) 2208.

32. P.A. Sava and J.A. Katzenellenbogen, J. Org. Chem., 46
    (1981) 329.

33. W. Adam and H.H Fick, J. Org. Chem., 43 (1978) 772.

34. G. Stork and L. Maldona, J. Amer. Chem. Soc., 93 (197)
    5286.

35. A.I. Meyers, D.R. Williams, G.W. Erickson, S. White and
    M. Druelinger, J. Amer. Chem. Soc., 103 (1981) 3081;
    A.I. Meyers, D.R. Williams, S. White, and G.W. Erickson, J.
    Amer. Chem. Soc., 103 (1981) 3088.

36. P. Helquist and M.S. Skekhani, J. Amer. Chem. Soc., 101
    (1979) 1057.

36a. H.J. Reich and S.K. Shaw, J. Amer. Chem. Soc., 97 (1975) 3250.

37. B.Harirchian and P. Magnus, J. Chem Soc., Chem Commun., (1977) 522.

38. U. Schollkopf and H. Beckhaus, Angew. Chemie (Int. Ed), 15 (1976) 50.

39. S. Raucher and G.A. Koolpe, J. Org. Chem., 43 (1978) 3794.

40. B. Ranger, H. Hugel, W. Wynkypiel and D. Seepach, Chem. Ber., 111 (1978) 2630.

41. (a) R.A. Olofson and C.M. Dougherty, J. Amer. Chem. Soc., 95 (1973) 581; (b) ibid., 95 (1973) 582.

42. G.N. Barber and R.A. Olofson, Tetrahedron Letters, (1976) 3783.

43. R.A. Olofson, K.D. Lotts and G.N. Barber, Tetrahedron Letters, (1976) 3381.

44. R.A. Olofson, D.H. Hoskins and K.D. Lotts, Tetrahedron Letters, (1978) 1677.

45. C. Burford, F. Cooke, E. Ehlinger and P. Magnus, J. Amer. Chem. Soc., 99 (1977) 4536.

46. R.L. Danheiser, C. Martinez-Davila, R.J. Auchus and J.T. Kadonaga, J. Amer. Chem Soc., 103 (1981) 2443.

47. K.L. Shepard, Tetrahedron Letters, (1975) 3371.

48. I. Flemming and T. Mah, J. Chem. Soc., Perkin I, (1975) 964.

49. C.J. Upton and P. Beak, J. Org. Chem., 40 (1975) 1094.

50. T. Tuschka, K. Naito, and B. Rickborn, J. Org. Chem., 48 (1983) 70.

51. C.H. Heathcock, C.T. Buse, W.A. Kleschick, M.C. Pirrung, J.E. Sohn and J. Lampe, J. Org. Chem., 45 (1980) 1066.

52. M.W. Rathke and A. Lindert, J. Amer. Chem. Soc., 93 (1971) 2318.

53. M.W. Rathke and D. Sullivan, Tetrahedron Letters, (1972) 4249.

54. G.H. Posner and G.L. Loomis, J. Chem. Soc., Chem. Commun., (1972) 892.

55. R.E. Ireland, R.H. Mueller and A.K. Willard, J. Amer. Chem. Soc., 98 (1976) 2868.

56. G. Stork, J.O. Gardner, R.K. Boeckman, Jr., and K.A. Parker, J. Amer Chem Soc., 95 (1973) 2014.

57. V. Wannegut and H. Niedersprum, Ber., 94 (1961) 1540.

58. E.H. Amonoo-Neizer, R.A. Shaw, D.O. Skovlin and B.C. Smith, J. Chem. Soc., (1965) 2997.

59. P. Deniff and D.I. Whiting, J. Chem. Soc., Chem. Commun., (1976) 712.

60. G. Stork and R.K. Boeckman, Jr., J. Amer. Chem Soc., 95 (1973) 2016.

61. G. Stork and J.F. Cohen, J. Amer. Chem. Soc., 96 (1974) 5272.

62. G. Stork, J.C. Depezay and J d'Anglo, Tetrahedron Letters, (1975) 389.

63. P.D. Magnus and T. Gallagher, J. Amer. Chem Soc., 105 (1983) 2086.

64. B. Martel and J.M. Hiriart, Synthesis, (1972) 201.

65. S. Arora and P. Binger, Synthesis, (1974) 801.

66. E. Piers, R.W. Britton, M.G. Geraghty, R.T. Kieziese and R.D. Smillie, Can. J. Chem., 53 (1975) 2827.

67. E.J. Corey and D. Seebach, J. Org. Chem., 31 (1966) 4097, for example.

68. R.B. Bates, W.A. Beavers, M.G. Green and J.H. Klein, J. Amer. Chem. Soc., 96 (1974) 5640; See also R. Bates, Vol. I, Chapter I, this Series.

69. (a) P. Beak and R.A. Brown, J. Org. Chem., 42 (1977) 1823; (b) ibid., 44 (1979) 4464; (c) M. Watanabe and V. Snieckus, J. Amer. Chem. Soc., 102 (1980) 1457; (d) M.P. Sibi and V. Snieckus, J. Org. Chem., 48 (1983) 1933 and references therein; (e) R.J. Mills and V. Snieckus, ibid., 48 (1983) 1565; (f) H.W. Gschwend and H.R. Rodriguez, Org. Reactions, 26 (1979) 1, for a general review of ortho melallation reactions.

70. P.L. Stotter and R.E. Hornish, J. Amer. Chem. Soc., 95 (1973) 4444.

71. (a) E. Ehlinger and P. Magnus, Tetrahedron Letters, 21 (1980) 11; (b) K.S. Kyler, M.A. Netzel, S. Arseniyadas and D.S. Watt, J. Org. Chem., 48, (1983) 383; (c) P.E. Bauer, K.S. Kyler and D.S. Watt, J. Org. Chem., 48 (1983) 34.

72. (a) D.A. Evans, G.C. Andrews and B. Buckwalter, J. Amer. Chem. Soc., 96 (1974) 5558; (b) W.C. Still and T.L. McDonald; ibid., 96 (1974) 5 562

73. F. Cooke and P. Magnus, J. Chem. Soc. Chem. Comm., (1977) 513.

74. (a) E.J. Corey and M.A. Tiuss, Tetrahedron Letters, 21
    (1980) 3535; (b) P. Magnus and G. Roy, J. Chem. Soc.,
    Chem. Commun., 1979, 822.

75. W.C. Still and C. Skreekumar   J. Amer. Chem. Soc.,
    102 (1980) 1201.

76. T. Cohen and J.R. Matz, J. Amer. Chem. Soc., 102 (1980)
    6900.

77. P. Magnus in "Organic Synthesis Today and Tomorrow",
    B.M. Trost, and C.R. Hutchinson, ed., Pergaman Press, 1980,
    p. 106; F. Cook, P. Magnus and G.L. Bundy, J. Chem Soc.,
    Chem Commun., (1978) 714.

78. (a) J.E. Baldwin, C.A. Hofle and O.W. Leven, Jr., J. Amer.
    Chem. Soc., 96 (1974) 7125; (b) U. Schollkopf and P. Hunsle,
    Ann., 763 (1972) 208.

79. R.K. Boekmann, Jr., and K.J. Bruza, Tetrahedron Letters,
    (1977) 4187.

80. H. Albrecht, G. Bonnet, D. Enders and G. Zimmermann, Tetra-
    hedron Letters, 21 (1980) 3175.

81. A.I. Meyers and W.B. Avila, Tetrahedron Letters, 21 (1980)
    3335.

82. M.R. Winkle and R.C. Ronald, J. Org. Chem., 47 (1982) 2101.

83. M. Braun and D. Seebach, Angew. Chem. Int. Ed. 13 (1974)
    277.

84. S.J. Branca and A.B. Smith, J. Amer. Chem. Soc., 100 (1978)
    7767.

85. T.H. Chan and W. Mychaijlowskij, Tetrahedron Letters, (1974)
    171.

86. R.J. Boatman, B.J. Whitlock and H.W. Whitlock, Jr., J. Amer.
    Chem. Soc., 99 (1977) 4822.

87. W.E. Parham, C.K. Bradsher and K.J. Edgar, J. Org. Chem.,
    46 (1981) 1057; W.E. Parham and Y.A. Sayed, J. Org. Chem.,
    39 (1974) 2051.

88. (a) D. Rajapaksa and R. Rodrigo, J. Amer. Chem. Soc., 103
    (1981) 6208; (b) E. Akgun, M.B. Glinski, K.L. Dhawan and T.
    Durst, J. Org. Chem., 46 (1981) 2730; (c) D.K. Bradsher
    and D.C. Reames, J. Org. Chem., 43 (1978) 1978.

89. (a) R.H. Wollenberg, Tetrahedron Lett., (1978) 717; (b)
    R.H. Wollenb rg, K.F. Albizati and R. Peries, J. Amer. Chem.
    Soc., 99 (1977) 7365; (c) J. Ficini, S. Falou, A.-M. Touzin

290

and J. d'Angelo, Tetrahedron Lett., (1977) 3589.

90. J.-P. Quintard, B. Ellisondo and M. Pereyre, J. Org. Chem., 48 (1983) 1559.

91. S. Halazy and A. Krief, Tetrahedron Letters, 21 (1980) 1997.

92. I Kuwajima, S. Hoshimo, T. Tanaka and M. Shimizu, Tetrahedron Letters, 21 (1980) 3209.

93. D. Seyferth and J. Pornet, J. Org. Chem., 45 (1980) 1721.

94. J.H. Edwards and F.J. McQuillans, J. Chem. Soc., Chem. Commun., (1977) 838.

95. M. Schlosser, H. Bosshardt, W. Walde and M. Stable, Angew. Chemie Int. Ed., 19 (1980) 303; M. Schlosser and G. Randschwalbe, J. Amer. Chem. Soc., 100 (1978) 3528.

96. D. Gange and P.D. Magnus, J. Amer. Chem. Soc., 100 (1978) 7746.

97. H.J. Reich and M.J. Kelly, J. Amer. Chem. Soc., 102 (1982) 1119.

98. R. Lett, S. Bory, B. Moreau, and A. Marquel, Bull. Soc. Chim. France, (1973) 2851.

99. B. Corbel and T. Durst, J. Org. Chem., 41 (1976) 3648.

100. B.M. Trost and D.E. Keely, J. Amer. Chem. Soc., 96 (1974) 1252; B.M. Trost, D. Keely and M.J. Bogdanowicz, ibid., 95 (1973) 3068.

101. (a) D.J. Ager, Tetrahedron Letters, 21 (1980) 4763; (b) P.J. Kocienski, ibid., 21 (1980) 1559; (c) D.J. Ager and R.C. Cookson, ibid., 21, (1980) 1677.

102. M.R. Binns, R.K. Haynes, T.L. Houston and W.R. Jackson, Tetrahedron Letters, 21, (1980) 573.

103. M. Hiramay, Tetrahedron Letters, 22 (1981) 1905.

104. P.W.K. Lau, and T.H. Chan, Tetrahedron Letters, (1978) 2383.

105. Y. Ito, M. Nakatsuka and T. Segusa, J. Amer. Chem. Soc., 103 (1981) 476; see also S. Djuric, T. Sarkar, and P. Magnus J. Amer. Chem. Soc., 102 (1980) 6885; Y. Ito, M. Nakatsuka and T. Saegusa, J. Amer. Chem. Soc., 104 (1983) 7609.

106. J. Eisch and James E. Galle, J. Amer. Chem Soc., 98 (1976) 4646.

107. A.D. Buss and S. Warren, J. Chem. Soc., Chem. Commun., (1981) 100.

108. C.H. Heathcock, E.F. Kleinman and E.S. Brinkley, J. Amer. Chem. Soc., 104 (1982) 1094; See also Chapter 4, this Volume.

109. D.A. Evans, C.H. Mitch, R.C. Thomas, D.M. Zimmerman and R.L. Robey, J. Amer. Chem. Soc. 102 (1980) 5955.

110. A.I. Meyers and W. Riecker, Tetrahedron Lett., (1982) 2091.

111. A.I. Meyers and K.K. Lutomski, J. Org. Chem., 44 (1979) 4464.

112. W.D. Slocum and C.A. Jennings, J. Org. Chem., 41 (1976) 3653.

113. A. Rusenko III, N.S. Mills and P. Morse, J. Org. Chem. 47 (1982) 5198.

114. T. Mukaiyama and K. Suzuki, Chem. Lett., (1980) 255.

115. M. Myashita, T. Kumazawa, Y. Yoshikoshi, J. Org. Chem., 45 (1980) 2945.

116. S.N. Huckin and L. Weiler, J. Amer. Chem. Soc., 96 (1974) 1082.

117. T.A. Bryson, J. Org. Chem., 38 (1973) 3428.

118. K. Hiroi, H. Mirura, K. Kotsuji and S. Sato, Chem. Lett., (1981) 559.

119. G.N. Barber and R.A. Olofson, J. Org. Chem., 43 (1978) 3015.

120. P. Beak and D.J. Kemf, J. Amer. Chem Soc., 102 (1980) 4550; J.J. Fih, and H.W. Gswend, J. Org. Chem., 45 (1980) 4259.

121. D. Seebach, Chem. and Ind., 17 (1978) 521.

122. M. Voyle and K.S. Watt, J. Org. Chem., 48 (1983) 470.

123. D. Seebach and H. Neumann, Tetrahedron Lett., (1976) 4839.

124. R.C. Ronald, J.M. Lansinger, T.S. Lillie and C.J. Wheeler, J. Org. Chem., 47 (1982) 2541.

125. F. Wudl and E. Aharon-Shalom, J. Amer. Chem. Soc., 104 (1982) 1154.

126. C.J. Kuwalski, A.E. Weber and K.W. Fields, J. Org. Chem., 47 (1982) 5088; C.J. Kowalski, M.L. O'Dowd, M.C. Burke and K.W. Fields, J. Amer. Chem. Soc., 102 (1980) 5411.

127. W. Fuhrer and H.W. Gschwend, J. Org. Chem., 44 (1979) 1133.

128. W.E. Parham, Y. Sayed and L.D. Jones, J. Org. Chem., 39 (1974) 2053.

129. G.W. Gribble and M.G. Saulnier, J. Org. Chem., 48 (1983) 603.

130. H.W. Gswend and A. Hamdan, J. Org. Chem., 40 (1975) 2008.